1. General cohomology theory and K-theor
4. Algebraic topology, J.F.ADAMS
5. Commutative algebra, J.T.KNIGHT
8. Integration and harmonic analysis on compact groups, R.E.EDWARDS
9. Elliptic functions and elliptic curves, P.DU VAL
10. Numerical ranges II, F.F.BONSALL & J.DUNCAN
11. New developments in topology, G.SEGAL (ed.)
12. Symposium on complex analysis, Canterbury, 1973, J.CLUNIE & W.K.HAYMAN (eds.)
13. Combinatorics: Proceedings of the British Combinatorial Conference 1973, T.P.McDONOUGH & V.C.MAVRON (eds.)
15. An introduction to topological groups, P.J.HIGGINS
16. Topics in finite groups, T.M.GAGEN
17. Differential germs and catastrophes, Th.BROCKER & L.LANDER
18. A geometric approach to homology theory, S.BUONCRISTIANO, C.P. BOURKE & B.J.SANDERSON
20. Sheaf theory, B.R.TENNISON
21. Automatic continuity of linear operators, A.M.SINCLAIR
23. Parallelisms of complete designs, P.J.CAMERON
24. The topology of Stiefel manifolds, I.M.JAMES
25. Lie groups and compact groups, J.F.PRICE
26. Transformation groups: Proceedings of the conference in the University of Newcastle-upon-Tyne, August 1976, C.KOSNIOWSKI
27. Skew field constructions, P.M.COHN
28. Brownian motion, Hardy spaces and bounded mean oscillations, K.E.PETERSEN
29. Pontryagin duality and the structure of locally compact Abelian groups, S.A.MORRIS
30. Interaction models, N.L.BIGGS
31. Continuous crossed products and type III von Neumann algebras, A.VAN DAELE
32. Uniform algebras and Jensen measures, T.W.GAMELIN
33. Permutation groups and combinatorial structures, N.L.BIGGS & A.T.WHITE
34. Representation theory of Lie groups, M.F. ATIYAH et al.
35. Trace ideals and their applications, B.SIMON
36. Homological group theory, C.T.C.WALL (ed.)
37. Partially ordered rings and semi-algebraic geometry, G.W.BRUMFIEL
38. Surveys in combinatorics, B.BOLLOBAS (ed.)
39. Affine sets and affine groups, D.G.NORTHCOTT
40. Introduction to Hp spaces, P.J.KOOSIS
41. Theory and applications of Hopf bifurcation, B.D.HASSARD, N.D.KAZARINOFF & Y-H.WAN
42. Topics in the theory of group presentations, D.L.JOHNSON
43. Graphs, codes and designs, P.J.CAMERON & J.H.VAN LINT
44. Z/2-homotopy theory, M.C.CRABB
45. Recursion theory: its generalisations and applications, F.R.DRAKE & S.S.WAINER (eds.)
46. p-adic analysis: a short course on recent work, N.KOBLITZ
47. Coding the Universe, A.BELLER, R.JENSEN & P.WELCH
48. Low-dimensional topology, R.BROWN & T.L.THICKSTUN (eds.)

49. Finite geometries and designs, P.CAMERON, J.W.P.HIRSCHFELD & D.R.HUGHES (eds.)
50. Commutator calculus and groups of homotopy classes, H.J.BAUES
51. Synthetic differential geometry, A. KOCK
52. Combinatorics, H.N.V.TEMPERLEY (ed.)
53. Singularity theory, V.I.ARNOLD
54. Markov processes and related problems of analysis, E.B.DYNKIN
55. Ordered permutation groups, A.M.W.GLASS
56. Journees arithmetiques 1980, J.V.ARMITAGE (ed.)
57. Techniques of geometric topology, R.A.FENN
58. Singularities of smooth functions and maps, J.MARTINET
59. Applicable differential geometry, F.A.E.PIRANI & M.CRAMPIN
60. Integrable systems, S.P.NOVIKOV et al.
61. The core model, A.DODD
62. Economics for mathematicians, J.W.S.CASSELS
63. Continuous semigroups in Banach algebras, A.M.SINCLAIR
64. Basic concepts of enriched category theory, G.M.KELLY
65. Several complex variables and complex manifolds I, M.J.FIELD
66. Several complex variables and complex manifolds II, M.J.FIELD
67. Classification problems in ergodic theory, W.PARRY & S.TUNCEL
68. Complex algebraic surfaces, A.BEAUVILLE
69. Representation theory, I.M.GELFAND et. al.
70. Stochastic differential equations on manifolds, K.D.ELWORTHY

London Mathematical Society Lecture Note Series : 69

Representation Theory

Selected Papers

I.M.GELFAND M.I.GRAEV
I.N.BERNSTEIN V.A.PONOMAREV
S.I.GELFAND A.M.VERSHIK

CAMBRIDGE UNIVERSITY PRESS

Cambridge

London New York New Rochelle

Melbourne Sydney

CAMBRIDGE UNIVERSITY PRESS
Cambridge, New York, Melbourne, Madrid, Cape Town, Singapore, São Paulo

Cambridge University Press
The Edinburgh Building, Cambridge CB2 8RU, UK

Published in the United States of America by Cambridge University Press, New York

www.cambridge.org
Information on this title: www.cambridge.org/9780521289818

First published 1982
Re-issued in this digitally printed version 2008

A catalogue record for this publication is available from the British Library

Library of Congress Catalogue Card Number: 82–4440

ISBN 978-0-521-28981-8 paperback

CONTENTS

TWO PAPERS ON REPRESENTATION THEORY

Graeme Segal

These two papers are devoted to the representation theory of two infinite dimensional Lie groups, the group $SL_2(\mathbf{R})^X$ of continuous maps from a space X into $SL_2(\mathbf{R})$, and the group Diff(X) of diffeomorphisms (with compact support) of a smooth manifold X. Almost nothing of a systematic kind is known about the representations of infinite dimensional groups, and the mathematical interest of studying these very natural examples hardly needs pointing out.

Nevertheless the stimulus to the work came from physics, and I shall try to indicate briefly how the representations arise there. Physicists encountered not the groups but their Lie algebras, the algebra $\mathfrak{g}^X$ of maps from X to the Lie algebra $\mathfrak{g}$ of $SL_2(\mathbf{R})$, and the algebra Vect(X) of vector fields on X. The space X is physical space $\mathbf{R}^3$. Choosing a basis in $\mathfrak{g}$, to represent $\mathfrak{g}^X$ is to associate linearly to each real-valued function f on $\mathbf{R}^3$ three operators $J_i(f)$ ($i = 1, 2, 3$), such that

$$[J_i(f), J_j(g)] = \sum_k c_{ijk} J_k(fg),$$

where c_{ijk} are the structural constants of $\mathfrak{g}$. In quantum field theory one writes $J_i(f)$ as $\int_{\mathbf{R}^3} f(x) j_i(x)\, dx$, where j_i is an operator-valued distribution. Then the relations to be satisfied are

$$[j_i(x), j_j(y)] = \sum_k c_{ijk}\, \delta(x-y) j_k(y) \qquad (*)$$

where δ is the Dirac delta-function.

Similarly, to represent Vect($\mathbf{R}^3$) is to associate operators $P(f)$ to vector-valued functions f: $\mathbf{R}^3 \to \mathbf{R}^3$ so that $[P(f), P(g)] = P(h)$, where

$$h = \sum_i \left(f_i \frac{\partial g}{\partial x_i} - g_i \frac{\partial f}{\partial x_i} \right).$$

Writing $P(f) = \sum_i \int f_i(x) p_i(x) dx$ this becomes

$$[p_i(x), p_j(y)] = \delta_i(x-y) p_j(y) - \delta_j(x-y) p_i(x), \qquad (**)$$

where $\delta_k = \partial\delta/\partial x_k$.

Operators with the properties of $j_i(x)$ and $p_i(x)$ arise commonly in quantum field theory in the guise of "current algebras". For example, if one has a complex scalar field given by operators $\psi(x)$ (for $x \in \mathbf{R}^3$) which satisfy either commutation or anticommutation relations of the form $[\psi^*(x), \psi(y)]_\pm = \delta(x-y)$, then the "current-like" operators $p_i(x)$ defined by

$$p_i(x) = \frac{1}{2}\left\{ \psi^*(x) \frac{\partial\psi(x)}{\partial x_i} - \frac{\partial\psi^*(x)}{\partial x_i} \psi(x) \right\}$$

satisfy (**). Similarly if one has an N-component field ψ satisfying $[\psi_\alpha^*(x), \psi_\beta(y)]_\pm = \delta_{\alpha\beta}\delta(x-y)$, and $\sigma_1, \ldots, \sigma_n$ are $N \times N$ matrices representing the generators of a Lie algebra $\mathfrak{g}$ then the operators $j_i(x) = \psi^*(x)\sigma_i\psi(x)$ satisfy (*). (These examples are taken from [3].)

In connection with the quantization of gauge fields it is also worth mentioning that, as we shall see below, the most natural representation of the group of all smooth automorphisms of a fibre bundle is its action on $L^2(E)$, where E is the space of connections ("gauge fields") in the bundle, endowed with a Gaussian measure.

Representations of the group $SL(2, \mathbf{R})^X$.

This paper is concerned with the construction of a single irreducible unitary representation of the group G^X of continuous maps from a space X equipped with a measure into the group $G = SL_2(\mathbf{R})$. (In this introduction I shall always think of G as $SU_{1,1}$, i.e. as the complex matrices $\begin{pmatrix} a & b \\ \bar{b} & \bar{a} \end{pmatrix}$ such that $|a|^2 - |b|^2 = 1$.)

An obvious way of obtaining an irreducible representation of G^X is to choose some point x of X and some irreducible representation of G by operators $\{U_g\}_{g \in G}$ on a Hilbert space H, and to make G^X act on H through the evaluation-map at x, i.e. to make $f \in G^X$ act on H by $U_{f(x)}$. This representation can be regarded as analogous to a "delta-function" at x. More generally, for any finite set of points $x_1, \ldots, x_n$ in X and corresponding irreducible representations $g \mapsto U_g^{(i)}$ of G on Hilbert spaces $H_1, \ldots, H_n$ one can make G act irreducibly on the tensor product $H_1 \otimes \cdots \otimes H_n$ by assigning to $f \in G^X$ the operator $U_{f(x_1)}^{(1)} \otimes \cdots \otimes U_{f(x_n)}^{(n)}$. The object of the paper is to generalize this construction and produce a representation on a "continuous tensor product" of a family of Hilbert spaces $\{H_x\}$ indexed by the points of X (and weighted by the measure on X). There is a simple criterion for deciding whether a representation is an acceptable solution of the problem, in view of the following remark. For any representation U of G^X and any continuous map $\phi: X \to X$ there is a twisted representation ϕ^*U given by $(\phi^*U)_f = U_{f\phi}$. The representation to be constructed ought to have the property that ϕ^*U is equivalent to U whenever ϕ is a measure-preserving homeomorphism of X, i.e. for each such ϕ there should be a unitary operator T such that $U_{f\phi} = T_\phi U_f T_\phi^{-1}$.

The paper describes six different constructions of the representation, but only three are essentially different. Of these, one, described in §4 of the paper, is extremely simple, but not very illuminating because it is a construction a posteriori. I shall deal with it first. For any group Γ and any cyclic unitary representation of Γ on a Hilbert space H with cyclic vector $\xi \in H$ ("cyclic" means that the vectors $U_\gamma \xi$, for all $\gamma \in \Gamma$, span a dense subspace of H) one can reconstruct the Hilbert space and the representation from the complex-valued function $\gamma \mapsto \Psi(\gamma) = \langle \xi, U_\gamma \xi \rangle$ on Γ. To see this, consider the abstract vector space H_0 whose basis is a collection of formal symbols $U_\gamma \xi$ indexed by $\gamma \in \Gamma$. An inner product can be introduced in H_0 by prescribing it on the basis elements:

$$\langle U_{\gamma_1} \xi, U_{\gamma_2} \xi \rangle = \Psi(\gamma_1^{-1} \gamma_2).$$

The group Γ has an obvious natural action on H_0, preserving the inner product. Then H is simply the Hilbert space completion of H_0. The function Ψ is called the *spherical function* of the representation corresponding to $\xi \in H$.

In our case the group $\Gamma = G^X$ has an abelian subgroup K^X, where $K = SO_2$ is the maximal compact subgroup of G, and it turns out that the desired representation H contains (up to a scalar multiple) a unique unit vector ξ invariant under K^X. The corresponding spherical function is easy to describe. The orbit of ξ can be identified with G^X/K^X, i.e. with the maps of X into G/K, which is the Lobachevskii plane. (I shall always think of G/K as the open unit disk in $\mathbf{C}$ with the Poincaré metric.) Given two maps $f_1, f_2 : X \to G/K$ the corresponding inner product is

$$\exp \int_X \log \operatorname{sech} \rho(f(x_1), f(x_2))\, dx,$$

where ρ is the G-invariant Lobachevskii or Poincaré metric on G/K. This means that the spherical function Ψ is given by

$$\Psi(f) = \exp \int_X \log \psi(f(x))\, dx,$$

where, if $g = \begin{pmatrix} a & b \\ \bar{b} & \bar{a} \end{pmatrix} \in G$, then $\psi(g) = |a|^{-1}$. To see that this construction does define a representation of G^X the only thing needing to be checked is that the inner product is positive. That is done in §4.2. But of course it is not clear from this point of view that the representation is irreducible.

A more illuminating construction of the representation is to realize the continuous tensor product as a limit of finite tensor products. To do this we actually represent the group of L^1 maps from X to G, i.e. the group obtained by completing the group of continuous maps in the L^1 metric (cf. §3.4). The L^1 maps contain as a dense subgroup the group of step-functions $X \to G$, and it is on the subgroup of step-functions that the representation is concretely defined.

If one is to form a limit from the tensor products of increasing numbers of vector spaces then the vector spaces must in some sense get "smaller". It happens that the group $SL_2(\mathbf{R})$ has the comparatively unusual property (cf. below) of possessing a family (called the "supplementary series") of irreducible representations H_λ (where $0 < \lambda < 1$) which do in a certain sense "tend to" the trivial one-dimensional representation as $\lambda \to 0$. Furthermore there is an isometric embedding $H_{\lambda+\mu} \to H_\lambda \otimes H_\mu$ whenever $\lambda + \mu < 1$. Now for any partition ν of X into parts $X_1, \ldots, X_n$ of measures $\lambda_1, \ldots, \lambda_n$ one can consider the group G_ν of those step-functions $X \to G$ which are constant on the steps X_i. The group G_ν acts on $\mathscr{H}_\nu = H_{\lambda_1} \otimes \cdots \otimes H_{\lambda_n}$; and when a partition ν' is a refinement of ν then $\mathscr{H}_\nu$ is naturally contained in $\mathscr{H}_{\nu'}$. Accordingly, the group $\bigcup_\nu G_\nu$ of all step-functions acts on $\bigcup_\nu \mathscr{H}_\nu$, and the desired representation is the completion of this.

The construction just outlined is carried out in §2 of the paper. A variant is described in §3, where the representations $\{H_\lambda\}$ of the supplementary series are replaced by another family $\{L_\lambda\}$ with analogous properties – the so-called "canonical" representations. These are cyclic but not irreducible, and L_λ contains H_λ as a summand. In terms of their spherical functions L_λ tends to H_λ as $\lambda \to 0$. The spherical function ψ_λ of L_λ is very simple, given by $\psi_\lambda(g) = |a|^{-\lambda}$ when $g = \begin{pmatrix} a & b \\ \bar{b} & \bar{a} \end{pmatrix}$. In other words, L_λ is spanned by vectors ξ_u indexed by u in the unit disk G/K, and $\langle \xi_u, \xi_{u'} \rangle = \operatorname{sech}^\lambda \rho(u, u')$. Notice that the size of the generating G-orbit $\{\xi_u\}$ in L_λ tends to 0 as $\lambda \to 0$.

The remaining constructions exploit a quite different idea, which is useful in other situations too, as we shall see. I shall explain it in general terms.

Gaussian measures on affine spaces

Suppose that a group Γ has an affine action on a real vector space E with an inner product; i.e. to each $\gamma \in \Gamma$ there corresponds a transformation of E of the form $v \mapsto T(\gamma)v + \beta(\gamma)$, where $T(\gamma): E \to E$ is linear and orthogonal, and $\beta(\gamma) \in E$. Then there is an induced unitary action of Γ on the space $L^2(E)$ of functions on E which are square-summable with respect to the standard Gaussian measure $e^{-\|v\|^2} dv$. Because this measure is not translation-invariant we have to define $U_\gamma: L^2(E) \to L^2(E)$ bv

$$(U_{\gamma^{-1}} f)(v) = \Phi_\gamma(v) f(T(\gamma)v + \beta(\gamma)),$$

where the factor

$$\Phi_\gamma(\nu) = e^{\frac{1}{2}\|v\|^2 - \frac{1}{2}\|T(\gamma)v + \beta(\gamma)\|^2} = e^{-\langle T(\gamma)v, \beta(\gamma)\rangle - \frac{1}{2}\|\beta(\gamma)\|^2}$$

is to achieve unitarity.

The importance of this construction is that the representation of Γ on $L^2(E)$ may be irreducible even when the underlying linear action on E by $\gamma \mapsto T(\gamma)$ is highly reducible. If the linear action T is given then the affine action is evidently described by the map $\beta: \Gamma \to E$. This is a "cocycle", i.e. $\beta(\gamma\gamma') = \beta(\gamma) + T(\gamma)\beta(\gamma')$, and it is easy to see that the affine space is precisely described up to isomorphism by the cohomology class of β in $H^1(\Gamma; E)$. One sometimes speaks of "twisting" the action of Γ on $L^2(E)$ by means of β.

Apart from the description just given there are two other useful ways of looking at $L^2(E)$. The first of these is as a "Fock space". For the Gaussian measure on E the polynomial functions are square-summable, and are dense in $L^2(E)$. So $L^2(E)$ can be identified with the Hilbert space completion of the symmetric algebra $S(E)$ of E. (A little care is necessary here: to make the natural inner product in $S(E)$ correspond to the Gaussian inner product in $L^2(E)$ one must identify $S^n(E)$ not with the homogeneous polynomials on E of degree n, but with the "generalized Hermite polynomials" of degree n.)

The other way of approaching $L^2(E)$ is to observe that it contains (and is spanned by) elements e^v for each $v \in E$, with the property that

$$\langle e^v, e^{v'}\rangle = e^{\langle v, v'\rangle}. \qquad \ldots (\dagger)$$

This means that $L^2(E)$ can be obtained from the abstract free vector space whose basis is a set of symbols $\{e^v\}_{v \in E}$ by completing it using the inner product defined by ($\dagger$). Better still, one can start with symbols ϵ_v and define

$$\langle \epsilon_v, \epsilon_{v'}\rangle = e^{-\frac{1}{2}\|v - v'\|^2}:$$

this makes it plain that the construction uses only the *affine* structure of E. (Of course $\epsilon_v = e^{-\frac{1}{2}\|v\|^2} e^v$.)

The group $SU_{1,1}$ acts on the circle S^1, and has a very natural affine action on the space E_λ of smooth measures on S^1 with integral λ. E_λ is a coset of the vector space E_0 of smooth measures with integral 0. The invariant norm in E_0 is given by

$$\|\alpha\|^2 = \sum_{n>0} \frac{1}{n} |a_n|^2 \text{ when } \alpha = \sum_{n \neq 0} a_n e^{in\theta} d\theta.$$

Then $L^2(E_\lambda)$ is the "canonical representation" L_{λ^2} mentioned above. (This is stated, not quite precisely, as Theorem (7.1) of the paper.) For if α_u is the Dirichlet measure on S^1 corresponding to u in the unit disk G/K (i.e. α_u is the transform of $d\theta$ by any element of G which takes 0 to u) then $\lambda\alpha_u \in E_\lambda$ and

$$\frac{1}{2}\| \lambda\alpha_u - \lambda\alpha_{u'}\|^2 = \lambda^2 \log \cosh \rho(u, u').$$

Returning to the group we are studying, G^X has an affine action (pointwise) on E_1^X, the space of maps $X \to E_1$. (Notice that the linear action of G^X on E_0^X is highly reducible.) The space $L^2(E_1^X)$ has the appropriate multiplicative property with respect to X: for if X is a disjoint union $X = X_1 \cup X_2$ then $E_1^{X_1 \cup X_2} = E_1^{X_1} \times E_1^{X_2}$ and $L^2(E_1^{X_1 \cup X_2}) = L^2(E_1^{X_1}) \otimes L^2(E_1^{X_2})$. In the paper the equivalence of the representations on $L_2(E_1^X)$ and on the continuous tensor product of §§2 and 3 is proved by calculating the spherical functions, but it is quite easy to give an explicit embedding of the continuous tensor product in $L^2(E_1^X)$. For if Y is a part of X with measure λ then the map $E_1^Y \to E_{\sqrt{\lambda}}$ given by $f \mapsto \lambda^{-\frac{1}{2}} \int_Y f$ is compatible with the Gaussian measures, so that $L^2(E_{\sqrt{\lambda}})$ is a subspace of $L^2(E_1^Y)$, and for any partition $X = X_1 \cup \ldots \cup X_n$ with $m(X_i) = \lambda_i$ we have

$$\begin{aligned} L_{\lambda_1} \otimes \cdots \otimes L_{\lambda_n} &= L^2(E_{\sqrt{\lambda_1}}) \otimes \cdots \otimes L^2(E_{\sqrt{\lambda_n}}) \\ &\subset L^2(E_1^{X_1}) \otimes \cdots \otimes L^2(E_1^{X_n}) \\ &= L^2(E_1^X). \end{aligned}$$

Indeed it is pointed out in [10] that for any affine space E the space $L^2(E^X)$ can always be interpreted as a continuous tensor product of copies of $L^2(E)$ indexed by the points of X.

The irreducibility of the representation can be seen very easily in the Fock version. For the cocycle β vanishes on the abelian subgroup K^X, and so under K^X the representation breaks up into its components $S^n(E_0^X)$, on which K^X acts just by multiplication operators. The characters of K^X which arise are all distinct, so the irreducibility of the representation follows from the fact that the vacuum vector is cyclic, which is easily proved (cf. §5.2).

The three approaches to $L^2(E_1^X)$ are described in §§5, 6 and 7 of the paper. In connection with §6 notice that to give an affine action of Γ on a vector space E is the same thing as to give a linear action on a vector space H together with an invariant linear form l: $H \to \mathbf{R}$ such that $l^{-1}(0) = E$: the affine space is then $l^{-1}(1)$. (There is no point, in §6, in considering functions f: $X \to H$ other than those satisfying $l(f(x)) = 1$, and the formulae become less cumbersome under that assumption.)

§5 describes the Fock space version, but not quite in the standard form. The space E_1 of measures on the circle can be identified (by Fourier series) with a space of maps $\mathbf{Z} \to \mathbf{C}$. Accordingly E_1^X is a space of maps $X \times \mathbf{Z} \to \mathbf{C}$, and the symmetric power $S^k(E_1^X)$ is a space of maps

$$\underbrace{X \times \ldots \times X}_{\leftarrow k \rightarrow} \times \underbrace{\mathbf{Z} \times \ldots \times \mathbf{Z}}_{\leftarrow k \rightarrow} \to \mathbf{C}$$

which are symmetric in the obvious sense. The effect of this point of view is to identify $L^2(E_1^X)$ with a space of functions on the free abelian group generated by the space X, i.e. on the space whose points are "virtual finite subsets" $\Sigma n_i x_i$ of X, with $n_i \in \mathbf{Z}$. This is intriguing, but whether it is more than a curiosity it is hard to say.

That concludes my account of the contents of the paper itself; but I shall mention some related matters. The most obvious question to ask is what class of groups G the method applies to. As it stands it evidently does not work for groups for which the trivial representation is isolated in the space of all irreducible representations. This excludes all compact groups, as for them the irreducible representations form a discrete set. The isolatedness of the trivial representation has been cleverly investigated by Kazhdan [5], who proved in particular that among semisimple groups the trivial representation is isolated if the group contains $\mathrm{SL}_3(\mathbf{R})$ as a subgroup. The only simple groups not excluded by Kazhdan's criteria are $\mathrm{SO}_{n,1}$ and $\mathrm{SU}_{n,1}$ – notice that $\mathrm{PSL}_2(\mathbf{R}) \cong \mathrm{SO}_{2,1}$ and $\mathrm{PSL}_2(\mathbf{C}) \cong \mathrm{SO}_{3,1}$. For these the method works just as for $\mathrm{SL}_2(\mathbf{R})$. (For example $\mathrm{SO}_{n,1}$ is the group of all conformal transformations of S^{n-1}, and the affine space E_λ used above can be replaced by the space of measures on S^{n-1} with integral λ.)

A class of groups for which the trivial representation is not isolated consists of the semidirect products $G \tilde{\times} V$, where G is a compact group with an orthogonal action on a real vector space V. Indeed if Ω is a G-orbit in V an element $g \in G$ acts naturally on $L^2(\Omega)$, and $v \in V$ can be made to act by multiplication by the function $e^{i\langle v, \omega\rangle}$. When the compact orbit Ω is close to the origin in V the representation $L^2(\Omega)$ is close to the trivial representation (in the sense of its spherical function). Furthermore $G \tilde{\times} V$ has an obvious affine action on V: the induced action on $L^2(V)$ is the direct integral of the irreducible representations $L^2(\Omega)$ for all orbits $\Omega \subset V$.

Thus the methods of the paper apply to all groups of the form $G \tilde{\times} V$. The importance of this is that it provides a way of constructing a representation of the group $(G^X)_{sm}$ of *smooth* maps from a manifold X to a compact group G. For a smooth map $f: X \to G$ induces a map of tangent bundles $Tf: TX \to TG$, and this can be regarded as a map which to each point $x \in X$ assigns a "1-jet" $j(x) \in J_x G = G \tilde{\times} (T_x^* X \otimes \mathfrak{g})$ where $\mathfrak{g}$ is the Lie algebra of G. As the groups $J_x G$ is of the form $G \tilde{\times} V$ the method of the paper provides a representation of the group Γ of bundle maps $TX \to TG$. (The fact that $J_x G$ depends on x, giving rise to a bundle of groups on X, is not important.) Γ contains $(G^X)_{sm}$ as a subgroup, and it turns out that the representation constructed remains irreducible when restricted to $(G^X)_{sm}$, at least when $\dim(X) \geqslant 4$. That is proved in the papers [1] and [2].

It is interesting to notice that the group of bundle maps Γ is just the semidirect product $(G^X)_{sm} \tilde{\times} \Omega^1(X;\ \mathfrak{g})$, where $\Omega^1(X;\ \mathfrak{g})$ is the space of 1-forms on X with values in $\mathfrak{g}$; and the associated affine space E is the space of connections in the trivial G-bundle on X. The fact that the space of the representation of $(G^X)_{sm}$ is $L^2(E)$ is, of course, suggestive from the point of view of gauge theories in physics.

Representations of the group of diffeomorphisms

This paper is devoted to the representation theory of the group Diff(X) of diffeomorphisms with compact support of a smooth manifold X. (A diffeomorphism has compact support if it is the identity outside a compact region.)

The most obvious unitary representation of Diff(X) is its natural action on $H = L^2(X)$, the space of square-summable $\frac{1}{2}$-densities on X. (By choosing a smooth measure m on X one can identify $L^2(X)$ with the usual space of functions f on X which are square-summable with respect to m. Then the action of a diffeomorphism ψ on f will be $f \mapsto \tilde{f}$, where

$$\tilde{f}(x) = J_\psi(x)^{\frac{1}{2}} f(\psi^{-1} x)$$

and $J_\psi(x) = dm(\psi^{-1} x)/dm(x)$. But it is worth noticing that $L^2(X)$ is canonically associated to X, and does not involve m.)

From H a whole class of irreducible representations of Diff(X) can be obtained by the well-known method introduced by Weyl to construct the representations of the general linear groups. For any integer n the symmetric group S_n acts on the n-fold tensor product $H^{\otimes n} = H \otimes \cdots \otimes H$ by permuting the factors, and the action commutes with that of Diff(X). It turns out that under Diff(X) $\times$ S_n the tensor product decomposes

$$H^{\otimes n} = \bigotimes_\rho V^\rho \otimes W_\rho,$$

where $\{W_\rho\}$ is the family of all irreducible representations of S_n, and V^ρ is a certain irreducible representation of Diff(X). More explicitly, V^ρ is the space of L^2 functions $\underbrace{X \times \ldots \times X}_{\leftarrow n \rightarrow} \rightarrow W_\rho$ which are equivariant with respect to S_n: thus it makes sense even when ρ is not irreducible, and $V^{\rho \oplus \rho'} \cong V^\rho \oplus V^{\rho'}$. The class of representations $\{V^\rho\}$, which were first studied by Kirillov, is closed under the tensor product: if ρ and σ are representations of S_n and S_m then $V^\rho \otimes V^\sigma \cong V^{\rho \cdot \sigma}$, where $\rho \cdot \sigma$ is the representation of S_{n+m} induced from $\rho \otimes \sigma$. All of this is explained in §1 of the paper.

It is then natural to ask, especially when X is not compact, whether new representations of Diff(X) can be constructed by forming some kind of infinite tensor product $H^{\otimes \infty}$ and decomposing it under the infinite symmetric group S^∞ of all permutations of $\{1, 2, 3, \ldots\}$. This question is the main subject of the paper, and it is considered in the following way.

The L^2 functions $X^n \rightarrow W_\rho$ are the same as those $\tilde{X}^n \rightarrow W_\rho$ where

$\widetilde{X}^n \subset X^n$ is the space of n-triples of *distinct* points. The symmetric group S_n acts on $\widetilde{X}^n$, and the quotient space is $B_X^{(n)}$, the space of n-point subsets of X. Diff(X) acts transitively on $B_X^{(n)}$, and there is a unique class of quasi-invariant measures on it. The representation V^ρ can be regarded as the space of sections of a vector bundle on $B_X^{(n)}$ whose fibre is W_ρ. An appropriate infinite analogue of $B_X^{(n)}$ is the space Γ_X of infinite "configurations" in X, i.e. the space of countable subsets γ of X such that $\gamma \cap K$ is finite for every compact subset K of X. This space, and the probability measures on it, play an important role in both statistical mechanics and probability theory. One can imagine the points of a configuration as molecules of a gas filling X, or as faulty telephones.

Diff(X) does not act transitively on Γ_X: two configurations are in the same orbit only if they coincide outside a compact region. Nevertheless one can define (in many ways) measures on Γ_X which are quasi-invariant and ergodic under Diff(X). For each such measure μ there is an irreducible representation U_μ of Diff(X) on $L^2(\Gamma_X; \mu)$. More generally, for each representation ρ of a finite symmetric group S_n there is an irreducible representation U_μ^ρ: it is the space of sections of the infinite dimensional vector bundle on Γ_X whose fibre is the representation H^ρ of S^∞ induced from the representation $\rho \otimes 1$ of $S_n \times S_n^\infty$. (S_n^∞ denotes the subgroup of permutations in S^∞ which leave $1, 2, \ldots, n$ fixed.) More explicitly, one can consider a covering space $\Gamma_{X,n}$ of Γ_X defined by

$$\Gamma_{X,n} = \{(\gamma; x_1, \ldots, x_n) \in \Gamma_X \times X^n : x_i \in \gamma \text{ for } i = 1, \ldots, n\}.$$

$\Gamma_{X,n}$ is locally homeomorphic to Γ_X, and therefore a measure μ on Γ_X defines a measure $\widetilde{\mu}$, the "Campbell measure", on $\Gamma_{X,n}$. The space of the representation U_μ^ρ is the space of maps $\Gamma_{X,n} \to W_\rho$ which are S_n-equivariant and square-summable for $\widetilde{\mu}$.

The simplest and most important measures on Γ_X are the Poisson measures μ_λ (parametrized by $\lambda > 0$), for which the measure of the set $\{\gamma \in \Gamma_X : \text{card}(\gamma \cap K) = n\}$ is $\left(\frac{\lambda m}{n!}\right)^n e^{-\lambda m}$, where m is the measure of K. More can be said about the representations $U_\lambda^\rho = U_{\mu_\lambda}^\rho$ in the Poisson case:

(i) They form a closed family under the tensor product, and have the following simple behaviour

(a) $U_\lambda^\rho \cong U_\lambda \otimes V^\rho$, and

(b) $U_\lambda \otimes U_{\lambda'} = U_{\lambda+\lambda'}$.

(ii) U_λ is what is called in statistical mechanics an "N/V limit". In other words, if X is the union of an expanding sequence $X_1 \subset X_2 \subset X_3 \subset \ldots$ of open relatively compact submanifolds such that X_N has volume $\lambda^{-1}N$ then $L^2(\Gamma_X; \mu_\lambda)$ is the limit as $N \to \infty$ of the spaces $L^2_{\text{sym}}((X_N)^N)$ of symmetric L^2 functions of N points in X_N. (This is explained in [4], [7], [8].)

(iii) U_λ has a more concrete realization as $L^2(E_\lambda)$, where E_λ is an affine space with a Gaussian measure (and an affine action of Diff(X)). E_λ is the space of $\frac{1}{2}$-densities f on X which are close to the standard Lebesgue $\frac{1}{2}$-density $f_\lambda = (\lambda\, dx)^{\frac{1}{2}}$ as $x \to \infty$, in the sense that $f - f_\lambda$ belongs to $H = L^2(X)$. This is an affine space associated to the vector space H and the cocycle β: $\mathrm{Diff}(X) \to H$ given by

$$\beta(\psi) = \lambda^{\frac{1}{2}}(J_\psi^{\frac{1}{2}} - 1),$$

where $J_\psi(x) = dm(\psi^{-1}x)/dm(x)$ as before. As we have seen when discussing the representations of G^X, $L^2(E_\lambda)$ can also be regarded as a Fock space $S(H) = \bigotimes_{n \geq 0} L^2_{\mathrm{sym}}(X^n)$, but with the natural action of Diff(X) twisted by the cocycle β. Because β vanishes on the subgroup Diff(X, m) of measure-preserving diffeomorphisms we see that in the Poisson case the representations associated to infinite configurations break up and give us nothing new when restricted to Diff(X, m).

In the paper the equivalence of $L^2(E_\lambda)$ and $L^2(\Gamma_X)$ is proved by considering the spherical functions, but it can also be described explicitly as a sequence of maps $L^2_{\mathrm{sym}}(X^n) \to L(\Gamma_X)$. In fact $L^2(X) \to L(\Gamma_X)$ takes $\lambda^{\frac{1}{2}} f$ to the function

$$\gamma \mapsto \sum_{X \in \gamma} f(x) - \lambda \int_X f(x)dx,$$

while $L^2_{\mathrm{sym}}(X \times X) \to L^2(\Gamma_X)$ takes λf to

$$\gamma \mapsto \sum_{x,y \in \gamma} f(x, y) - 2\lambda \sum_{x \in \gamma} \int_X f(x, y)dy + \lambda^2 \int\int_{X \times X} f(x, y)dxdy,$$

and so on.

The fact that there is a Gaussian realization of the representation is closely connected with the property of the Poisson measure μ_λ called "infinite divisibility". The latter means that if X is the disjoint union of two pieces X_1 and X_2, so that $\Gamma_X = \Gamma_{X_1} \times \Gamma_{X_2}$ up to sets of measure zero, then $\mu_\lambda = \mu_\lambda^{(1)} \times \mu_\lambda^{(2)}$, where $\mu_\lambda^{(i)}$ is the projection of μ_λ on Γ_{X_i}. This implies that when the representation U_λ of Diff(X) is restricted to the subgroup Diff(X_1) $\times$ Diff(X_2) it becomes $U_\lambda^{(X_1)} \otimes U_\lambda^{(X_2)}$, a property which must certainly be possessed by a construction of the type of $L^2(E_\lambda)$.

The reader may at first be confused by the fact that the affine action on $L^2(E_\lambda)$ used in this paper is the Fourier transform of the natural one used in the paper on G^X. Perhaps it is worth pointing out explicitly that if a group G acts orthogonally on a real vector space H with an inner product, and β: $G \to H$ is a cocycle, and $L^2(H)$ is formed using the standard Gaussian measure, then the following two unitary actions of G on $L^2(H)$ are unitarily equivalent

(a) $g \mapsto A_g$, where $(A_g\phi)(h) = e^{\frac{1}{2}\|h\|^2 - \frac{1}{2}\|h - \beta(g)\|^2}\ \phi(g^{-1}(h - \beta(g)))$,

(b) $g \mapsto B_g$, where $(B_g\phi)(h) = e^{i\langle\beta(g),h\rangle}\,\phi(g^{-1}h)$.

The automorphism of $L^2(H)$ relating them is characterized by

$$e^{\langle a,h\rangle-\frac{1}{2}\|a\|^2} \longleftrightarrow e^{i\langle a,h\rangle}$$

for all $a \in H$. The important thing to notice about it is that it takes polynomials to polynomials.

I shall conclude this account by drawing attention to the matters treated rather sketchily in Appendix 2, as I think they are interesting and deserve to be investigated further. The representations we obtained from Γ_X were constructed from a particularly simple family $\{H^\rho\}$ of irreducible unitary representations of the uncountable discrete group S^∞. But the group which seems more obviously relevant – because a diffeomorphism with compact support can move only finitely many points of a configuration – is the countable group S_∞ of the permutations of the natural numbers which leave almost all fixed. The representations H^ρ restrict to irreducible representations of S_∞; but most representations of S_∞, notably the one-dimensional sign representation, do not extend to S^∞. (There is a natural compact convex set of primary representations of S_∞ which has been elegantly described by Thoma [9]. It is the family of all primary representations which admit a finite trace. It contains the trivial representation, the sign representation, and the regular representation. All members are of type II except for the two one-dimensional representations.) Menikoff [8] has constructed a representation of Diff(X) corresponding to the sign representation of S_∞ as an N/V limit of the fermionic space $L^2_{\text{skew}}((S_N)^N)$ of antisymmetric functions of N particles in X_N.

Can one construct a representation of Diff(X) corresponding to any unitary representation H of S_∞? A possible method is described in Appendix 2. Let us choose an arbitrary rule for ordering the points of each configuration $\gamma \in \Gamma_X$. This gives us a map $s\colon \Gamma_X \to \widetilde{X}^\infty$ (where $\widetilde{X}^\infty$ is the space of ordered configurations), which clearly cannot be continuous. We require of the ordering only "correctness": if γ and γ' differ only in a compact region then the sequences $s(\gamma)$ and $s(\gamma')$ are required to coincide after finitely many terms. Consider the subspace $\Delta_s = S_\infty \cdot s(\Gamma_X)$ of $\widetilde{X}^\infty$. It is invariant under Diff(X) $\times$ S_∞, and $\Delta_s/S_\infty = \Gamma_X$. It was proved in §2.3 of the paper that for any quasi-invariant ergodic measure μ on Γ_X there is a quasi-invariant ergodic measure $\widetilde{\mu}$ on Δ_s. Then the space of S_∞-invariant maps $\Delta_s \to H$ which are square-summable with respect to $\widetilde{\mu}$ affords a unitary representation of Diff(X) associated to (μ, H, s). The extent of its dependence on the arbitrary and inexplicit choice of s is rather unclear, as is its relation to the N/V limit of the physicists. But the method does seem to produce, at least, a large supply of type II representations of Diff(X).

Vershik and Kerov [11] have proved that Thoma's family of representations can be obtained as limits of finite-dimensional representations of the finite symmetric groups. (One associates representations of S_{n+1} to representations of S_n by induction.) I imagine that this description should permit one both to construct the corresponding representations of Diff(X) as N/V limits, and to describe them in terms of a Gaussian measure.

Note. The definition of the topology of Γ_X given in the paper does not seem quite correct. One method of obtaining it is as follows.

The topology on the space B_X of finite configuration in X is obvious and uncontroversial. For a connected open manifold X the connected components of B_X are the $B_X^{(n)}$ for $n = 0, 1, 2, \ldots$ If Y is an open relatively compact submanifold of X let us topologize $\Gamma_{\bar{Y}}$ as a quotient space of B_X. Then if Y is connected so is $\Gamma_{\bar{Y}}$. Now define Γ_X as $\varprojlim \Gamma_{\bar{Y}}$, where Y runs through all open relatively compact submanifolds of X. This means that a configuration moves continuously precisely when it appears to move continuously to every observer with a bounded field of vision.

An alternative definition is: Γ_X has the coarsest topology such that $\tilde{f}: \Gamma_X \to \mathbf{R}$ is continuous for every continuous function $f: X \to \mathbf{R}$ with compact support, where $\tilde{f}(\gamma) = \sum_{X \in \gamma} f(x)$.

It is easy to see (cf. [6]) that Γ_X is metrizable, separable, and complete. On the other hand if $\tilde{X}^\infty$ is given the product topology then the map $\tilde{X}^\infty \to \Gamma_X$ is not continuous, and I do not see how to obtain Γ_X as a quotient space from a sensible topology on $\tilde{X}^\infty$. (It does not seem, however, that the topology on $\tilde{X}^\infty$ plays a significant role in the paper.)

In conclusion, notice that the fundamental group of Γ_X is S^∞; but of course Γ_X is not locally simply connected. The map $\Gamma_{X,n} \to \Gamma_X$, for example, is a local homeomorphism, but not a locally trivial fibration.

References

[1] I. M. Gelfand, M. I. Graev and A. M. Veršik, Representations of the group of smooth mappings of a manifold X into a compact Lie group. Compositio Math., **35** (1977), 299–334.

[2] I. M. Gelfand, M. I. Graev and A. M. Veršik, Representations of the group of functions taking values in a compact Lie group. Compositio Math., **42** (1981), 217–243.

[3] G. Goldin, Non-relativistic current algebras as unitary representations of groups. J. Math. Phys., 12 (1971), 462–488.

[4] G. Goldin, K. J. Grodnik, R. Powers and D. Sharp, Non-relativistic current algebras in the N/V limit. J. Math. Phys., **15** (1974), 88–100.

[5] D. A. Kazhdan, The connection of the dual space of a group with the structure of its closed subgroups. Functsional. Anal. i Prilozhen **1** (1967), 71–74. = Functional Anal. Appl. **1** (1967), 63–66.

[6] K. Matthes, J. Kerstan and J. Mecke, Infinitely divisible point processes. John Wiley, 1978.

[7] R. Menikoff, The hamiltonian and generating functional for a non-relativistic local current algebra. J. Math. Phys., **15** (1974), 1138–1152.

[8] R. Menikoff, Generating functionals determining representations of a non-relativistic local current algebra in the N/V limit. J. Math. Phys., **15** (1974), 1394–1408.
[9] E. Thoma, Die unzerlegboren, positiv-definiten Klassenfunktionen der abzählbar unendlichen symmetrischen Gruppe. Math. Z., **85** (1964), 40–61.
[10] A. M. Vershik, I. M. Gelfand and M. I. Graev, Irreducible representations of the group G^X and cohomology. Functsional. Anal. i Prilozhen., 8 (1974), 67–69. = Functional Anal. Appl., 8 (1974), 151–153.
[11] A. M. Vershik and S. V. Kerov, Characters and factor-representations of the infinite symmetric group. Doklady AN SSSR **257** (1981), 1037–1040.

REPRESENTATIONS OF THE GROUP SL (2, R), WHERE R IS A RING OF FUNCTIONS

Dedicated to
Andreĭ Nikolaevich Kolmogorov

A. M. Vershik, I. M. Gel'fand, and M. I. Graev

We obtain a construction of the irreducible unitary representations of the group of continuous transformations $X \to G$, where X is a compact space with a measure m and $G = PSL(2, \mathbf{R})$, that commute with transformations in X preserving m.

This construction is the starting point for a non-commutative theory of generalized functions (distributions). On the other hand, this approach makes it possible to treat the representations of the group of currents investigated by Streater, Araki, Parthasarathy, and Schmidt from a single point of view.

Contents

Introduction

One stimulus to the present work was the desire to extend the theory of generalized functions to the non-commutative case. Let us explain what we have in mind.

Let $\mathbf{R}$ be the real line, X a compact manifold, and $f(x)$ an infinitely differentiable function on X with values in $\mathbf{R}$, that is, a mapping $X \to \mathbf{R}$. A group structure arises naturally on the set of functions $f(x)$, which we denote by $\mathbf{R}^X$. Irreducible unitary representations of this group are defined by the formula $f(x) \mapsto e^{il(f)}$, where l is a linear functional in the space of "test" functions $f(x)$. Thus, to each generalized function (distribution) there corresponds an *irreducible* representation of $\mathbf{R}^X$. If we replace $\mathbf{R}$ by any other Lie group G, then it is natural to ask for the construction of *irreducible* unitary representations of the group G^X, regarded as a natural non-commutative analogue to the theory of distributions. Such an attempt was made in [1], §3.

However, our progress was only partial. We succeeded in defining distributions with support at a single point or at a finite number of points (for the group $SU(2)$) – analogues to the delta function and its derivatives; we were also able to introduce the concept of a derivative and show that δ' is the derivative of δ. The work came to a halt because we did not succeed in introducing the concept of an integral, without which the theory of generalized functions cannot go on.

The problem of constructing an integral for G^X can be stated as follows. Suppose that an X measure m is given. We have to find *irreducible* unitary representations of G^X that go over into equivalent ones under transformations of X preserving m. Reducible representations of this kind can be constructed without special difficulty. However, even the case $G = \mathbf{R}$ indicates that for our purposes reducible representations are unsuitable.

For a long time it was not clear to the authors whether such irreducible representations exist for semisimple groups G. Finally we succeeded in constructing such representations for a number of semisimple groups, namely, groups in which the identity representation is not isolated in the set of all irreducible unitary representations.

In this paper we analyze in detail only the case of the group $SL(2, \mathbf{R})$. The fact is, as experience with representation theory shows, that an understanding of any new situation is impossible without a preliminary study of the group SL_2 from all points of view.

We have performed the construction of the integral several times, each time from a somewhat different standpoint. The order in which we have written down the various constructions corresponds more or less to the order in which we thought them out. The first construction proceeds from a very simple idea: to obtain the multiplicative integral as the limit of a tensor product of representations, each member of the product being a closer approximation to the identity representation than the last, more precisely, to the point of the representation space to which the identity representation is attached.

From the last few sections it is clear that this representation can also be interpreted in terms of the cocycles of Streater, Araki, Parthasarathy and Schmidt (more precisely, it is not the 1-cocycles that play the fundamental role, but rather the reducible representations from Ext^1). The proof of the irreducibility of these representations is a new feature in our constructions.

At the end of §6 we construct two other projective unitary representations of the group $(PSL(2, \mathbf{R}))^X$.

The construction of the integral for all other groups G^X for which G satisfies the condition that the identity representation is not isolated in the set of all irreducible representations will be presented elsewhere.

The integral constructed in this paper provides us with a constructive representation of the group G^X, which in the terminology of mathematical physics is the group of currents of G.

Thus, this paper can also be regarded as a survey, from a somewhat different standpoint, of work on the representations of the group of currents.

Representations of the group of currents have been widely studied by a number of authors (Streater, Araki, Parthasarathy and Schmidt). See [4], [7]–[11], [13], [14], [15], and the further literature cited in the survey papers [6] and [12].

§1. Some information on the representations of the group of real 2 X 2 matrices

1. Representations of the supplementary series. We consider here the group $G = PSL(2, \mathbf{R})$ of real matrices $g = \begin{pmatrix} \alpha & \beta \\ \gamma & \delta \end{pmatrix}$ with determinant 1 in which g and $-g$ are identified. This group is known to be isomorphic to the group of complex matrices of the form $\begin{pmatrix} \alpha & \beta \\ \bar{\beta} & \bar{\alpha} \end{pmatrix}$, where $|\alpha|^2 - |\beta|^2 = 1$ and g and $-g$ are again identified. In what follows we use either the first or the second definition of G, as convenient.

Let G be given in the second form.

We introduce the space K of continuous infinitely differentiable functions on the unit circle $|\zeta| = 1$ in the complex plane. With each real number λ in the interval $0 < \lambda < 1$ we associate a representation T_λ of G in K.

$$\left(T_\lambda \begin{pmatrix} \alpha & \beta \\ \bar{\beta} & \bar{\alpha} \end{pmatrix} f\right)(\zeta) = f\left(\frac{\alpha\zeta + \bar{\beta}}{\beta\zeta + \bar{\alpha}}\right) |\beta\zeta + \bar{\alpha}|^{\lambda - 2}. \tag{1}$$

In K there is a positive definite Hermitian form $(f_1, f_2)_\lambda$ that is invariant under the operators T_λ:[1)]

$$(f_1, f_2)_\lambda = \frac{\Gamma\left(1 - \frac{\lambda}{2}\right)}{4\sqrt{\pi}\,\Gamma\left(\frac{1-\lambda}{2}\right)} \int_0^{2\pi}\int_0^{2\pi} \left|\sin\frac{\varphi_1 - \varphi_2}{2}\right|^{-\lambda} f_1(e^{i\varphi_1})\overline{f_2(e^{i\varphi_2})}\, d\varphi_1\, d\varphi_2 \tag{2}$$

We denote by H_λ the completion of K in the norm $\|f\|_\lambda^2 = (f, f)_\lambda$.

---

1) The numerical factor is chosen so that $(\mathbf{1}, \mathbf{1})_\lambda = 1$, where $\mathbf{1}$ is the function identically equal to unity.

It is evident that the T_λ can be extended to unitary operators in H_λ.

DEFINITION. A unitary representation in the Hilbert space H_λ, as defined by (1), is called a representation T_λ of the supplementary series of G.

It is known that all representations T_λ, $0 < \lambda < 1$, of the supplementary series are irreducible and pairwise inequivalent.[1)]

It is sometimes convenient to specify the representations of the supplementary series in another manner. Let G be given in the first form. For each λ, $0 < \lambda < 1$, we introduce the space $\mathscr{D}_\lambda$ of continuous real functions such that $f(x) = O(|x|^{\lambda-2})$ as $x \to \pm\infty$. In $\mathscr{D}_\lambda$ we introduce the positive definite Hermitian form $(f_1, f_2)_\lambda$:

$$(3) \qquad (f_1, f_2)_\lambda = \int_{-\infty}^{+\infty}\int_{-\infty}^{+\infty} |x_1 - x_2|^{-\lambda} f_1(x_1)\overline{f_2(x_2)}\, dx_1\, dx_2.$$

A representation of the supplementary series acts in the Hilbert space obtained by completing $\mathscr{D}_\lambda$ in the norm $\|f\|_\lambda^2 = (f, f)_\lambda$. The representation operators have the following form:

$$(4) \qquad \left(T_\lambda\begin{pmatrix}\alpha & \beta\\ \gamma & \delta\end{pmatrix} f\right)(x) = f\left(\frac{\alpha x+\gamma}{\beta x+\delta}\right)|\beta x+\delta|^{\lambda-2}.$$

2. Canonical representations of G. Some unitary representations, which we call canonical, of the group G of 2×2 real matrices play an important role in our work. These very pretty representations of the matrix group are interesting for their own sake. We present two methods of specifying the canonical representations.

2a. THE FIRST METHOD. We specify G in the second form. Further, we let $K \subset G$ be the maximal compact subgroup in G consisting of the matrices of the form $\begin{pmatrix} e^{it} & 0 \\ 0 & e^{-it}\end{pmatrix}$.

Of fundamental importance for the first specification of the canonical representations is the function

$$\psi(g) = \left(\frac{4}{\operatorname{Sp} gg^* + 2}\right)^{1/2}.$$

THEOREM 1.1. *For any* $\lambda > 0$ *the function* $\psi^\lambda(g)$ *is positive definite on* G *and constant on the double cosets of* K.

PROOF. The fact that $\psi^\lambda(g)$ is constant on the double cosets of K is obvious. That it is positive definite is a consequence of the following two lemmas.

---

1) (1) defines a unitary representation T_λ also in the interval $1 < \lambda < 2$. The scalar product $(f_1, f_2)_\lambda$ in the space of this representation is defined as the analytic continuation of the function of λ defined in the domain $\operatorname{Re}\lambda < 1$ by the convergent integral (2).

The representations T_λ and $T_{2-\lambda}$, $0 < \lambda < 2$, are known to be equivalent; hence we can always restrict our attention to the interval $0 < \lambda < 1$.

We denote by $\phi_\lambda(g)$ the zonal spherical function of the representation T_λ of the supplementary series with respect to K, normalized so that $\phi_\lambda(e) = 1$.

LEMMA 1.1. *The function $\phi_\lambda(g)$ is continuous and differentiable in λ at $\lambda = 0$. Also $\lim_{\lambda\to 0} \phi_\lambda(g) = 1$ and*

$$\lim_{\lambda\to 0} \frac{d\varphi_\lambda(g)}{d\lambda} = \ln \psi(g).$$

The proof follows easily from the explicit form of $\phi_\lambda(g)$:

$$\varphi_\lambda(g) = P_{-\lambda/2}\left(\frac{1}{2}\operatorname{Sp}(gg^*)\right),$$

where

$$P_{-\lambda/2}(x) = \frac{1}{\pi}\int_0^\pi (x + \sqrt{(x^2 - 1} \cos \phi))^{-\lambda/2} d\phi$$

is the Legendre function.

LEMMA 1.2. *The function*

$$\psi^\mu(g) = \exp\left(\mu \frac{d\varphi_\lambda(g)}{d\lambda}\bigg|_{\lambda=0}\right)$$

is positive definite for $\mu > 0$.

PROOF. By Lemma 1.1, $\frac{d\varphi_\lambda(g)}{d\lambda}\Big|_{\lambda=0} = \lim_{\lambda\to 0} \frac{\varphi_\lambda(g)-1}{\lambda}$. It is evident that $\frac{\varphi_\lambda(g)-1}{\lambda}$ is conditionally positive definite, that is, $\sum_{i,j} \frac{\varphi_\lambda(g_i g_j^{-1})-1}{\lambda}\xi_i\bar{\xi}_j \geqslant 0$ under the condition that $\sum \xi_i = 0$. It then follows that $\exp\left(\mu\frac{\varphi_\lambda(g)-1}{\lambda}\right)$ is positive definite for $\mu > 0$. Since the positive definite functions form a weakly closed set, the limit $\exp\left(\mu \frac{d(\varphi_\lambda(g)-1)}{d\lambda}\Big|_{\lambda=0}\right) = \lim_{\lambda\to 0} \exp\left(\mu\frac{\varphi_\lambda(g)-1}{\lambda}\right)$ is positive definite.

DEFINITION. The unitary representation of G defined by the positive definite function $\psi^\lambda(g)$, $\lambda > 0$, is called *canonical.* A cyclic vector ξ_λ in the space of a canonical representation for which $(T(g)\xi_\lambda, \xi_\lambda) = \psi^\lambda(g)$ is called *canonical.*

Let us construct a canonical representation. We consider the space $Y = K\backslash G$, which is a Lobachevskii plane. Let y_0 be the point of Y that corresponds to the coset of the identity element. We define the kernel $\Psi^\lambda(y_1, y_2)$, where $\lambda > 0$, on the Lobachevskii plane by the formula

$$\Psi^\lambda(y_1, y_2) = \psi^\lambda(g_1 g_2^{-1}),$$

where $y_1 = y_0 g_1$, $y_2 = y_0 g_2$. By Theorem 1.1 this kernel is positive definite.

We consider the space of all finite continuous functions on Y. We denote by L_λ the completion of this space in the norm:

$$\|f\|^2 = \int \Psi^\lambda(y_1, y_2) f(y_1)\overline{f(y_2)}\, dy_1\, dy_2,$$

where dy is an invariant measure on Y.

A canonical representation of G is defined by operators in L_λ of the form

$$(T(g)f)(y) = f(yg).$$

(That the operators $T(g)$ are unitary and form a representation of G is obvious.)

THEOREM 1.2. *If $\lambda > 1$, then a canonical representation in L_λ splits into a direct integral over the representations of the principal continuous series of G. If $0 < \lambda < 1$, then*

$$L_\lambda = H_\lambda \oplus L_\lambda^0,$$

where H_λ is the space of the representation T_λ of the supplementary series, and L_λ^0 splits into representations of the principal continuous series only.

PROOF. It suffices to verify that $\psi^\lambda(g)$ can be expanded in zonal spherical functions of the corresponding irreducible representations. We may limit our attention to the matrices $g = \begin{pmatrix} \cosh t & \sinh t \\ \sinh t & \cosh t \end{pmatrix}$. For these matrices we have $\psi^\lambda(g) = \left(\frac{2}{\cosh 2t + 1}\right)^{\lambda/2}$. Furthermore, we know that the zonal spherical functions of the representations of the principal continuous series have the following form:

$$\varphi_{1+i\rho}(g) = P_{\frac{-1-i\rho}{2}}(\cosh 2t),$$

where $P_{\frac{-1-i\rho}{2}}$ is the Legendre function.

Let $\lambda > 1$; then $\left(\frac{2}{x+1}\right)^{\lambda/2}$ is square integrable on $[1, \infty)$ and can therefore be expanded in an integral of functions $P_{\frac{-1-i\rho}{2}}(x)$ (the Fock-Mehler transform). Thus, we have

$$\psi^\lambda(g) = \int_0^\infty a_\lambda(\rho)\,\varphi_{1+i\rho}(g)\,d\rho.$$

The coefficients $a_\lambda(\rho)$ in this expression can be calculated by the inversion formula for the Fock-Mehler transform; we obtain

$$a_\lambda(\rho) = \frac{1}{2\pi^2}\,\frac{\Gamma\left(\lambda - \frac{1}{2} - i\rho\right)\Gamma\left(\lambda - \frac{1}{2} + i\rho\right)}{\Gamma^2(\lambda)}\,\rho\tanh \pi\rho$$

Now let $0 < \lambda < 1$. It is known that the zonal spherical function $\phi_\lambda(g) = P_{-\lambda/2}(\cosh 2t)$ of the representation T_λ of the supplementary series has the following asymptotic form:

$$P_{-\lambda/2}(x) = \frac{2^{-\lambda/2}\,\Gamma\left(\frac{1-\lambda}{2}\right)}{\sqrt{\pi}\,\Gamma\left(1-\frac{\lambda}{2}\right)}\,x^{-\lambda/2} + O\left(\frac{1}{x}\right) \quad \text{as } x \to \infty.$$

It follows that the function

$$\left(\frac{2}{x+1}\right)^{\lambda/2} - \frac{\sqrt{\pi}\, 2^{\lambda}\Gamma\left(1-\frac{\lambda}{2}\right)}{\Gamma\left(\frac{1-\lambda}{2}\right)} P_{-\lambda/2}(x)$$

is square integrable on $[1, \infty)$ and can therefore be expanded in an integral of functions $P_{\frac{-1-i\rho}{2}}(x)$.

In what follows we say that the canonical representation L_λ for $0 < \lambda < 1$ is congruent to the representation H_λ of the supplementary series modulo representations of the principal series.

The explicit separation of the component of the supplementary series in L_λ will be carried out a little later (see Theorem 1.3).

In conclusion we give another two expressions for the kernel $\Psi^\lambda(y_1, y_2) = \psi^\lambda(g_1 g_2^{-1})$, where $y_1 = y_0 g_1$, $y_2 = y_0 g_2$.

From the definition of $\psi(g)$ it follows easily that

$$\Psi^\lambda(y_1, y_2) = \cosh^{-\lambda}\rho(y_1, y_2),$$

where $\rho(y_1, y_2)$ is the invariant metric on the Lobachevskii plane.

We suppose further that the Lobachevskii plane Y is realized as the interior of the unit disk $|z| < 1$ in the complex plane, where G acts by fractional linear transformations: $z \to \frac{\alpha z+\bar{\beta}}{\beta z+\bar{\alpha}}$. It is easy to verify that in this realization Ψ^λ has the following form:

$$\Psi^\lambda(z_1, z_2) = \left[\frac{(1-|z_1|^2)(1-|z_2|^2)}{|1-z_1\bar{z}_2|^2}\right]^{\lambda/2}.$$

We observe that the invariant measure on the unit disk is $(1-|z|^2)^{-2}dz\, d\bar{z}$. Thus, in the realization on the unit disk the norm in the space of the canonical representation has the following form:

$$(5) \quad \|f\|^2 =$$
$$= \int \left[\frac{(1-|z_1|^2)(1-|z_2|^2)}{|1-z_1\bar{z}_2|^2}\right]^{\lambda/2} (1-|z_1|^2)^{-2}(1-|z_2|^2)^{-2} f(z_1)\overline{f(z_2)}\, dz_1\, d\bar{z}_1\, dz_2\, d\bar{z}_2.$$

2b. A SECOND METHOD OF SPECIFYING A CANONICAL REPRESENTATION. Suppose that the Lobachevskii plane is realized as the interior of the unit disk $|z| < 1$. Then the norm in the space L_λ of a canonical representation is given by (5). If we now go over from the functions $f(z)$ to the functions $(1 - |z|^2)^{\frac{1}{2}\lambda-2}f(z)$, then we obtain a new and very convenient realization of a canonical representation.

In this realization the space L_λ of a canonical representation is the completion of the space of finite (that is, vanishing close to the boundary) continuous functions in the unit disk $|z| < 1$ with respect to the norm

(6) $$\|f\|^2 = \int\limits_{|z_1|<1,\ |z_2|<1} |1-z_1\bar{z}_2|^{-\lambda} f(z_1)\overline{f(z_2)}\,dz_1\,d\bar{z}_1\,dz_2\,d\bar{z}_2.$$

The representation operators act according to the formula

(7) $$\left(T\begin{pmatrix}\alpha & \beta\\ \bar{\beta} & \bar{\alpha}\end{pmatrix} f\right)(z) = f\left(\frac{\alpha z+\bar{\beta}}{\beta z+\bar{\alpha}}\right)|\beta z+\bar{\alpha}|^{\lambda-4}.$$

We note that for finite continuous functions $f(z)$ in the disk $|z| < 1$ the norm $\|f\|^2$ can be written in the following convenient form:

(8) $$\|f\|^2 = \sum_{m,\,n=0}^{\infty} c_{mn}(\lambda)\left|\int\limits_{|z|<1} f(z)\,z^m\bar{z}^n\,dz\,d\bar{z}\right|^2,$$

where

$$c_{mn}(\lambda) = \frac{\Gamma\left(\frac{\lambda}{2}+m\right)\Gamma\left(\frac{\lambda}{2}+n\right)}{m!\,n!\left(\Gamma\left(\frac{\lambda}{2}\right)\right)^2}.$$

In particular, if f depends only on the modulus r, then the norm $\|f\|^2$ takes the following simple form:

$$\|f\|^2 = 2\pi\sum_{n=0}^{\infty} c_{nn}(\lambda)\left|\int_0^1 f(r)\,r^{2n+1}\,dr\right|^2.$$

To derive (8) it is sufficient to represent the kernel $|1 - z_1\bar{z}_2|^{-\lambda}$ as the product $(1 - z_1\bar{z}_2)^{-\lambda/2}(1 - \bar{z}_1 z_2)^{-\lambda/2}$, and then to expand each factor in a binomial series.

This space is very interesting. We shall see now that it contains a large store of generalized functions.

We consider the space K of test functions $\phi(z)$ that are continuous and infinitely differentiable in the closed disk $|z| \leqslant 1$. Let $l(\phi)$ be a generalized function, that is, a linear functional on K. From l we construct a new functional in the space K_0 of functions $f(z)$ that are infinitely differentiable and finite in the disk $|z| < 1$ (that is, vanish near the boundary):

$$(l, f) = \sum_{m,\,n=0}^{\infty} c_{mn}(\lambda)\,l(z^m\bar{z}^n)\int\limits_{|z|<1} f(z)\,z^m\bar{z}^n\,dz\,d\bar{z}.$$

If this series converges absolutely and $|(l, f)| \leqslant c\,\|f\|$, then the functional (l, f) can be extended to a continuous linear functional in L_λ and therefore specifies an element l in L_λ.

We claim that the delta function $\xi_\lambda = \delta(z)$ concentrated at the point $z = 0$ belongs to L_λ. In fact, since $\xi_\lambda(1) = 1$ and $\xi_\lambda(z^m\bar{z}^n) = 0$ for $m + n > 0$, we have

(9) $$(\xi_\lambda, f) = (\delta, f) = \int f(z)\,dz\,d\bar{z}.$$

Consequently, by the definition of the norm in L_λ, $|(\xi_\lambda, f)| \leqslant \|f\|$ and hence $\delta(z) \in L_\lambda$.

It can be shown that ξ_λ is a canonical vector in L_λ, that is,

$$(T(g)\xi_\lambda, \xi_\lambda) = \psi^\lambda(g).$$

This is easily derived from (9) if we replace f by $T(g)f_n$, where f_n is a sequence of test functions converging to ξ_λ, and then proceed to the limit.[1]

It is not difficult to verify that L_λ contains not only $\delta(z)$, but also all its derivatives, $\delta^{(m,n)}(z) = \dfrac{\partial^{m+n}\delta(z)}{\partial z^m\,\partial\bar{z}^n}$. In particular, all derivatives $\delta^{(2n+1)}(r) = \dfrac{d^{2n+1}\delta(r)}{dr^{2n+1}}$ lie in the subspace of L_λ consisting of functions[2] that depend only on the modulus r.

We now look at the generalized functions $l = a(z)\delta(1 - |z|^2)$, where $a(z)$ is a continuous function on the circle $|z| = 1$ (that is,

$$l(\varphi) = \frac{1}{2}\int_0^{2\pi} a(e^{it})\,\varphi(e^{it})\,dt)$$

---

[1] We can obtain this result formally by substituting in (9)

$$f(z) = T(g)\,\xi_\lambda = \delta\left(\frac{\alpha z + \bar{\beta}}{\beta z + \bar{\alpha}}\right)|\beta z + \bar{\alpha}|^{\lambda-4} = \delta\left(z + \frac{\bar{\beta}}{\alpha}\right)|\alpha|^{-\lambda}.$$

[2] The functions $\delta^{(m,n)}(z)$ form an orthogonal basis in L_λ. Thus, each element l of this space can be written in the form

$$l = \sum_{m,n=0}^{\infty} b_{mn}\frac{\partial^{m+n}\delta(z)}{\partial z^m\,\partial\bar{z}^n},$$

with

$$\|l\|^2 = \sum_{m,n=0}^{\infty} (m!\,n!)^2\,c_{mn}(\lambda)\,|b_{mn}|^2.$$

In particular, in the subspace of functions depending only on the modulus r there is an orthogonal basis consisting of the functions $\delta^{(2n+1)}(r) = \dfrac{d^{2n+1}\delta(r)}{dr^{2n+1}}$. Thus, any element of this subspace can be written in the form $l = \sum_{n=0}^{\infty} b_n \dfrac{d^{2n+1}\delta(r)}{dr^{2n+1}}$, with

$$\|l\|^2 = \sum_{n=0}^{\infty} [(2n+1)!]^2\,c_{mn}(\lambda)\,|b|^2.$$

LEMMA 1.3. *The functions* $l = a(z)\,\delta(1-|z|^2)$ *belong to the space* L_λ *for* $0<\lambda<1$.

PROOF. Since

$$l(z^m\bar{z}^n) = \frac{1}{2}\int_0^{2\pi} a(e^{it})\, e^{i(m-n)t}\, dt = \alpha_{m-n},$$

we have

$$|(l, f)| \leqslant \sum_{m,n=0}^{\infty} c_{mn}(\lambda)\,|\alpha_{m-n}|\left|\int_{|z|<1} f(z)\, z^m\bar{z}^n\, dz\, d\bar{z}\right|.$$

Hence, by the Cauchy inequality

$$|(l, f)| \leqslant \Big(\sum_{m,n=0} c_{mn}(\lambda)\,|\alpha_{m-n}|^2\Big)^{1/2}\,\|f\|.$$

Thus, it remains to prove the convergence of the series

$$\sum_{m,n=0}^{\infty} c_{mn}(\lambda)\,|\alpha_{m-n}|^2 = \sum_{k=-\infty}^{+\infty} |\alpha_k|^2\Big(\sum_{m-n=k} c_{mn}(\lambda)\Big). \tag{10}$$

To do so we use the following estimate for the coefficients $c_{mn}(\lambda) = \dfrac{\Gamma\left(\frac{\lambda}{2}+m\right)\Gamma\left(\frac{\lambda}{2}+n\right)}{m!\,n!\,(\Gamma(\lambda/2))^2}$:

$$c_{mn}(\lambda) \leqslant C m^{\frac{\lambda}{2}-1} n^{\frac{\lambda}{2}-1}.$$

If follows from this estimate that for $0<\lambda<1$ the series $\sum_{n=0}^{\infty} c_{nn}(\lambda)$ converges. On the other hand, it is not hard to see that $\sum_{m-n=k} c_{mn}(\lambda) \leqslant \sum_{n=0}^{\infty} c_{nn}(\lambda)$ and, therefore, $\sum_{m-n=k} c_{mn}(\lambda) \leqslant C_1$, where C_1 does not depend on k. Since $\sum_{k=0}^{\infty} |\alpha_k|^2$ converges, the convergence of the series (10) follows.[1)]

[1)] We note that the function $\delta(z)$ can also be obtained by a limit passage. Let us consider the sequence of functions $f_n \in L_\lambda$ of the form

$$f_n(z) = \begin{cases} c_n & \text{for } |z| < \frac{1}{n}, \\ 0 & \text{for } \frac{1}{n} < |z| < 1, \end{cases}$$

where $c_n > 0$ is defined by the condition $\|f_n\| = 1$. It is easy to verify that this is a fundamental sequence in L_λ and that its limit is equal to $\delta(z)$. Similarly, we can obtain $\delta(1-|z|^2)$ as the limit of a fundamental sequence of functions of the form

$$f_n(z) = \begin{cases} c_n, & \text{for } \frac{n-1}{n} < |z| < \frac{n}{n+1}, \\ 0 & \text{if } |z| < \frac{n-1}{n} \quad \text{or} \quad \frac{n}{n+1} < |z| < 1 \end{cases}$$

where c_n is determined from the relation $\int f_n(z)\, dz\, d\bar{z} = \pi$.

We state without proof two simple propositions on the functions $\xi_\lambda = \delta(z)$ and $l = f(z)\,\delta(1 - |z|^2)$.

LEMMA 1.4. *Let* $l_1 = f_1(z)\,\delta(1 - |z|^2)$, $l_2 = f_2(z)\,\delta(1 - |z|^2)$, *then*

$$(11)\qquad 1)\ (l_1, l_2)_\lambda = 2^{-\lambda-2}\int_0^{2\pi}\int_0^{2\pi}\left|\sin\frac{\varphi_1-\varphi_2}{2}\right|^{-\lambda} f_1(e^{i\varphi_1})\,\overline{f_2(e^{i\varphi_2})}\,d\varphi_1\,d\varphi_2.$$

$$(12)\qquad 2)\ (\delta(z),\ \delta(1-|z|^2))_\lambda = \pi.$$

From (11), it follows, in particular, that

$$(13)\qquad \|\delta(1-|z|^2)\|^2 = 2^{-\lambda}\pi^{3/2}\,\frac{\Gamma\left(\frac{1-\lambda}{2}\right)}{\Gamma\left(1-\frac{\lambda}{2}\right)}.$$

LEMMA 1.5. *The representation operator* $T(g)$, *where* $g = \begin{pmatrix}\alpha & \beta\\ \bar\beta & \bar\alpha\end{pmatrix}$, *takes the function* $f(z)\delta(1 - |z|^2)$ *into* $f\left(\frac{\alpha z+\bar\beta}{\beta z+\bar\alpha}\right)|\beta z + \bar\alpha|^{\lambda-2}\delta(1 - |z|^2)$.

The next result follows immediately from Lemmas 1.4 and 1.5.

THEOREM 1.3. *For* $0 < \lambda < 1$ *the subspace* $H_\lambda \subset L_\lambda$ *generated by the functions* $f(z)\,\delta(1 - |z|^2)$ *is invariant and the representation of G acting in it is the representation* T_λ *of the supplementary series.*

In what follows we find it useful to know the projection of the canonical vector ξ_λ onto H_λ. This projection is obviously invariant under the maximal compact subgroup and is consequently proportional to $\delta(1 - |z|^2)$. Let

$$\eta_\lambda = c(\lambda)\delta(1 - |z|^2), \qquad \|\eta_\lambda\| = 1,$$

where $c(\lambda) = \|\delta(1 - |z|^2)\|^{-1}$. Then, according to Lemma 1.4, 2) we have $(\xi_\lambda, \eta_\lambda) = \pi c(\lambda)$; consequently the projection of the canonical vector ξ_λ onto the subspace H_λ is equal to $\pi c(\lambda)\eta_\lambda$, where $c^2(\lambda) = \|\delta(1 - |z|^2)\|^{-2} = 2^\lambda\pi^{-3/2}\,\Gamma(1 - \lambda/2)/\Gamma(\frac{1}{2}(1 - \lambda))$.

The next result follows easily from standard estimates for $\Gamma(\lambda)$.

LEMMA 1.6. $(\xi_\lambda, \eta_\lambda) = 1 + O(\lambda^2)$ *as* $\lambda \to 0$.

COROLLARY. *As* $\lambda \to 0$, *the distance from the canonical vector* ξ_λ *in* L_λ *to* H_λ *is* $O(\lambda^2)$.

3. Theorems on tensor products of representations of G. THEOREM 1.4. *Let* T_{λ_1} *and* T_{λ_2} *be two representations of the supplementary series. If* $\lambda_1 + \lambda_2 < 1$, *then* $T_{\lambda_1} \otimes T_{\lambda_2} = T_{\lambda_1+\lambda_2} \oplus T$, *where* T *splits into representations of the principal continuous and the discrete series only.*

If $\lambda_1 + \lambda_2 \geqslant 1$, *then* $T_{\lambda_1} \otimes T_{\lambda_2}$ *splits into representations of the principal continuous and the discrete series only.*

The tensor product of two irreducible unitary representations of which at least one belongs to the principal continuous or the discrete series splits into representations of the principal continuous and the discrete series only.

For the proof see [12].

We now let $G_n = \underbrace{G \times \ldots \times G}_{n}$. Since G is of type 1, it is standard knowledge that any irreducible unitary representation T of G_n can be obtained as follows. We are given irreducible unitary representations $T^{(1)}, \ldots, T^{(n)}$ of G, acting in Hilbert spaces $H_1, \ldots, H_n$, respectively. A representation T of G_n acts in the tensor product $H_1 \otimes \ldots \otimes H_n$ according to the following formula:

$$(14) \qquad T(g_1, \ldots, g_n)(\xi_1 \otimes \ldots \otimes \xi_n) = (T^{(1)}(g_1)\xi_1) \otimes \ldots \otimes (T^{(n)}(g_n)\xi_n).$$

Here two representations T' and T'' of G_n are equivalent if and only if all the corresponding representations $T'^{(i)}$ and $T''^{(i)}$ of G are equivalent, $i = 1, \ldots, n$.

We say that a representation T of G_n of the form (14) is *purely of the supplementary series* if all the $T^{(i)}$ are representations of G of the supplementary series. In this case, if $T^{(i)} = T_{\lambda_i}$, $i = 1, \ldots, n$, then the corresponding representations of the group G_n are denoted by $T_{\lambda_1, \ldots, \lambda_n}$, and the representation space by $H_{\lambda_1, \ldots, \lambda_n}$.

The next theorem follows from the one stated above.

THEOREM 1.5. *Let $T_{\lambda'_1, \ldots, \lambda'_n}$ and $T_{\lambda''_1, \ldots, \lambda''_n}$ be two representations of G_n purely of the supplementary series: and let $\lambda'_i + \lambda''_i < 1$, $i = 1, \ldots, n$. Then $T_{\lambda'_1, \ldots, \lambda'_n} \otimes T_{\lambda''_1, \ldots, \lambda''_n} = T_{\lambda'_1 + \lambda''_1, \ldots, \lambda'_n + \lambda''_n} \oplus T$, where in the decomposition of T into irreducible representations there are no representations purely of the supplementary series.*

In what follows we find it useful to specify explicitly an embedding of $H_{\lambda_1 + \lambda_2} (\lambda_1 + \lambda_2 < 1)$ in the tensor product $H_{\lambda_1} \otimes H_{\lambda_2}$.

Let G be defined in the first form. Then $H_{\lambda_1 + \lambda_2}$ is the completion of the space of finite continuous real functions $f(x)$ with the norm

$$\| f \|^2 = \int_{-\infty}^{+\infty} \int_{-\infty}^{+\infty} | x - x' |^{-\lambda_1 - \lambda_2} f(x) \overline{f(x')} \, dx \, dx'.$$

Now $H_{\lambda_1} \otimes H_{\lambda_2}$ is the completion of the space of finite continuous functions $F(x_1, x_2)$ of two variables with the norm

$$\| F \|^2 = \int | x_1 - x'_1 |^{-\lambda_1} | x_2 - x'_2 |^{-\lambda_2} F(x_1, x_2) \overline{F(x'_1, x'_2)} \, dx_1 \, dx_2 \, dx'_1 \, dx'_2.$$

In $H_{\lambda_1} \otimes H_{\lambda_2}$ there are many generalized functions. The precise meaning of this statement is the following.

Let $l(\tilde{F})$ be a linear functional on the space of infinitely differentiable functions $\tilde{F}(x_1, x_2)$ such that $| \tilde{F}(x_1, x_2) | < C(1 + x_1^2)^{-\lambda_1/2} (1 + x_2^2)^{-\lambda_2/2}$. By means of l we construct the following linear functional $\tilde{l}$ on the space of finite infinitely differentiable functions $F(x_1, x_2)$: $(\tilde{l}, F) = l(\tilde{F})$, where

$$\tilde{F}(x_1, x_2) = \int_{-\infty}^{+\infty} \int_{-\infty}^{+\infty} | x_1 - x'_1 |^{-\lambda_1} | x_2 - x'_2 |^{-\lambda_2} F(x'_1, x'_2) \, dx'_1 \, dx'_2.$$

If the functional $(\tilde{l}, F)$ is defined and continuous in the norm of $H_{\lambda_1} \otimes H_{\lambda_2}$, then we identify the generalized function l with the vector $\tilde{l} \in H_{\lambda_1} \otimes H_{\lambda_2}$.

LEMMA 1.7. *If* $\lambda_1 > 0$, $\lambda_2 > 0$, $\lambda_1 + \lambda_2 < 1$ *and* l *is a generalized function having the form* $l = f(x_1)\,\delta\,(x_1 - x_2)$ ($f(x)$ *finite*), *that is,*

$$l(F) = \int_{-\infty}^{+\infty} F(x, x) f(x)\, dx,$$

then the functional $(\tilde{l}, F)$ *is continuous in the norm of* $H_{\lambda_1} \otimes H_{\lambda_2}$.

This lemma is proved by standard calculations involving the Fourier transform, and we omit the proof. Next we can establish the following lemma.

LEMMA 1.8. *Let* $\tilde{l}_1$ *and* $\tilde{l}_2$ *be defined by the generalized functions* $l_1 = f_1(x_1)\,\delta\,(x_1 - x_2)$ *and* $l_2 = f_2(x_1)\,\delta\,(x_1 - x_2)$. *Then*

$$(l_1, l_2) = \int_{-\infty}^{+\infty} \int_{-\infty}^{+\infty} |x_1 - x_2|^{-\lambda_1 - \lambda_2} f_1(x_1) f_2(x_2) dx_1 dx_2 .$$

THEOREM 1.6. *If* $\lambda_1 > 0$, $\lambda_2 > 0$, $\lambda_1 + \lambda_2 < 1$, *then the mapping defines an isometric embedding of* $H_{\lambda_1 + \lambda_2}$ *in* $H_{\lambda_1} \otimes H_{\lambda_2}$, *consistent with the action of* G *on these spaces.*

This theorem follows immediately from the lemmas stated above.

We need a somewhat more general theorem, which can be proved similarly.

THEOREM 1.7. *Let* $\lambda_1, \lambda_2, \ldots, \lambda_k > 0$ and $\lambda_1 + \ldots + \lambda_k < 1$. *Then the mapping*

$$f(x) \longmapsto \delta(x_1 - x_k, \ldots, x_{k-1} - x_k) f(x_k)$$

defines an isometric embedding

$$H_{\lambda_1 + \cdots + \lambda_k} \to H_{\lambda_1} \otimes \ldots \otimes H_{\lambda_k},$$

consistent with the action of G.

The meaning of the concepts and mappings introduced is the same as that explained earlier for $k = 2$.

The mappings indicated above are consistent; namely, if

$$\lambda = \sum_i \lambda_i, \quad \lambda_i = \sum_j \lambda_{ij}, \quad \lambda_{ij} > 0, \quad \lambda < 1,$$

then the mapping

$$H_\lambda \to \bigotimes_{i,\, j} H_{\lambda_{ij}}$$

is the composition of the mappings

$$H_\lambda \to \bigotimes_i H_{\lambda_i} \text{ and } H_{\lambda_i} \to \bigotimes_j H_{\lambda_{ij}}.$$

§2. Construction of the multiplicative integral of representations of $G = PSL(2, \mathbf{R})$.

Let $G = PSL(2, \mathbf{R})$ and let X be a compact topological space with a given measure m. We define a group operation on the set of functions

$g: X \to G$ as pointwise multiplication: $(g_1g_2)(x) = g_1(x)g_2(x)$. We define G^X as the group of all continuous functions $g: X \to G$ with the topology of uniform convergence.

We give here a construction of an irreducible unitary representation of the group G^X, which we call the *multiplicative integral of representations of G.*

Following the definition of the integral as closely as possible, we replace G^X by the group of step functions and define an integral on it as a tensor product of representations. As we decrease the length of the intervals of subdivision and simultaneously allow the parameter on which the representations in the tensor product depend to approach a certain limit, we obtain a representation of G^X, which we call the integral of representations. It is remarkable that the integral of representations is an irreducible representation.

We now proceed to precise definitions.

1. Definition of the group G^0. For every Borel subset $X' \subset X$ we denote by $G_{X'}$ the group of functions $g: X \to G$ that are constant on X' and equal to 1 on the complement of X'. It is obvious that there exists a natural isomorphism $G_{X'} \cong G$.

A partition ν: $X = \cup X_i$ of X into finitely many disjoint Borel subsets is called *admissible.*

On the set of admissible partitions we define an ordering, setting $\nu_1 < \nu_2$ if ν_2 is a refinement of ν_1. It is obvious that the set of admissible partitions is directed (that is, for any ν_1 and ν_2 there exists a ν such that $\nu_1 \subset \nu$ and $\nu_2 < \nu$).

For any admissible partition ν: $X = \bigcup_{i=1}^{k} X_i$ we denote by G_ν the group of functions $g: X \to G$ that are constant on each of the subsets X_i. It is obvious that

$$G_\nu = G_{X_1} \times \ldots \times G_{X_k}.$$

Observe that for $\nu_1 < \nu_2$ there is a natural embedding: $G_{\nu_1} \to G_{\nu_2}$. We define the group of step functions G^0 as the inductive limit of the G_ν:

$$G^0 = \lim_{\longrightarrow} G_\nu.$$

In this section we construct a representation of G^0. We make the transition from this representation to a representation of G^X in §3.

2. Construction of a representation of G^0. Let m be a positive finite measure on X, defined on all Borel subsets of X. We always assume that m is countably additive.

Let us consider the Hilbert spaces in which the representations T_λ of the supplementary series act. In §1 we denoted these spaces with the action of G defined on them by H_λ, $0 < \lambda < 1$. Next, we denote by H_0 the one-dimensional space in which the identity representation of G acts.

Let ν: $X = \bigcup_{i=1}^{k} X_i$ be an arbitrary admissible partition such that

$\lambda_i = m(X_i) < 1$. We set

$$\mathcal{H}_\nu = H_{\lambda_1} \otimes \dots \otimes H_{\lambda_k}.$$

In $\mathcal{H}_\nu$ we define a representation of the group

$$G_\nu = G_{X_1} \times \dots \times G_{X_k},$$

supposing that $G_{X_i} \cong G$ ($i = 1, \dots, k$) acts in H_{λ_i}.

The representation of G_ν in $\mathcal{H}_\nu$ is irreducible (see §1.3); we have agreed to call such representations purely of the supplementary series.

Note that since $G_{\nu_1} \subset G_{\nu_2}$ for $\nu_1 < \nu_2$, a representation of each of the groups $G_{\nu'}$, $\nu' < \nu$, is also defined in $\mathcal{H}_\nu$.

LEMMA 2.1. *If $\nu_1 < \nu_2$, then $\mathcal{H}_{\nu_2}$ splits into the direct sum of subspaces invariant under G_{ν_1}: $\mathcal{H}_{\nu_2} = \mathcal{H}_{\nu_1} \oplus \mathcal{H}'$ where $\mathcal{H}'$ does not contain invariant subspaces in which a representation of G_{ν_1} purely of the supplementary series acts.*

PROOF. Let $\nu_2 > \nu_1$, that is, ν_1: $X = \bigcup_{i=1}^{k} X_i$, ν_2: $X = \bigcup_{i,j} X_{ij}$, where $X_i = \bigcup_j X_{ij}$. We set $\lambda_i = m(X_i)$, $\lambda_{ij} = m(X_{ij})$; thus, $\mathcal{H}_{\nu_1} = \bigotimes_i H_{\lambda_i}$, $\mathcal{H}_{\nu_2} = \bigotimes_{i,j} H_{\lambda_{ij}}$. We also set $\mathcal{H}^i_{\nu_2} = \bigotimes_j H_{\lambda_{ij}}$; then $\mathcal{H}_{\nu_2} = \bigotimes_i \mathcal{H}^i_{\nu_2}$. It is evident that G_{X_i} acts diagonally in $\mathcal{H}^i_{\nu_2} = \bigotimes_j H_{\lambda_{ij}}$ (that is, acts simultaneously on each factor $H_{\lambda_{ij}}$). Thus, the representation of $G_{X_i} \cong G$ in $\mathcal{H}^i_{\nu_2}$ is a tensor product of representations $T_{\lambda_{ij}}$ of the supplementary series.

From this it follows that $\mathcal{H}^i_{\nu_2} = H_{\lambda_i} \oplus H^i_{\nu_2}$, $\lambda_i = \sum_j \lambda_{ij}$, where H_{λ_i} is the space in which the representation T_{λ_i} of the supplementary series acts, and $H^i_{\lambda_2}$ splits only into representations of the principal, the continuous, and the supplementary series (see §1.3). Forming the tensor product of the spaces $\mathcal{H}^i_{\nu_2}$ and bearing in mind $\bigotimes_i H_{\lambda_i} = \mathcal{H}_{\nu_1}$ and $G_{X_1} \times \dots \times G_{X_k} = G_{\nu_1}$, we obtain: $\mathcal{H}_{\nu_2} = \mathcal{H}_{\nu_1} \oplus \mathcal{H}'$, where $\mathcal{H}'$ does not contain representations purely of the supplementary series.

THEOREM 2.1. *There exist morphisms of Hilbert spaces*

$$j_{\nu_2\nu_1}: \mathcal{H}_{\nu_1} \to \mathcal{H}_{\nu_2},$$

defined for each pair $\nu_1 < \nu_2$ of admissible partitions of X satisfying the following conditions:

1) *$j_{\nu_2\nu_1}$ commutes with the action of G_{ν_1} in $\mathcal{H}_{\nu_1}$ and $\mathcal{H}_{\nu_2}$;*

2) *$j_{\nu_3\nu_2} \circ j_{\nu_2\nu_1} = j_{\nu_3\nu_1}$ for any $\nu_1 < \nu_2 < \nu_3$.*

These morphisms are determined uniquely to within factors $c_{\nu_2\nu_1}$ ($|c_{\nu_2\nu_1}| = 1$).

PROOF. From Lemma 2.1 it follows that for each pair $\nu_1 < \nu_2$ there exists a morphism $j_{\nu_2\nu_1}: \mathcal{H}_{\nu_1} \to \mathcal{H}_{\nu_2}$ that commutes with the action of G_{ν_1}, and that this morphism is uniquely determined to within a factor. We claim that the morphism $j_{\nu_2\nu_1}$ can be chosen so that condition 2) is satisfied.

Let $\nu_2 > \nu_1$, that is, ν_1: $X = \bigcup_{i=1}^{k} X_i$, ν_2: $X = \bigcup_{i,j} X_{ij}$, where $X_i = \bigcup_j X_{ij}$. We set $\lambda_i = m(X_i)$, $\lambda_{ij} = m(X_{ij})$; thus, $\mathcal{H}_{\nu_1} = \bigotimes_{i=1}^{k} H_{\lambda_i}$, $\mathcal{H}_{\nu_2} = \bigotimes_{i,j} H_{\lambda_{ij}}$.

For each $i = 1, 2, \ldots, k$ we define a mapping $H_{\lambda_i} \to \bigotimes_{i,j} H_{\lambda_{ij}}$, $\lambda_i = \sum_j \lambda_{ij}$, compatible with the action of $G_{X_i} \cong G$, just as this was done in §1.3. These mappings induce the mapping

$$j_{\nu_2\nu_1} : \mathcal{H}_{\nu_1} \to \mathcal{H}_{\nu_2},$$

which is compatible with the action of G_ν in these spaces.

From the definition of the mappings $H_{\lambda_i} \to \bigotimes_j H_{\lambda_{ij}}$ it follows easily that the mappings $j_{\nu_2\nu_1}$ so defined satisfy the compatibility requirement 2) of the Theorem.

REMARK. Another method of specifying the compatibility of the system of morphisms $j_{\nu_2\nu_1}$ will be given in §3.

DEFINITION. We assign to all possible pairs $\nu_1 < \nu_2$ of admissible partitions of X the morphisms $j_{\nu_2\nu_1} : \mathcal{H}_{\nu_1} \to \mathcal{H}_{\nu_2}$, which commute with the action of G_{ν_1} in $\mathcal{H}_{\nu_1}$ and $\mathcal{H}_{\nu_2}$ and satisfy the compatibility condition $j_{\nu_3\nu_2} \circ j_{\nu_2\nu_1} = j_{\nu_3\nu_1}$ for $\nu_1 < \nu_2 < \nu_3$. We introduce the space

$$\mathcal{H}^0 = \underrightarrow{\lim}\, \mathcal{H}_\nu,$$

and let $\mathcal{H}$ be the completion of $\mathcal{H}^0$ in the norm defined in $\mathcal{H}^0$.

Then there is a natural way of defining a unitary representation U of G^0 in $\mathcal{H}$.

(Specifically, we have the natural embeddings $G_\nu \hookrightarrow G^0$, $\mathcal{H}_\nu \hookrightarrow \mathcal{H}^0$. Let $\tilde{g} \in G^0$ and $\xi \in \mathcal{H}^0$. Then there exists an admissible partition ν such that $\tilde{g} \in G_\nu$, $\xi \in \mathcal{H}_\nu$, and we set $U_{\tilde{g}}\xi = T(\tilde{g})\xi$, where T is a representation operator of G_ν in $\mathcal{H}_\nu$. It is evident that this definition does not depend on the choice of ν. The unitary operator $U_{\tilde{g}}$ on $\mathcal{H}^0$ so constructed can be extended by continuity from $\mathcal{H}^0$ to its completion $\mathcal{H}$.)

LEMMA 2.2. *The representation $U_{\tilde{g}}$ does not depend on the choice of the morphisms $j_{\nu_2\nu_1}$.*

PROOF. Let $j'_{\nu_2\nu_1} : \mathcal{H}_{\nu_1} \to \mathcal{H}_{\nu_2}$ be another system of morphisms satisfying the conditions of the definition, and let $\mathcal{H}'^0 = \underrightarrow{\lim}\, \mathcal{H}_\nu$ be the inductive limit constructed with respect to this system of morphisms. We claim that the representations of G^0 in $\mathcal{H}^0$ and $\mathcal{H}'^0$ are equivalent.

By Theorem 2.1. we have $j'_{\nu_2\nu_1} = c_{\nu_2\nu_1} j_{\nu_2\nu_1}$, where $c_{\nu_2\nu_1}$ are numerical factors satisfying $c_{\nu_3\nu_2} c_{\nu_2\nu_1} = c_{\nu_3\nu_1}$ for $\nu_1 < \nu_2 < \nu_3$. It follows from this condition that $c_{\nu_2\nu_1} = c_{\nu_2\nu_0} c^{-1}_{\nu_1\nu_0}$, where ν_0 is a fixed admissible partition ($\nu_0 < \nu_1 < \nu_2$). For $\nu > \nu_0$ we specify isomorphisms of the spaces $\mathcal{H}_\nu \to \mathcal{H}_\nu$ in the following manner: $\xi \mapsto c_{\nu\nu_0}\xi$. Since $j'_{\nu_2\nu_1} = c_{\nu_2\nu_0} c^{-1}_{\nu_1\nu_0} j_{\nu_2\nu_1}$, they take

$j_{\nu_2\nu_1}$ into $j'_{\nu_2\nu_1}$ and consequently induce an isomorphism of the spaces $\mathcal{H}^0$ and $\mathcal{H}'^0$ compatible with the action of G^0.

THEOREM 2.2. *The representation U of G^0 in $\mathcal{H}$ is irreducible.*

PROOF. Let ν: $X = \bigcup_{i=1}^{k} X_i$ be an admissible partition such that $\lambda_i = m(X_i) < 1$. We restrict the representation of G^0 in $\mathcal{H}$ to G_ν.

An irreducible representation of G_ν acts in $\mathcal{H}_\nu = \bigotimes_{i=1}^{k} H_{\lambda_i}$. By Lemma 2.1 $\mathcal{H}_\nu$ occurs with multiplicity 1 in each $\mathcal{H}_{\nu'}$ for $\nu' > \nu$. Since $\mathcal{H}_\nu$ is irreducible, it follows that it occurs with multiplicity[1] in the whole space $\mathcal{H}$.

We now suppose that $\mathcal{H}$ splits into the direct sum of invariant subspaces, $\mathcal{H} = \mathcal{H}' \oplus \mathcal{H}''$. Then $\mathcal{H}_\nu$ is contained in one of the summands, for example, in $\mathcal{H}'$. Now let $\nu' > \nu$. Since $\mathcal{H}_{\nu'} \supset \mathcal{H}_\nu$ and an irreducible representation of $G_{\nu'}$ acts in $\mathcal{H}_{\nu'}$, we have $\mathcal{H}_{\nu'} \subset \mathcal{H}'$. Consequently, $\mathcal{H}'$ contains all the subspaces $\mathcal{H}_{\nu'}$, $\nu' > \nu$, and therefore coincides with $\mathcal{H}$. This completes the proof.

THEOREM 2.3. *Let m_1 and m_2 be two positive measures on X, $U^{(1)}$ and $U^{(2)}$ representations of G^0 defined on these measures. If $m_1 \neq m_2$, then the representations $U^{(1)}$ and $U^{(2)}$ are inequivalent.*

PROOF. We denote by $\mathcal{H}^{(1)}$ and $\mathcal{H}^{(2)}$ the representation spaces of $U^{(1)}$ and $U^{(2)}$. Since $m_1 \neq m_2$, there exists an admissible partition ν: $X = \bigcup_{i=1}^{k} X_i$ such that $\lambda_i^{(1)} = m_1(X_i) < 1$, $\lambda_i^{(2)} = m_2(X_i) < 1$ for $i = 1, \ldots, k$ and $m_1(X_i) \neq m_2(X_i)$ at least for one i.

We claim that the representations of $G_\nu \subset G^0$ in $\mathcal{H}^{(1)}$ and $\mathcal{H}^{(2)}$ are inequivalent. It then follows that the representations of the whole group G^0 in these spaces a fortiori are inequivalent.

Let us first consider the spaces

$$\mathcal{H}_\nu^{(1)} = \bigotimes_{i=1}^{k} H_{\lambda_i^{(1)}} \text{ and } \mathcal{H}_\nu^{(2)} = \bigotimes_{i=1}^{k} H_{\lambda_i^{(2)}},$$

in which the irreducible representations of G_ν act. Since $\lambda_i^1 \neq \lambda_i^2$ for at least one i, the representations of G_ν in $\mathcal{H}_\nu^{(1)}$ and $\mathcal{H}_\nu^{(2)}$ are inequivalent.

Furthermore, $\mathcal{H}_{\nu'}^{(2)} = \mathcal{H}_\nu^{(2)} \oplus \mathcal{H}'_\nu$, for every $\nu' > \nu$, and $\mathcal{H}'_\nu$ does not contain representations purely of the supplementary series of G_ν, therefore does not contain representations equivalent to $\mathcal{H}_\nu^{(1)}$ (see Lemma 2.1). Consequently the whole space $\mathcal{H}_{\nu'}^{(2)}$ does not contain representations equivalent to $\mathcal{H}_\nu^{(1)}$. But then $\mathcal{H}^{(2)}$ also does not contain representations of G_ν equivalent to $\mathcal{H}_\nu^{(1)}$. Since obviously $\mathcal{H}_\nu^{(1)} \subset \mathcal{H}^{(1)}$, the representations of G_ν in $\mathcal{H}^{(1)}$ and $\mathcal{H}^{(2)}$ are inequivalent.

§3. Another construction of the multiplicative integral of representations of $G = PSL(2, \mathbf{R})$.

The concept of the multiplicative integral of representations of $G = PSL(2, \mathbf{R})$ introduced in §2 can also be obtained starting out from the canonical representations of G. Here we explain this second method. It is surprising that although the representations in the product are significantly more "massive", their product turns out to be the same as before.

1. Construction of the representation. As before, let X be a compact topological space on which a positive finite measure m is given, defined on all Borel subsets and countably additive.

We consider the canonical representations of G in the Hilbert spaces L_λ introduced in §1.2. Next we denote by L_0 the one-dimensional space in which the identity representation T_0 of G acts.

We recall that in each space L_λ we have fixed a cyclic vector ξ_λ which we have called canonical. For this vector

$$(T(g)\xi_\lambda, \xi_\lambda) = \psi^\lambda(g),$$

where $\psi(g)$ is the function defined in §1.2.

With each admissible partition ν: $X = \bigcup_{i=1}^{k} X_i$, we associate a Hilbert space

$$\mathscr{L}_\nu = L_{\lambda_1} \otimes \ldots \otimes L_{\lambda_k},$$

where $\lambda_i = m(X_i)$. We define in $\mathscr{L}_\nu$ a unitary representation of

$$G_\nu = G_{X_1} \otimes \ldots \otimes G_{X_k},$$

assuming that each group $G_{X_i} \cong G$ acts in L_{λ_i} in accordance with the corresponding canonical representation, and trivially on the remaining factors L_{λ_j}, $j \neq i$. We observe that since $G_{\nu_1} \subset G_{\nu_2}$ for $\nu_1 < \nu_2$, an action of each of the groups $G_{\nu'}$, $\nu' < \nu$, is also defined in $\mathscr{L}_\nu$.

For each admissible partition ν: $X = \bigcup_{i=1}^{k} X_i$ we specify a vector $\xi_\nu \in \mathscr{L}_\nu$:

$$\xi_\nu = \xi_{\lambda_1} \otimes \ldots \otimes \xi_{\lambda_k},$$

where ξ_{λ_i} is the canonical vector in L_{λ_i}. It is obvious that ξ_ν is a cyclic vector in $\mathscr{L}_\nu$.

LEMMA 3.1. *For any pair of partitions $\nu_1 < \nu_2$ the mapping $\xi_{\nu_1} \mapsto \xi_{\nu_2}$ can be extended to a morphism*

$$j_{\nu_2\nu_1} : \mathscr{L}_{\nu_1} \to \mathscr{L}_{\nu_2},$$

which commutes with the action of G_{ν_1}.

PROOF. It is sufficient to verify that

$$(T(g_{\nu_1})\xi_{\nu_1}, \xi_{\nu_1})_{\mathscr{L}_{\nu_1}} = (T(g_{\nu_1})\xi_{\nu_2}, \xi_{\nu_2})_{\mathscr{L}_{\nu_2}}$$

for any $g_{\nu_1} \in G_{\nu_1}$, where the parentheses denote the scalar product in the corresponding space.

According to hypothesis we have ν_1: $X = \bigcup_{i=1}^{k} X_i$, ν_2: $X = \bigcup_{i,j} X_{ij}$, where $X_i = \bigcup_j X_{ij}$. Let $g_{\nu_1} \in G_{\nu_2}$, that is, $g_{\nu_1} = g_1 \ldots g_n$, where $g_i \in G_{X_i} \cong G$. Then

$$(1) \qquad (T(g_{\nu_1})\xi_{\nu_1}, \xi_{\nu_1})_{\mathscr{L}_{\nu_1}} = \coprod_{i=1}^{k} (T(g_i)\xi_{\lambda_i}, \xi_{\lambda_i})_{L_i} = \coprod_{i=1}^{k} \psi^{\lambda_i}(g_i),$$

where $\lambda_i = m(X_i)$. Similarly, let $g_{\nu_2} \in G_{\nu_2}$, that is, $g_{\nu_2} = \coprod_{i,j} g_{ij}$, where $g_{ij} \in G_{X_{ij}} \cong G$. Then

$$(2) \qquad (T(g_{\nu_2})\xi_{\nu_2}, \xi_{\nu_2})_{\mathscr{L}_{\nu_2}} = \coprod_{i,j} \psi^{\lambda_{ij}}(g_{ij}),$$

where $\lambda_{ij} = m(X_{ij})$.

If now $g_{\nu_2} = g_{\nu_1}$, this means that $g_{ij} = g_i$ for every $i = 1, \ldots, k$. In addition, since $\lambda_i = \sum_j \lambda_{ij}$, for any $i = 1, \ldots, k$ we have $\coprod_j \psi^{\lambda_{ij}}(g_{ij}) = \psi^{\lambda_i}(g_i)$. Consequently the expressions (1) and (2) are the same and the lemma is proved.

It is obvious that the morphisms $j_{\nu_2\nu_1}$ satisfy the compatibility condition $j_{\nu_3\nu_2} \circ j_{\nu_2\nu_1} = j_{\nu_3\nu_1}$ for $\nu_1 < \nu_2 < \nu_3$.

DEFINITION. We denote by $\mathscr{L}^0$ the inductive limit of the spaces $\mathscr{L}_\nu$, $\mathscr{L}^0 = \varinjlim \mathscr{L}_\nu$ and by $\mathscr{L}$, the completion of $\mathscr{L}^0$ in the norm defined in $\mathscr{L}^0$.

There is a natural way of defining in $\mathscr{L}$ a unitary representation of the group $G^0 = \varinjlim G_\nu$, which we denote by $U_{\tilde{g}}$.

2. Definition of a vacuum vector in $\mathscr{L}$. Let K be a maximal compact subgroup of G, and let $K^0 \subset G^0$ be the subgroup of step functions on X with values in K. Vectors in $\mathscr{L}$ of unit norm that are invariant under K^0 are called *vacuum* vectors.

We claim that a vacuum vector exists in $\mathscr{L}$.

Let j_ν be the natural mapping $\mathscr{L}_\nu \to \mathscr{L}$. It follows from the definition of ξ_ν that their images $j_\nu \xi_\nu$ in $\mathscr{L}$ coincide, that is, there exists a vector $\xi_0 \in \mathscr{L}$ such that $\xi_0 = j_\nu \xi_\nu$ for any admissible partition ν.

Since each vector $\xi_\nu \in \mathscr{L}_\nu$ is invariant under $K_\nu \subset G_\nu$, it follows that $\xi_0 = \varinjlim \xi_\nu$ is invariant under $K^0 = \varinjlim K_\nu$, so that it is a vacuum vector in $\mathscr{L}$.

In addition, since each $\xi_\nu \in \mathscr{L}_\nu$ is a cyclic vector in $\mathscr{L}_\nu$ under G_ν, $\xi_0 = \varinjlim \xi_\nu$ is a cyclic vector in $\mathscr{L}$ under $G^0 = \varinjlim G_\nu$.

In §3.3 we shall prove the uniqueness of the vacuum vector in $\mathscr{L}$ (Theorem 3.2).

Let us calculate the spherical function of the representation $U_{\tilde{g}}$:

$$\Psi(\tilde{g}) = (U_{\tilde{g}}\xi_0, \xi_0)_{\mathscr{L}},$$

where $\tilde{g} = g(\cdot) \in G^0$.

LEMMA 3.2.

$$\Psi(\tilde{g}) = \exp\left(\int_X \ln \psi(g(x))\, dm(x)\right),$$

where $\psi(g)$ is the function on G introduced in §1.2.

PROOF. For any $\tilde{g} = g(\cdot) \in G^0$ we can find an admissible partition ν: $X = \bigcup_{i=1}^{k} X_i$ of X such that $\tilde{g} \in G_\nu$, that is, $\tilde{g} = g_1 \ldots g_k$, where $g_i \in G_{X_i} \cong G$. But then, setting $\lambda_i = m(X_i)$, we have

$$(U_{\tilde{g}}\xi_0, \xi_0)_{\mathscr{L}} = (T(\tilde{g})\xi_\nu, \xi_\nu)_{\mathscr{L}_\nu} = \prod_{i=1}^{k} (T(g_i)\xi_{\lambda_i}, \xi_{\lambda_i})_{L_{\lambda_i}} =$$

$$= \prod_{i=1}^{k} \psi^{\lambda_i}(g_i) = \exp\left(\sum_{i=1}^{k} \ln(\psi(g_i))\, m(X_i)\right) =$$

$$= \exp\left(\int \ln(\psi(g(x))\, dm(x)\right).$$

3. Equivalence of the two constructions of the representations of G^0. We claim that the unitary representation of G^0 in $\mathscr{L}$ constructed here is equivalent to the representation in $\mathscr{H}$ constructed in §2.

LEMMA 3.3. *There is an embedding morphism $\mathscr{H} \to \mathscr{L}$ that is compatible with the action of G^0.*

PROOF. In §1.2 we have shown that for $0 < \lambda < 1$ the canonical representation L_λ of G is congruent to the representation H_λ of the supplementary series modulo representations of the principal continuous series; in other words, $L_\lambda = H_\lambda \oplus L'_\lambda$, where L'_λ can be expanded in an integral over representations of the principal continuous series.

Hence it follows that the representation space $\mathscr{L}_\nu = \bigotimes_{i=1}^{k} L_{\lambda_i}$ of $G_\nu = G_{X_1} \times \ldots \times G_{X_k}$, where ν: $X = \bigcup_{i=1}^{k} X_i$, $\lambda_i = m(X_i) < 1$, splits into a direct sum of invariant subspaces:

$$\mathscr{L}_\nu = \mathscr{H}_\nu \oplus \mathscr{L}'_\nu,$$

where $\mathscr{H}_\nu = \bigotimes_{i=1}^{k} H_{\lambda_i}$, and $\mathscr{L}'_\nu$ for any $\nu' < \nu$ does not contain representations purely of the supplementary series of $G_{\nu'}$.

This also shows that under the morphism $j_{\nu_2\nu_1}: \mathscr{L}_{\nu_1} \to \mathscr{L}_{\nu_2}$, $\nu_1 < \nu_2$, the subspace $\mathscr{H}_{\nu_1} \subset \mathscr{L}_{\nu_1}$ is mapped into $\mathscr{H}_{\nu_2} \subset \mathscr{L}_{\nu_2}$. Consequently, $\mathscr{L}^0 = \varinjlim \mathscr{L}_\nu$ contains $\mathscr{H}^0 = \varinjlim \mathscr{H}_\nu$. Going over to the completions, we see that $\mathscr{L}$ contains $\mathscr{H}$.

REMARK. We have incidentally constructed a compatible system of morphisms $j_{\nu_2\nu_1}: \mathscr{H}_{\nu_1} \to \mathscr{H}_{\nu_2}$, $\nu_1 < \nu_2$, that commute with the action of G_{ν_1}.

In §2 we have given another method of specifying such a compatible system of morphisms. However, the method given there has the advantage that it does not make use of the concept of a vacuum vector and does not depend on the choice of a maximal compact subgroup.

THEOREM 3.1. *The morphism $\mathcal{H} \to \mathcal{L}$ defined above is an isomorphism. Thus, the representations of G^0 in $\mathcal{H}$ and $\mathcal{L}$ are equivalent.*

PROOF. Since the vacuum vector $\xi_0 \in \mathcal{L}$ is cyclic in $\mathcal{L}$, it is sufficient for us to verify that ξ belongs to $\mathcal{H}$. To do this we find the projection of ξ_0 onto each $\mathcal{H}_\nu$.

Let ν: $X = \bigcup_{i=1}^{k} X_i$ be an arbitrary admissible partition such that $\lambda_i = m(X_i) < 1$. Then we represent ξ_0 as an element of $\mathcal{L}_\nu = \bigotimes_{i=1}^{k} L_{\lambda_i}$ in the form $\xi_0 = \bigotimes_{i=1}^{k} \xi_{\lambda_i}$, where ξ_{λ_i} is the canonical vector in L_{λ_i}.

According to §1.2b the projection of $\xi_\lambda \in L_\lambda$ onto $H_\lambda \subset L_\lambda$ is equal to $c_\lambda \eta_\lambda$, where η_λ is the fixed unit vector of H_λ explicitly constructed there; here $1 - c_\lambda = O(\lambda^2)$ as $\lambda \to 0$.

Hence it follows that the projection of $\xi_0 = \bigotimes_{i=1}^{k} \xi_{\lambda_i}$ onto $\mathcal{H}_\nu = \bigotimes_{i=1}^{k} H_{\lambda_i}$ is equal to

$$\eta'_\nu = \left(\prod_{i=1}^{k} c_{\lambda_i}\right) \eta_\nu,$$

where $\eta_\nu = \bigotimes_{i=1}^{k} \eta_{\lambda_i}$ is a vector of unit norm.

From the estimate $c_\lambda = 1 + O(\lambda^2)$ it follows that for an indefinite refinement of ν, as $\max \lambda_i$ tends to zero, the norm of η'_ν tends to 1, hence that η'_ν itself tends to ξ_0, and the theorem is proved.

THEOREM 3.2. *The vacuum vector in $\mathcal{L}$ is uniquely determined to within a factor.*

PROOF. Let ξ'_0 and ξ''_0 be two vacuum vectors in $\mathcal{L}$, and let η'_ν and η''_ν be their projections onto $\mathcal{H}_\nu = \bigotimes_{i=1}^{k} H_{\lambda_i}$, where ν: $X = \bigcup_{i=1}^{k} X_i$, $\lambda_i = m(X_i) < 1$. Since ξ'_0 and ξ''_0 are invariant under $K_\nu = K_{X_1} \times \ldots \times K_{X_k} \subset G_\nu$, their projections η'_ν and η''_ν onto $\mathcal{H}_\nu$ have the same property. But in $\mathcal{H}_\nu$ there is, to within a factor, only one vector that is invariant under K_ν; consequently, η'_ν and η''_ν are proportional.

On the other hand, since $\mathcal{H}^0 = \varinjlim \mathcal{H}_\nu$ is everywhere dense in $\mathcal{L}$ (by Theorem 3.1), the vectors η'_ν and η''_ν converge, respectively, to ξ'_0 and ξ''_0 when ν is refined indefinitely provided that $\max \lambda_i \to 0$. Consequently, since η'_ν and η''_ν are proportional, so are the limit vectors ξ'_0 and ξ''_0. This completes the proof.

REMARK. We emphasize that the vacuum vector ξ_0 is contained in each subspace $\mathcal{L}_\nu$ whereas it is not contained in any of the subspaces $\mathcal{H}_\nu$.

4. A representation of G^X. So far we have constructed a representation $U_{\tilde{g}}$ of the group G^0 of step functions $X \to G$. We now show how a representation of the group G^X of continuous functions on $X \to G$ can be defined in terms of this representation. Namely, we claim that the representation $U_{\tilde{g}}$ of G^0 (second construction) can be extended to a representation of a complete metric group containing both G^0 and G^X as everywhere dense subgroups. This then defines an irreducible unitary representation of G^X.

For simplicity we assume further that the support of m is the whole space X.

We first construct a certain metric on G^0. Let $\rho(y_1, y_2)$ be the invariant metric on the Lobachevskii plane $Y = K\backslash G$. We define on G a metric $d(g_1, g_2)$ invariant under right translations and such that

$$d(g_1, g_2) \geqslant \rho(y_0 g_1, y_0 g_2)$$

for any $g_1, g_2 \in G$, where y_0 is the point on the Lobachevskii plane Y that corresponds to the unit coset. (Such a metric exists; for example, we may set $d(g_1, g_2) = \rho(y_0 g_1, y_0 g_2) + \rho(y_1 g_1, y_1 g_2)$, where $y_1 \neq y_0$.) We now introduce a metric δ on the group G^0 of step functions, setting

$$\delta(g_1(\cdot), g_2(\cdot)) = \int d(g_1(x), g_2(x)) dm(x).$$

Completing G^0 in this metric we obtain a complete metric group $\bar{G}^X$, consisting of all m-measurable functions $g(\cdot)$ for which

$$\int d(g(x), e)\, dm(x) < \infty.$$

Observe that the completion of G^0 in the metric δ contains, in particular, the group G^X of continuous functions; G^X is everywhere dense in this completion.

We claim that the representation $U_{\tilde{g}}$ of G^0 constructed above can be extended to a representation of $\bar{G}^X$.

For this purpose we consider the functional Ψ on G^0 introduced earlier:

$$\Psi(\tilde{g}) = (U_{\tilde{g}} \xi_0, \xi_0)_{\mathscr{L}},$$

where $\xi_0 \in \mathscr{L}$ is a vacuum vector. It was shown above that

$$\Psi(\tilde{g}) = \exp\left(\int \ln \psi(g(x))\, dm(x) \right). \tag{3}$$

LEMMA 3.4. *The functional $\Psi(\tilde{g})$ can be extended from G^0 to a continuous functional on the whole group $\bar{G}^X$.*

PROOF. It is sufficient to verify that $\Psi(\tilde{g})$ in (3) is defined and continuous on the whole group $\bar{G}^X$.

We use the following expression for $\psi(g)$ introduced in §1.2a: $\psi(g) = \cosh^{-1}\rho(y_0 g, g)$.

From this expression it follows that $|\ln \psi(g)| \leqslant \rho(y_0 g, y_0) \leqslant d(g, e)$, therefore the integral $\int \ln \psi(g(x))dm(x)$ converges absolutely.

The continuity of $\Psi(\tilde{g})$ follows immediately from the following estimate:

$$\left| \int \ln \psi(g_1(x))dm(x) - \int \ln \psi(g_2(x))dm(x) \right| \leqslant \delta(g_1(\cdot), g_2(\cdot)).$$

We shall prove this inequality. We set $\tau_i(x) = \rho(y_0 g_i(x), y_0)$ and use the bound $\left| \ln \frac{\cosh \tau_1}{\cosh \tau_2} \right| \leqslant \left| \tau_1 - \tau_2 \right|$. We have

$$\left| \int \ln \psi(g_1(x))dm(x) - \int \ln \psi(g_2(x))dm(x) \right| \leqslant \int |\ln \psi(g_1(x)) - \ln \psi(g_2(x))| dm(x) =$$

$$= \int \left| \ln \frac{\cosh \tau_1(x)}{\cosh \tau_2(x)} \right| dm(x) \leqslant \int |\tau_1(x) - \tau_2(x)| dm(x) \leqslant \int \rho(y_0 g_1(x), y_0 g_2(x)) dm(x) \leqslant$$

$$\leqslant \int d(g_1(x), g_2(x))dm(x) = \delta(g_1(\cdot), g_2(\cdot)).$$

THEOREM 3.3. *The representation $U_{\tilde{g}}$ of G^0 can be extended by continuity to a unitary representation of $\tilde{G}^X$.*

This follows immediately from the preceding lemma and the following proposition.

LEMMA 3.5. *Let G be a topological group satisfying the first axiom of countability, $G^0 \subset G$ a subgroup everywhere dense in G, and T a continuous unitary representation of G^0 in a Hilbert space H with a cyclic vector ξ. Further, let $\Phi(g) = (T(g)\xi, \xi)$, $g \in G^0$. If $\Phi(\cdot)$ can be extended to a continuous function on G, then the representation T of G^0 can be extended by continuity to a unitary representation of G.*

PROOF OF THE LEMMA. We first show that for any $\eta_1, \eta_2 \in H$ the function $\Phi_{\eta_1\eta_2}(g) = (T(g)\eta_1, \eta_2)$ can be extended by continuity from G^0 to G.

Let H_0 be the space consisting of finite linear combinations of vectors $T(g)\xi$, $g \in G^0$. Since ξ is a cyclic vector, H_0 is everywhere dense in H.

It is evident that if $\eta_1, \eta_2 \in H_0$, then $\Phi_{\eta_1\eta_2}(\cdot)$ can be extended by continuity from G^0 to G. For if $\eta_1 = \Sigma a_i T(g_i')\xi$, $\eta_2 = \Sigma b_j T(g_j'')\xi$, then

$$\Phi_{\eta_1\eta_2}(g) = \sum a_i \bar{b}_j \Phi(g_j''^{-1} g g_i').$$

We now let η_1, η_2 be arbitrary; without loss of generality we may suppose that $\|\eta_1\| = \|\eta_2\| = 1$. For any $\eta_1', \eta_2' \in H$, $\|\eta_1'\| = \|\eta_2'\| = 1$, we have

$$|\Phi_{\eta_1\eta_2}(g) - \Phi_{\eta_1'\eta_2'}(g)| \leqslant \|\eta_1 - \eta_1'\| + \|\eta_2 - \eta_2'\|.$$

Hence the family $\Phi_{\eta_1\eta_2}(g)$ is equicontinuous in η_1, η_2, and since it can be extended to all $g \in G$ for an everywhere dense set of vectors η_1, η_2, this proves that $\Phi_{\eta_1\eta_2}(\cdot)$ can be extended to G for all η_1, η_2.

We claim that the operators $T(g)$, $g \in G$, so obtained are unitary, that is,

$(T(g)\xi, T(g)\xi) = (\xi, \xi)$ for any $\xi \in H$. Let $\{g_n\}$ be a sequence of elements of G^0 that converges to g.

For any $\varepsilon > 0$ there exists an N such that for $m, n > N$

$$|(T(g_m)\xi, T(g_n)\xi) - (\xi, \xi)| < \varepsilon \tag{4}$$

(since $(T(g_m)\xi, T(g_n)\xi) = (T(g_n g_m)\xi, \xi)$, hence converges to (ξ, ξ)).

On the other hand, we can find m and n, greater than N, such that

$$|(T(g_m)\xi, T(g_n)\xi) - (T(g)\xi, T(g_n)\xi)| + \\ + |(T(g)\xi, T(g_n)\xi) - (T(g)\xi, T(g)\xi)| < \varepsilon. \tag{5}$$

Comparing (4) and (5), we see that

$$|(T(g)\xi, T(g)\xi) - (\xi, \xi)| < 2\varepsilon,$$

which proves that $T(g)$ is unitary.

From the fact that the $T(g)$ are unitary and weakly convergent it follows that $\|T(g_n)\xi - T(g)\xi\| \to 0$, as $g_n \to g$, $g_n \in G^0$. From this it follows automatically that $T(g_1)T(g_2)\xi = T(g_1 g_2)\xi$ for all $g_1, g_2 \in G$.

Thus, we have constructed an irreducible unitary representation of the complete metric group $\bar{G}^X$ in $\mathcal{L}$. Restricting it to the everywhere dense subgroup G^X of $\bar{G}^X$, we obtain the required irreducible unitary representation of G^X in $\mathcal{L}$.

Since G^X is dense in $\bar{G}^X$, the vacuum vector ξ_0 is also cyclic relative to G^X, and for every $\tilde{g} \in G^X$ we have

$$(U_{\tilde{g}}\xi_0, \xi_0) = \exp\left(\int \ln \psi(g(x))dm(x)\right). \tag{6}$$

We see that the metric δ does not figure in (6). Hence our representation of G^X does not depend on the choice of the metric δ in the construction.

5. The representation $U_{\tilde{g}}$ commutes with the transformations of X that preserve the measure m. We consider continuous transformations $\sigma: x \mapsto x^\sigma$ of X. They induce automorphisms of G^X:

$$\tilde{g} = g(\cdot) \mapsto \tilde{g}^\sigma = g^\sigma(\cdot), \quad \text{where } g^\sigma(x) = g(x^{\sigma^{-1}}).$$

If $U_{\tilde{g}}$ is the representation of G^X constructed in this section and σ is an arbitrary continuous transformation of X, then we can define a new representation $U_{\tilde{g}}^\sigma$ of G^X by setting $U_{\tilde{g}}^\sigma = U_{\tilde{g}^\sigma}$.

THEOREM 3.4. *The representation $U_{\tilde{g}}^\sigma$ is equivalent to the representation of G^X defined in terms of the measure m^σ on X, where $m^\sigma(X') = m(X'^\sigma)$ for any measurable subset $X' \subset X$. In particular, if σ preserves m, then $U_{\tilde{g}}^\sigma$ and $U_{\tilde{g}}$ are equivalent.*

PROOF. Let $U'_{\tilde{g}}$ be the representation of G^X defined in terms of the measure m^σ on X. We compare the spherical functions $(U'_{\tilde{g}}\xi_0, \xi_0)$ and $(U_{\tilde{g}}^\sigma\xi_0, \xi_0)$, where ξ_0 is the vacuum vector. On the one hand,

$$(U'_{\tilde{g}}\xi_0, \xi_0) = \exp\left(\int \ln\cosh^{-1}\psi(g(x))dm^\sigma(x)\right).$$

On the other hand,

$$(U^\sigma_{\tilde{g}}\xi_0, \xi_0) = (U_{\tilde{g}^\sigma}\xi_0, \xi_0) = \exp\left(\int \ln\cosh^{-1}\psi(g(x^{\sigma^{-1}}))dm(x)\right) =$$

$$= \exp\left(\int \ln\cosh^{-1}\psi(g(x))dm^\sigma(x)\right).$$

Thus, $(U'_{\tilde{g}}\xi_0, \xi_0) = (U^\sigma_{\tilde{g}}\xi_0, \xi_0)$ for any $\tilde{g} \in G^X$, hence $U'_{\tilde{g}}$ and $U^\sigma_{\tilde{g}}$ are equivalent.

6. Invariant definition of a canonical representation. In §1 a canonical representation of G was defined constructively. We wish to demonstrate that its connection with the representation of G^X in $\mathcal{L}$ we have constructed is not accidental.

There is a natural embedding of G in G^X. When we restrict our representation of G^X in $\mathcal{L}$ to G, we obtain a certain representation of G. We look at the vacuum vector ξ_0 in $\mathcal{L}$, that is, the vector ξ_0, $\|\xi_0\| = 1$, that is invariant under K^X. We consider the minimal G-invariant subspace that contains ξ_0 and denote it by L_λ. It follows from the construction performed in §3.1 that L_λ *is a canonical representation of* G^X with $\lambda = m(X)$. We draw attention to the following interesting fact.

If in $\mathcal{L}$ we consider the restriction of the representation of G^X to G, naturally embedded in G^X, then for $\lambda = m(X) < 1$ there is precisely one representation of the supplementary series H_λ, with $\lambda = m(X)$, that occurs as a discrete component in the decomposition. The orthogonal complement to this space splits into representations only of the principal continuous and the discrete series. For $m(X) > 1$ this representation of the supplementary series is absent. This follows from the construction of the representation of G^X carried out in §2.

§4. A representation of G^X associated with the Lobachevskii plane.

Here we give an explicit form of the multiplicative integral of representations of $G = PSL(2, \mathbf{R})$.

1. Construction of a representation of G^X. Let X be a compact topological space, m a positive finite measure on X defined on all Borel subsets and countably additive. For simplicity we assume that the support of m is the whole space X.

Let Y be a Lobachevskiĭ plane on which the group of motions G acts transitively. We consider the set Y^X of all continuous mappings $\tilde{y} = y(\cdot)$: $X \to Y$. We introduce the linear space $\mathcal{H}^0$, whose elements are formal finite linear combinations of such mappings:

$$\sum \lambda_i \circ \tilde{y}_i, \quad \lambda_i \in \mathbf{C}, \quad \tilde{y}_i \in Y^X.$$

In other words, $\mathcal{H}^0$ is a free linear space over $\mathbf{C}$ with Y^X as a set of generators.

We introduce a scalar product in $\mathcal{H}^0$. Let $\rho(y_1, y_2)$ be the invariant metric on the Lobachevskiĭ plane. For any pair of mappings in Y^X, $\tilde{y}_1 = y_1(\cdot)$ and $\tilde{y}_2 = y_2(\cdot)$ we set

$$(\tilde{y}_1, \tilde{y}_2) = \exp\left(\int \ln \cosh^{-1} \rho(y_1(x), y_2(x)) dm(x)\right)$$

and then extend this scalar product by linearity to the whole space $\mathcal{H}^0$. The Hermitian form so defined on $\mathcal{H}^0$ is positive definite (for a proof see the end of §4.2 below). Let $\mathcal{H}$ be the completion of $\mathcal{H}^0$ in the norm $\|\xi\|^2 = (\xi, \xi)$.

We define a unitary representation $U_{\tilde{g}}$ of G^X in $\mathcal{H}$. For this purpose we observe first that an action of G^X on the set Y^X of continuous mappings $X \to Y$ is naturally defined. Namely, an element $\tilde{g} = g(\cdot) \in G^X$ takes $\tilde{y} = y(\cdot)$ into $\tilde{y}\tilde{g} = y_1(\cdot)$, where $y_1(x) = y(x)g(x)$.

We assign to each $\tilde{g} \in G^X$ the following operator $U_{\tilde{g}}$ in $\mathcal{H}^0$:

$$U_{\tilde{g}}\left(\sum \lambda_i \circ \tilde{y}_i\right) = \sum \lambda_i \circ (\tilde{y}_i \tilde{g}^{-1}).$$

THEOREM 4.1. *The operators $U_{\tilde{g}}$ are unitary on $\mathcal{H}^0$ and form a representation of G^X.*

PROOF. The fact that the operators $U_{\tilde{g}}$ form a representation is obvious. That they are unitary follows immediately from the invariance of $\rho(y_1, y_2)$ on Y.

Since the operators $U_{\tilde{g}}$ are unitary on $\mathcal{H}^0$, they can be extended to unitary operators in the whole space $\mathcal{H}$. So we have constructed a unitary representation of G^X in $\mathcal{H}$.

2. Realization in the unit disk. We provide explicit expressions for the scalar product in $\mathcal{H}^0$ and for the operator $U_{\tilde{g}}$ when Y is realized as the interior of the unit disk $|z| < 1$.

Let G be given as the group of matrices $g = \begin{pmatrix} \alpha & \beta \\ \bar{\beta} & \bar{\alpha} \end{pmatrix}$, let Y be the interior of the unit disk $|z| < 1$, and let G act in the unit disk in the following manner: $z \to zg^{-1} = \frac{\alpha z + \beta}{\bar{\beta} z + \bar{\alpha}}$.

Then $\mathcal{H}^0$ is the space of finite formal linear combinations

$$\sum \lambda_i \circ z_i(\cdot),$$

where $z(\cdot)$ are continuous mappings of X into the unit disk $|z| < 1$. For a pair of mappings $z_1(\cdot)$ and $z_2(\cdot)$ the scalar product in $\mathcal{H}^0$ has the following form:

$$(z_1(\cdot), z_2(\cdot)) = \exp \int \ln \left(\frac{(1 - |z_1(x)|^2)(1 - |z_2(x)|^2)}{|1 - z_1(x)\overline{z_2(x)}|^2} \right)^{1/2} dm(x). \tag{1}$$

The representation operator $U_{\tilde{g}}$, $\tilde{g} = \begin{pmatrix} \alpha(\cdot) & \beta(\cdot) \\ \overline{\beta(\cdot)} & \overline{\alpha(\cdot)} \end{pmatrix}$, takes $z(\cdot)$ into $z(\cdot)\tilde{g}^{-1} = z_1(\cdot)$, where

$$z_1(x) = \frac{\alpha(x) z(x) + \beta(x)}{\overline{\beta(x)} z(x) + \overline{\alpha(x)}}. \tag{2}$$

We indicate another convenient realization of the representation (1). (It can be obtained from the first by the transformation $z(\cdot) \to \lambda(z(\cdot)) \circ z(\cdot)$, where $\lambda(z(\cdot)) = \exp \int \ln (1 - |z(x)|^2)^{-1} dm(x)$.)

In this realization, as before, the elements of $\mathcal{H}^0$ are formal finite linear combinations of continuous transformations of X into the unit disc $|z| < 1$:

$$\sum \lambda_i \circ z_i(\cdot);$$

but the scalar product has the simpler form:

$$(3) \qquad (z_1(\cdot), z_2(\cdot)) = \exp \int \ln |1 - z_1(x)\overline{z_2(x)}|^{-1} dm(x).$$

The representation operator $U_{\tilde{g}}$ is given by the formula:

$$U_{\tilde{g}} z(\cdot) = \exp\Big(\int \ln |\overline{\beta(x)} z(x) + \overline{\alpha(x)}|^{-1} dm(x)\Big) \circ z_1(\cdot),$$

where $z_1(x)$ is defined by (2).

In conclusion we verify that the Hermitian form introduced in $\mathcal{H}^0$ is positive definite. It is simplest to confirm this for the Hermitian form given by (3).

We introduce the following notation:

$$f_i(x, n) = \begin{cases} z_i^n(x) & \text{for } n > 0, \\ \overline{z_i^{|n|}(x)} & \text{for } n < 0, \end{cases}$$

$$F_i(x_1, \ldots, x_k; n_1, \ldots, n_k) = \prod_{s=1}^{k} f_i(x_s, n_s), \text{ where } i = 1, 2.$$

It is not hard to check that the scalar product (3) can be represented in the following form:

$$(4) \quad (z_1(\cdot), z_2(\cdot)) = \sum_{k=0}^{\infty} \sum_{n_1, \ldots, n_k \, (n_i \neq 0)} \frac{2^{-k}}{k!} \frac{1}{|n_1 \ldots n_k|} \times$$

$$\times \int F_1(x_1, \ldots, x_k; n_1, \ldots, n_k) \overline{F_2(x_1, \ldots, x_k; n_1, \ldots, n_k)} \, dm(x_1) \ldots dm(x_k).$$

To obtain this expression from (3) we have to expand first the function $\ln |1 - z_1(x)\overline{z_2(x)}|^{-1}$ in a series:

$$\ln |1 - z_1(x)\overline{z_2(x)}|^{-1} = \frac{1}{2} \sum_{n \neq 0} \frac{1}{|n|} \int f_1(x, n) \overline{f_2(x, n)} \, dm(x).$$

Then we expand $\exp u$ in a power series, where

$$u = \frac{1}{2} \sum_{n \neq 0} \frac{1}{|n|} \int f_1(x, n) \overline{f_2(x, n)} \, dm(x),$$

and obtain the required expression (4).

It is evident that each term in (4) gives a positive definite Hermitian form on $\mathcal{H}^0$; consequently, the Hermitian form given by (4) for any pair of mappings $z_1(\cdot)$ and $z_2(\cdot)$ is positive definite.

3. Equivalence of the representation constructed here with the preceding ones.

THEOREM 4.2. *The representation $U_{\tilde{g}}$ of G^X in $\mathcal{H}$ is equivalent* *to the representation constructed in* §3. *Hence it follows, in particular, that* $U_{\tilde{g}}$ *is irreducible.*

Let y_0 be the point of the Lobachevskii plane $Y = K\backslash G$ (where K is a fixed maximal compact subgroup) corresponding to the unit coset. We denote by $\tilde{y}_0 = y_0(\cdot)$ the mapping that takes X into y_0. The vector $\tilde{y}_0$ belongs to $\mathcal{H}$, and it is clear that

$$U_{\tilde{g}}\tilde{y}_0 = \tilde{y}_0$$

for every $\tilde{g} \in K^X$. Thus $\tilde{y}_0$ is a vacuum vector in $\mathcal{H}$.

LEMMA 4.1. *The vector* $\tilde{y}_0$ *is cyclic in* $\mathcal{H}$.

PROOF. It is sufficient to verify that as $\tilde{g}$ ranges over G^X, $U_{\tilde{g}}\tilde{y}_0$ ranges over the whole of Y^X.

It is known that the natural fibration $G \to Y = K\backslash G$ is trivial, hence there exists a continuous cross section $s\colon Y \to G$. Now s induces the mapping $Y^X \to G^X$ under which $\tilde{y} = y(\cdot) \in Y^X$ goes into $\tilde{g} = g(\cdot) \in G^X$, where $g(x) = s[y(x)]$. It is also clear that $y(\cdot) = \tilde{y}_0 g(\cdot)$. This completes the proof.

Let us find the spherical function $(U_{\tilde{g}}\tilde{y}_0, \tilde{y}_0)$, where $\tilde{y}_0$ is a vacuum vector. Since $U_{\tilde{g}}\tilde{y}_0 = \tilde{y}_0\tilde{g}^{-1}$, we obtain by the formula for the scalar product in $\mathcal{H}$

$$(U_{\tilde{g}}\tilde{y}_0, \tilde{y}_0) = \exp \int \ln \cosh^{-1} \rho\,(y_0 g^{-1}(x), y_0)\, dm(x) = \\ = \exp \int \ln \cosh^{-1} \rho\,(y_0 g(x), y_0)\, dm(x) = \exp \int \ln \psi\,(g(x))\, dm(x).$$

We proceed now to the proof of Theorem 4.2. In §3 the representation $U_{\tilde{g}}$ of G^X was defined in the Hilbert space $\mathcal{L}$ with the cyclic vacuum vector ξ_0. It was also established that $(U_{\tilde{g}}\xi_0, \xi_0) = \exp \int \ln \psi(g(x))dm(x)$.

So we see that $(U_{\tilde{g}}\xi_0, \xi_0)_{\mathcal{L}} = (U_{\tilde{g}}\tilde{y}_0, \tilde{y}_0)_{\mathcal{H}}$. Since the vectors ξ_0 and $\tilde{y}_0$ are cyclic in their respective spaces, it follows that the mapping $\xi_0 \mapsto \tilde{y}_0$ can be extended to an isomorphism $\mathcal{L} \to \mathcal{H}$ that commutes with the action of G^X in $\mathcal{L}$ and $\mathcal{H}$. This proves Theorem 4.2.

§5. A representation of G^X associated with a maximal compact group $K \subset G$

1. Construction of a representation of G. We take G to be the group of matrices $\begin{pmatrix} \alpha & \beta \\ \bar{\beta} & \bar{\alpha} \end{pmatrix}$, $|\alpha|^2 - |\beta|^2 = 1$. As before, let X be a compact topological space with positive finite measure m. For simplicity we assume that the support of m is the whole space X and that $m(X) = 1$. Henceforth we write dx instead of $dm(x)$.

Although the method of construction that we use here for the representation of G^X is cumbersome, it has the advantage that all the formulae can be written out explicitly and completely and are to some extent analogous to the expression of representations of the rotation group by means of spherical functions.

We suggest that on first reading the reader should omit the simple but tedious proof in the second half of §5.1 of the fact that the formulae gives a unitary representation.

Formulae for representations of the Lie algebra of G^X are given in two forms at the end of this section.

We introduce the Hilbert space $\mathcal{H}$ whose elements are all the sequences

$$F = (f_0, f_1, \ldots, f_k, \ldots),$$

where $f_0 \in \mathbf{C}$, and f_k for $k > 0$ are functions $\underbrace{X \times \ldots \times X}_{k} \times \underbrace{\mathbf{Z} \times \ldots \times \mathbf{Z}}_{k} \to \mathbf{C}$, satisfying the following conditions:[1]

1) $f_k(x_1, \ldots, x_k; n_1, \ldots, n_k)$ is symmetric with respect to permutations of the pairs (x_i, n_i), (x_j, n_j);

2) $f_k(x_1, \ldots, x_k; n_1, \ldots, n_k)|_{n_i=0} =$

$f_{k-1}(x_1, \ldots, \hat{x}_i, \ldots, x_k; n_1, \ldots, \hat{n}_i, \ldots, n_k)$ $(i = 1, \ldots, k)$;

(1) 3) $\|F\|^2 = |f_0|^2 +$

$$+\sum_{k=1}^{\infty} \sum_{\substack{n_1, \ldots, n_k \\ (n_i \neq 0)}} \frac{1}{k!} \frac{1}{|n_1 \ldots n_k|} \int |f_k(x_1, \ldots, x_k; n_1, \ldots, n_k)|^2 \, dx_1 \ldots dx_k < \infty.$$

REMARK. Nothing would be changed in the definition of $\mathcal{H}$ if we were to assume that all the integral indices n_i are non-zero. Then, of course, condition 2) is unnecessary, and the norm, as before, is given by (1).

We construct a unitary representation of G^X in $\mathcal{H}$. First we introduce on G functions $P_{mn}(g)$ and $p_n(g)$. We define $P_{mn}(g)$ for $n \geqslant 0$ as the coefficient of z^m in the power series expansion of $\left(\frac{\alpha z+\beta}{\bar{\beta} z+\bar{\alpha}}\right)^n$, where $g = \begin{pmatrix} \alpha & \beta \\ \bar{\beta} & \bar{\alpha} \end{pmatrix}$. Thus.

$$\left(\frac{\alpha z+\beta}{\bar{\beta} z+\bar{\alpha}}\right)^n = \sum_m P_{mn}(g) z^m, \qquad n \geqslant 0.$$

For $n \leqslant 0$ we define $P_{mn}(g)$ by:

$$\left(\frac{\bar{\alpha}\bar{z}+\bar{\beta}}{\beta \bar{z}+\alpha}\right)^{|n|} = \sum_m P_{-m,n}(g) \bar{z}^m, \qquad n \leqslant 0.$$

---

[1] $\hat{x}_i$, $\hat{n}_i$ indicate that the corresponding variables are omitted.

From this definition it follows that 1) $P_{-m,-n}(g) = P_{mn}(g)$; 2) $P_{mn}(g) = 0$ if $mn < 0$; 3) $P_{m0}(g) = 1$ for $m = 0$ and $P_{m0}(g) = 0$ for $m \neq 0$; 4) $P_{0n}(g) = \left(\frac{\beta}{\bar{\alpha}}\right)^n$ for $n > 0$ and $P_{0n}(g) = \left(\frac{\bar{\beta}}{\alpha}\right)^{|n|}$ for $n < 0$.

Next we set $p_n(g) = \left(\frac{\bar{\beta}}{\alpha}\right)^n$ for $n > 0$, $p_n(g) = \left(\frac{\beta}{\alpha}\right)^{|n|}$ for $n < 0$, $p_0(g) = 1$.

Let $\tilde{g} = \begin{pmatrix} \alpha(\cdot) & \beta(\cdot) \\ \bar{\beta}(\cdot) & \bar{\alpha}(\cdot) \end{pmatrix}$ be an arbitrary element of G^X. We associate with it the operator $U_{\tilde{g}}$ in $\mathcal{H}$ that is given by the formula

$$U_{\tilde{g}}\{f_k\} = \{\varphi_k\},$$

where

$$\varphi_k(x_1, \ldots, x_k; n_1, \ldots, n_k) = \Psi(\tilde{g}) \times \tag{2}$$

$$\times \sum_{m_1, \ldots, m_k} \left\{ \prod_{i=1}^{k} P_{m_i n_i}(g(x_i)) \times \sum_{s=0}^{\infty} \sum_{\substack{l_1, \ldots, l_s \\ (l_i \neq 0)}} \frac{(-1)^{l_1 + \ldots + l_s}}{s! \, | l_1 \ldots l_s |} \times \right.$$

$$\left. \times \int \prod_{j=1}^{s} p_{l_j}(g(t_j)) f_{k+s}(x_1, \ldots, x_k, t_1, \ldots, t_s; m_1, \ldots, m_k, l_1, \ldots, l_s) \, dt_1 \ldots dt_s \right\},$$

and

$$\Psi(\tilde{g}) = \exp \int \ln \psi^2(g(x)) \, dx. \tag{3}$$

Here $\psi(g)$ denotes $|\alpha|^{-1}$.

THEOREM 5.1. *The operators $U_{\tilde{g}}$ are unitary and form a representation of G^X, that is,*

$$(U_{\tilde{g}} F_1, U_{\tilde{g}} F_2) = (F_1, F_2)$$

for any $\tilde{g} \in G^X$ and $F_1, F_2 \in \mathcal{H}$ and

$$U_{\tilde{g}_1} U_{\tilde{g}_2} F = U_{\tilde{g}_1 \tilde{g}_2} F$$

for any $\tilde{g}_1, \tilde{g}_2 \in G^X$ and $F \in \mathcal{H}$.

We verify these relations for the vectors F of a certain space $\mathcal{H}^0$ everywhere dense in $\mathcal{H}$, which we now introduce.

We denote by M the set of sequences of the form

$$F = (1, f_1, f_2, \ldots, f_k, \ldots),$$

where

$$f_k(x_1, \ldots, x_k; n_1, \ldots, n_k) = u(x_1, n_1) \ldots u(x_k, n_k),$$

and $u(x, n)$ is a function continuous in x for any fixed n such that $u(x, 0) = 1$ and

$$(4)\qquad \sum_{n\neq 0}\frac{1}{|n|}\int |u(x,n)|^2\,dx<\infty.$$

We denote by $\mathcal{H}^0$ the space of all finite linear combinations of elements of M.

We must verify that $\mathcal{H}^0\subset\mathcal{H}$. Now it is evident that the vectors $F\in M$ satisfy conditions 1) and 2) in the definition of $\mathcal{H}$. Furthermore, if

$$(5)\qquad F=(1,\,u(x,n),\,\ldots,\,u(x_1,n),\,\ldots,\,u(x_k,n_k),\,\ldots)$$

is a vector of M, then its norm $\|F\|$ can be represented in the form:

$$(6)\qquad \|F\|^2=\exp\Big(\sum_{n\neq 0}\frac{1}{|n|}\int|u(x,n)|^2\,dx\Big).$$

Consequently, by (4), F also satisfies condition 3), and hence $F\in\mathcal{H}$. We observe that if $F\in M$, that is, if it has the form (4), then the expression for $U_{\tilde g}F$ reduces to the following simple form:

$$(7)\qquad U_{\tilde g}F=\lambda(\tilde g,u)(1,\,v(x,n),\,\ldots,\,v(x_1,n_1),\ldots,v(x_k,n_k),\ldots),$$

where

$$(8)\qquad \lambda(\tilde g,u)=\Psi(\tilde g)\exp\Big(\sum_{l\neq 0}\frac{(-1)^l}{|l|}\int p_l(g(x))\,u(x,l)\,dx\Big),$$

$$(9)\qquad v(x,n)=\sum_m P_{mn}(g(x))\,u(x,m).$$

LEMMA 5.1. *The space $\mathcal{H}^0$ is everywhere dense in $\mathcal{H}$.*

PROOF. We assume that all the indices n_i are non-zero (see the Remark on p. 115). Suppose that $\mathcal{H}^0$ is not dense in $\mathcal{H}$, hence that there exists a non-zero vector $F=\{f_k^0\}$, orthogonal to $\mathcal{H}^0$. We consider in $\mathcal{H}$ the vectors of the form $F_\lambda=\{\lambda^k f_k\}$, where $f_0=1$, $f_k(x_1,\ldots,x_k;\,n_1,\ldots,n_k)=$ $=u(x_1,n_1)\ldots u(x_k,n_k)$ for $k>0$, $u(x,n)$ is a continuous function, and λ is an arbitrary number.

Since $F_\lambda\in\mathcal{H}^0$, we have $(F_\lambda,F)=0$, that is,

$$(10)\qquad \sum_{k=0}^{\infty}\frac{\lambda^k}{k!}\Big(\sum_{\substack{n_1,\ldots,n_k\\(n_i\neq 0)}}|n_1\ldots n_k|^{-1}\times$$
$$\times\int f_k(x_1,\ldots,x_k;\,n_1,\ldots,n_k)\,\overline{f_k^0(x_1,\ldots,x_k;\,n_1,\ldots,n_k)}\,dx_1\ldots dx_k\Big)=0$$

for any λ. Hence it follows that for every $k=0,1,\ldots$ we have the relation

$$\sum_{\substack{n_1,\ldots,n_k\\(n_i\neq 0)}}|n_1\ldots n_k|^{-1}\times$$
$$\times\int f_k(x_1,\ldots,x_k;\,n_1,\ldots,n_k)\overline{f_k^0(x_1,\ldots,x_k;\,n_1,\ldots,n_k)}\,dx_1\ldots dx_k=0,$$

or

$$(11) \qquad \sum_{\substack{n_1, \ldots, n_k \\ (n_i \neq 0)}} |n_1 \ldots n_k|^{-1} \times$$
$$\times \int u(x_1, n_1) \ldots u(x_k, n_k) f_k^0(x_1, \ldots, x_k; n_1, \ldots, n_k)\, dx_1 \ldots dx_k = 0$$

for any continuous function $u(x, n)$ such that

$$\sum_{n \neq 0} \frac{1}{|n|} \int |u(x, n)|^2 \, dx < \infty.$$

Since $f_k^0(x_1, \ldots, x_k; n_1, \ldots, n_k)$ is symmetric under permutations of the pairs (x_i, n_i) and (x_j, n_j), it follows from (11) that $f_k^0 \equiv 0$ $(k = 0, 1, \ldots)$. Thus, $F = 0$, in contradiction to the hypothesis.

To prove Theorem 5.1 we need certain relations for the functions $P_{mn}(g)$ and $p_l(g)$:

a) $\dfrac{(-1)^n}{|n|} P_{mn}(g) = \dfrac{(-1)^m}{|m|} P_{nm}(g)$ for $m \neq 0$, $n \neq 0$;

b) for any compact subset $V \subset G$ there exist constants $C > 0$ and r, $0 < r < 1$, such that $|P_{mn}(g)| < Cr^{|m|+|n|}$;

c) $\sum_{m'} P_{m'n}(g_1) P_{mm'}(g_2) = P_{mn}(g_1 g_2)$;

d)

$$\sum_{l \neq 0} \frac{(-1)^l}{|l|} p_l(g_1) P_{ml}(g_2) = \begin{cases} \dfrac{(-1)^m}{|m|} (p_m(g_1 g_2) - p_m(g_2)) & \text{for } m \neq 0, \\ -2 \ln \dfrac{\psi(g_1)\psi(g_2)}{\psi(g_1 g_2)} & \text{for } m = 0; \end{cases}$$

e)

$$\sum_{n \neq 0} \frac{1}{|n|} P_{mn}(g) \overline{P_{m'n}(g)} = \begin{cases} |m|^{-1} \delta_{mm'} & \text{for } m \neq 0,\ m' \neq 0, \\ -(-1)^m |m|^{-1} p_m(g) & \text{for } m \neq 0,\ m' = 0, \\ -(-1)^{m'} |m'|^{-1} \overline{p_{m'}(g)} & \text{for } m = 0,\ m' \neq 0, \\ -4 \ln \psi(g) & \text{for } m = m' = 0 \end{cases}$$

($\delta_{mm'}$ is the Kroneker delta).[1])

LEMMA 5.2. *The function* $\lambda(\tilde{g}, u)$ *defined by* (8) *satisfies the following functional relation:*

$$(12) \qquad \lambda(\tilde{g}_1, v)\lambda(\tilde{g}_1, u) = \lambda(\tilde{g}_1 \tilde{g}_2, u),$$

where

$$v(x, n) = \sum_m P_{mn}(g_2(x)) u(x, m).$$

---

1) We can derive a) from the relation $P_{mn}(g) = \dfrac{1}{2\pi i} \displaystyle\int_{|z|=1} \left(\dfrac{\alpha z + \beta}{\bar{\beta} z + \bar{\alpha}} \right)^n z^{-m-1}\, dz$, $n > 0$; the bound b) for $P_{mn}(g)$ follows from the fact that the radius of convergence of the series $\sum_m P_{mn}(g) z^m$ is greater than 1; the relation c) follows immediately from the definition of the functions P_{mn}; d) follows easily from the definition of P_{mn} and p_l and a).

PROOF. It follows from the definition of λ that

$$\lambda(\tilde{g}_1, v) = \Psi(\tilde{g}_1) \exp\left(\sum_{l\neq 0} \frac{(-1)^l}{|l|} \int p_l(g_1(x))\, v(x, l)\, dx\right) =$$

$$= \Psi(\tilde{g}_1) \exp\left(\sum_{l\neq 0} \sum_{m} \frac{(-1)^l}{|l|} \int p_l(g_1(x))\, P_{ml}(g_2(x))\, u(x, m)\, dx\right).$$

We sum over l, apply d),[1] and obtain

$$\lambda(\tilde{g}_1, v) = \Psi(\tilde{g}_1) \exp\left\{\sum_{m\neq 0} \frac{(-1)^m}{|m|} \int p_m((g_1g_2)(x))\, u(x, m)\, dx - \right.$$

$$\left. - \sum_{m\neq 0} \frac{(-1)^m}{|m|} \int p_m(g_2(x))\, u(x, m)\, dx\right\} \frac{\Psi(\tilde{g}_1\tilde{g}_2)}{\Psi(\tilde{g}_1)\Psi(\tilde{g}_2)} = \frac{\lambda(\tilde{g}_1\tilde{g}_2, u)}{\lambda(\tilde{g}_2, u)}.$$

LEMMA 5.3. *$U_{\tilde{g}_1} U_{\tilde{g}_2} = U_{\tilde{g}_1\tilde{g}_2} F$ for any $F \in M$ and any $\tilde{g}_1, \tilde{g}_2 \in G^X$.*

PROOF. Let

$$F = (1, u(x, n), \ldots, u(x_1, n_1), \ldots, u(x_k, n_k), \ldots).$$

Then

$$U_{\tilde{g}_2}F = \lambda(\tilde{g}_2, u)(1, v(x, n), \ldots, v(x_1, n_1) \ldots v(x_k, n_k), \ldots),$$

where

$$v(x, n) = \sum_m P_{mn}(g_2(x))u(x, m);$$

$$U_{\tilde{g}_1}U_{\tilde{g}_2}F = \lambda(\tilde{g}_1, v)\lambda(\tilde{g}_2, u)(1, w(x, n), \ldots, w(x_1, n_1), \ldots, w(x_k, n_k), \ldots),$$

where

$$w(x, n) = \sum_{m'} P_{m'n}(g_1(x))\, v(x, m') = \sum_{m', m} P_{m'n}(g_1(x))\, P_{mm'}(g_2(x))\, u(x, m).$$

It follows from Lemma 5.2 that $\lambda(\tilde{g}_1, v)\lambda(\tilde{g}_2, u) = \lambda(\tilde{g}_1\tilde{g}_2, u)$. On the other hand, by c) for $P_{mn}(g)$, we have $w(x, n) = \sum_m P_{mn}((g_1g_2)(x))u(x, m)$.

Consequently, $U_{\tilde{g}_1}U_{\tilde{g}_2} F = U_{\tilde{g}_1\tilde{g}_2} F$.

COROLLARY. *The operators $U_{\tilde{g}}$ form a representation of G^X in $\mathcal{H}^0$.*

LEMMA. 5.4. *$(U_{\tilde{g}}F_1, U_{\tilde{g}}F_2) = (F_1, F_2)$ for any $F_1, F_2 \in M$* and *$\tilde{g} \in G^X$.*

PROOF. Let

$$F_1 = (1, u_1(x, n), \ldots, u_1(x_1, n_1), \ldots, u_1(x_k, n_k), \ldots),$$
$$F_2 = (1, u_2(x, n), \ldots, u_2(x_1, n_1), \ldots, u_2(x_k, n_k), \ldots).$$

Then

$$(F_1, F_2) = \exp\left(\sum_{n\neq 0} \frac{1}{|n|} \int u_1(x, n)\, \overline{u_2(x, n)}\, dx\right).$$

On the other hand, using the expression (7) for $U_{\tilde{g}}F$ we have

$$(U_{\tilde{g}}F_1, U_{\tilde{g}}F_2) = \exp\left\{4 \int \ln \psi(g(x))\, dx + \right.$$

$$+ \sum_{l\neq 0} \frac{(-1)^l}{|l|} \int p_l(g(x))\, u_1(x, l)\, dx + \sum_{l\neq 0} \frac{(-1)^l}{|l|} \int \overline{p_l(g(x))\, u_2(x, l)}\, dx +$$

$$\left. + \sum_{n\neq 0} \frac{1}{|n|} \sum_{m, m'} \int P_{mn}(g(x))\, \overline{P_{m'n}(g(x))}\, u_1(x, m)\, \overline{u_2(x, m')}\, dx\right\}.$$

---

[1] Reversal of the order of the summations is permissible in view of b).

In the last expression we sum over n under the exponential sign, then use e) and $u_i(x, 0) = 1$, and obtain $(U_{\tilde{g}}F_1, U_{\tilde{g}}F_2) = (F_1, F_2)$ after some elementary simplifications.

COROLLARY. *The $U_{\tilde{g}}$ are unitary operators in $\mathscr{H}^0$.*

2. Irreducibility of the representation $U_{\tilde{g}}$. The representation operators $U_{\tilde{g}}$ assume a specially simple form when restricted to the subgroup of matrices

$$\tilde{k} = \begin{pmatrix} e^{i\varphi(\cdot)/2} & 0 \\ 0 & e^{-i\varphi(\cdot)/2} \end{pmatrix}.$$

Namely,

$$U_{\tilde{k}}\{f_k\} = \{f'_k\}, \tag{13}$$

where $f'_k(x_1, \ldots, x_k;\, n_1, \ldots, n_k) = \exp\left(i\sum_{s=1}^{k} n_s\varphi(x_s)\right) f_k(x_1, \ldots, x_k;\, n_1, \ldots, n_k)$.

From this expression it is clear that *the family of commuting operators $U_{\tilde{k}}$ has a simple spectrum in $\mathscr{H}$. The vacuum vector in $\mathscr{H}$ is*

$$\xi_0 = (1, f_1(x, n), \ldots, f_k(x_1, \ldots, x_k;\, n_1, \ldots, n_k), \ldots),$$

where $f_k(x_1, \ldots, x_k;\, n_1, \ldots, n_k) = 0$ if $|n_1| + \ldots + |n_k| > 0$.

THEOREM 5.2. *The representation $U_{\tilde{g}}$ of G^X in $\mathscr{H}$ is irreducible.*

PROOF. Let A be a bounded operator in $\mathscr{H}$ that commutes with all the operators $U_{\tilde{g}}$, in particular, with the $U_{\tilde{k}}$ of the form (13). Since the family of operators $U_{\tilde{k}}$ has a simple spectrum, an operator A that commutes with them has the form

$$A\{f_k\} = \{a_k f_k\}, \tag{14}$$

where $a_k(x_1, \ldots, x_k;\, n_1, \ldots, n_1)$ are measurable functions (satisfying the same relations as the f_k). Let ξ_0 be the vacuum vector in $\mathscr{H}$ defined above. It follows from (14) that $A\xi_0 = a_0\xi_0$.

We apply to ξ_0, the operator $U_{\tilde{g}}$, $\tilde{g} \in G$, where

$$\tilde{g} = \begin{pmatrix} \cosh\tau & \sinh\tau \\ \sinh\tau & \cosh\tau \end{pmatrix}, \quad \tau \neq 0,$$

and τ does not depend on x. We obtain

$$U_{\tilde{g}}\xi_0 = \{f_k\},$$

where $f_k(x_1, \ldots, x_k;\, n_1, \ldots, n_k) = \cosh^{-1}\tau \prod_{i=1}^{k} \tanh^{|n_i|}\tau$.

From $U_{\tilde{g}}A\xi_0 = AU_{\tilde{g}}\xi_0$ it follows that $a_0\{f_k\} = \{a_k f_k\}$, hence $a_0 f_k = a_k f_k$. Consequently, since the f_k do not vanish, $a_k = a_0$ for every k, that is, A is a multiple of the unit operator.

3. Equivalence of the representation $U_{\tilde{g}}$ to representations constructed earlier. THEOREM 5.3. *The representation $U_{\tilde{g}}$ of G^X in $\mathscr{H}$ is equivalent to* *the representations constructed in the preceding sections.*

PROOF. It follows easily from the definition of $U_{\tilde{g}}$ and the scalar product in $\mathcal{H}$ that the spherical function corresponding to the vacuum vector ξ_0 has the form

$$(U_{\tilde{g}}\xi_0, \xi_0) = \exp\left(2\int \ln \psi(g(x))dx\right).$$

So we see that this spherical function coincides (with suitable agreement of measures on X) with the spherical functions of the representations of G^X constructed in the preceding sections. The theorem follows immediately from this fact and the irreducibility of $U_{\tilde{g}}$.

4. Infinitesimal formulae for the representation. We give formulae for the representation operators of the Lie algebra of G^X in $\mathcal{H}$ (that is, of the algebra $\mathfrak{G}^X$ of continuous mappings $X \to \mathfrak{G}$ of X to the Lie algebra $\mathfrak{G}$ of G with the natural commutation relations).

We take the following matrices as generators of $\mathfrak{G}$:

$$a_{\tilde{\varphi}}^0 = \begin{pmatrix} i\frac{\varphi(\cdot)}{2} & 0 \\ 0 & -i\frac{\varphi(\cdot)}{2} \end{pmatrix}, \quad a_{\tilde{\tau}} = \begin{pmatrix} 0 & \tau(\cdot) \\ \tau(\cdot) & 0 \end{pmatrix}, \quad a_{i\tilde{\tau}} = \begin{pmatrix} 0 & i\tau(\cdot) \\ -i\tau(\cdot) & 0 \end{pmatrix},$$

where $\varphi(\cdot)$, $\tau(\cdot)$ are continuous mappings $X \to \mathbf{R}$. We denote the Lie operators in $\mathcal{H}$ corresponding to these elements by $A_{\tilde{\varphi}}^0$, $A_{\tilde{\tau}}$ and $A_{i\tilde{\tau}}$

Expressions for $A_{\tilde{\varphi}}^0$, $A_{\tilde{\tau}}$, and $A_{i\tilde{\tau}}$ are easily obtained from the formula (1) for the representation operators $U_{\tilde{g}}$ of G^X. Namely,

$$(A_{\tilde{\varphi}}^0 f)_k(x_1, \ldots, x_k; n_1, \ldots, n_k) = i\left(\sum_{s=1} n_s\varphi(x_s)\right) f_k(x_1, \ldots, x_k; n_1, \ldots, n_k);$$

$$\begin{aligned}(A_{\tilde{\tau}}f)_k(x_1, \ldots, x_k; n_1, \ldots, n_k) = \\ = -\sum_{s=1}^{k} n_s\tau(x_s)(f_k(x_1, \ldots, x_k; n_1, \ldots, n_s+1, \ldots, n_k) - \\ - f_k(x_1, \ldots, x_k; n_1, \ldots, n_s-1, \ldots, n_k)) - \\ - \int \tau(t)(f_{k+1}(x_1, \ldots, x_k, t; n_1, \ldots, n_k, 1) + \\ + f_{k+1}(x_1, \ldots, x_k, t; n_1, \ldots, n_k, -1))\,dt,\end{aligned}$$

$$\begin{aligned}(A_{i\tilde{\tau}}f)_k(x_1, \ldots, x_k; n_1, \ldots, n_k) = \\ = i\sum_{s=0}^{k} n_s\tau(x_s)(f_k(x_1, \ldots, x_k, n_1, \ldots, n_s+1, \ldots, n_k) + \\ + f_k(x_1, \ldots, x_k, n_1, \ldots, n_s-1, \ldots, n_k)) + \\ + i\int \tau(t)(f_{k+1}(x_1, \ldots, x_k, t; n_1, \ldots, n_k, 1) - \\ - f_{k+1}(x_1, \ldots, x_k, t; n_1, \ldots, n_k, -1))\,dt.\end{aligned}$$

It is convenient to go from $A_{\tilde{\tau}}$ and $A_{i\tilde{\tau}}$ to

$$A_{\tilde{\tau}}^{+} = \frac{1}{2}(A_{\tilde{\tau}} + iA_{i\tilde{\tau}}), \quad A_{\tilde{\tau}}^{-} = \frac{1}{2}(A_{\tilde{\tau}} - iA_{i\tilde{\tau}}).$$

The operators A_τ^+ and A_τ^- act in the following manner:

$$(A_\tau^+ f)_k(x_1, \ldots, x_k; n_1, \ldots, n_k) =$$

$$= -\sum_{s=1}^{k} n_s \tau(x_s) f_k(x_1, \ldots, x_k; n_1, \ldots, n_s+1, \ldots, n_k) -$$

$$- \int \tau(t) f_{k+1}(x_1, \ldots, x_k, t; n_1, \ldots, n_k, 1)\, dt,$$

$$(A_\tau^- f)_k(x_1, \ldots, x_k; n_1, \ldots, n_k) =$$

$$= \sum_{s=1}^{k} n_s \tau(x_s) f_k(x_1, \ldots, x_k; n_1, \ldots, n_s - 1, \ldots, n_k) -$$

$$- \int \tau(t) f_{k+1}(x_1, \ldots, x_k, t; n_1, \ldots, n_k, -1)\, dt.$$

Since A_ϕ^0, A_τ^+, and A_τ^- form a representation of the algebra $\mathfrak{G}^X$, the same commutation relations hold for them as for the corresponding elements of the Lie algebra $\mathfrak{G}^X$, namely,

$$[A_\tau^+, A_\varphi^0] = -iA_{\tau\varphi}^+, \quad [A_\tau^-, A_\varphi^0] = iA_{\tau\varphi}^-, \quad [A_{\tau_1}^+, A_{\tau_2}^-] = -2iA_{\tau_1\tau_2}^0.$$

5. Another method of realizing the operators A_ϕ^0, A_τ^+ and A_τ^-. The method of realization proposed here seems very interesting to us. We specify the elements of $\mathscr{H}$ not as sequences of functions of $x_1, \ldots, x_k$, $n_1, \ldots, n_k$, but as sequences of functions of the x-parameters alone.

Let us consider, for example, the function $f_3(x_1, x_2, x_3; 2, 1, -4)$. We assign to it the function $f_{3,4}(x_1, x_1, x_2; y_1, y_1, y_1, y_1)$. More generally, if a function $f_k(x_1, \ldots, x_k; n_1, \ldots, n_k)$ is given, we first discard the zeros among the numbers $n_1, \ldots, n_k$. We then pick out the positive numbers among the n_i and denote their sum by m and the sum of the absolute values of the negative n_i by n. x_1 is then repeated $|n_1|$ times, x_2 is repeated $|n_2|$ times, etc. Because of the symmetry of f in the pairs (x_i, n_i) we can write down first all the arguments x_i with positive n_i, then all those with negative n_i; we denote the resulting function by

$$f_{mn}(\underbrace{x_1, \ldots, x_1}_{|n_1|}, \ldots, \underbrace{x_k, \ldots, x_k}_{|n_k|}).$$

Now we give a precise definition that does not depend on these arguments.

We introduce the space $\mathscr{H}^0$, whose elements are infinite sequences $f = \{f_{mn}\}_{m,n=0,1,\ldots}$, *where* $f_0 \in \mathbf{C}$, $f_{mn}\colon \underbrace{X \times \ldots \times X}_{m+n} \to \mathbf{C}$ *for* $m + n > 0$,

satisfying the following conditions:

1) the functions $f_{mn}(x_1, \ldots, x_m; y_1, \ldots, y_n)$ are continuous;

2) the functions $f_{mn}(x_1, \ldots, x_m; y_1, \ldots, y_n)$ are symmetric in the first m arguments and in the last n arguments:

$$3) \quad \|f\|^2 = \sum_{m,\,n=0}^{\infty} \sum \frac{1}{(k+l)!\,|m|\,|n|} \int |f_{mn}(\underbrace{x_1, \ldots, x_1}_{m_1}, \ldots, \underbrace{x_k, \ldots, x_k}_{m_k};$$

$$\underbrace{y_1, \ldots, y_1}_{n_1}, \ldots, \underbrace{y_l, \ldots, y_l}_{n_l})|^2\, dx_1 \ldots dx_k\, dy_1 \ldots dy_l < \infty,$$

where the inner sum is taken over all partitions $m = m_1 + \ldots + m_k$, $n = n_1 + \ldots + n_k$, and $|m| = m_1, \ldots, m_k$, $|n| = n_1, \ldots, n_k$. (For $m = 0$ we take $|m| = 1$.)

An isometric correspondence between $\mathscr{H}^0$ and the previous space is given by

$$\{f_{mn}\} \longmapsto \{f_k\},$$

where

$$f_k(x_1, \ldots, x_p, y_1, \ldots, y_q;\ m_1, \ldots, m_p,\ n_1, \ldots, n_q) =$$

$$= f_{mn}(\underbrace{x_1, \ldots, x_1}_{m_1}, \ldots, \underbrace{x_p, \ldots, x_p}_{m_p};\ \underbrace{y_1, \ldots, y_1}_{|n_1|}, \ldots, \underbrace{y_q, \ldots, y_q}_{|n_q|})$$

$$(p+q=k;\ m_i > 0;\ n_i < 0,\ m = m_1 + \ldots + m_p,\ n = |n_1| + \ldots + |n_q|).$$

Now A^0_{φ}, A^+_{τ}, and A^-_{τ} act in $\mathscr{H}^0$ according to the formulae:

$$(A^0_{\varphi} f)_{mn}(x_1, \ldots, x_m;\ y_1, \ldots, y_n) =$$

$$= i\left(\sum_{s=1}^{m} \varphi(y_s) - \sum_{s=1}^{n} \varphi(y_s)\right) f_{mn}(x_1, \ldots, x_m;\ y_1, \ldots, y_n),$$

$$(A^+_{\tau} f)_{mn}(x_1, \ldots, x_m;\ y_1, \ldots, y_n) =$$

$$= -\sum_{s=1}^{m} \tau(x_s) f_{m+1,\,n}(x_1, \ldots, x_m, x_s;\ y_1, \ldots, y_n) +$$

$$+ \sum_{s=1}^{n} \tau(y_s) f_{m,\,n-1}(x_1, \ldots, x_m;\ y_1, \ldots, \hat{y}_s, \ldots, y_n) -$$

$$- \int \tau(t) f_{m+1,\,n}(x_1, \ldots, x_m, t;\ y_1, \ldots, y_n)\, dt,$$

$$(A^-_{\tau} f)_{mn}(x_1, \ldots, x_m;\ y_1, \ldots, y_n) =$$

$$= \sum_{s=1}^{m} \tau(x_s) f_{m-1,\,n}(x_1, \ldots, \hat{x}_s, \ldots, x_m;\ y_1, \ldots, y_n) -$$

$$- \sum_{s=1}^{n} \tau(y_s) f_{m,\,n+1}(x_1, \ldots, x_m;\ y_1, \ldots, y_n, y_s) -$$

$$- \int \tau(t) f_{m,\,n+1}(x_1, \ldots, x_m;\ y_1, \ldots, y_n, t)\, dt.$$

§6. Another method of constructing a representation of G^X

1. The general construction. a) *Construction of the representation space.* Let G be an arbitrary topological group and H a linear topological space in which a representation $T(g)$ of G is defined. Further, Let X be a compact topological space with a positive finite measure m. As before, we assume that m is a countably additive, non-negative Borel measure whose support is X.

We suppose that in H there is a linear functional l $(l \neq 0)$ invariant under $T(g)$. Then we construct a representation $U_{\tilde{g}}$ of the group G^X of continuous mappings $X \to G$ from the representation $T(g)$ of G and the functional l.

We denote by H^X the set of all continuous mappings $\tilde{f} = f(\cdot)\colon X \to H$ such that $l(f(x))$ does not depend on x and $l(f(x)) = 0$.

We introduce a new linear space $\mathcal{H}^0$ whose elements are formal finite sums of elements of H^X:

(1) $$\tilde{f}_1 \dotplus \ldots \dotplus \tilde{f}_n.$$

Here we set $\lambda_1 \tilde{f} \dotplus \lambda_2 \tilde{f} = (\lambda_1 + \lambda_2)\tilde{f}$ if $\lambda_1 + \lambda_2 \neq 0$, and $\tilde{f} \dotplus (-\tilde{f}) = 0$. We emphasize that if $f_1(x)$ and $f_2(x)$ are not proportional, then $\tilde{f} = \tilde{f}_1 + \tilde{f}_2$ and $\tilde{f}_1 \dotplus \tilde{f}_2$ are regarded as distinct elements.

In $\mathcal{H}^0$ operations of addition and multiplication by a factor $\lambda \in \mathbf{C}$ are defined in the natural way. Namely, the product of $\tilde{f} = f(\cdot)$ by $\lambda \in \mathbf{C}$ is defined as $\lambda \circ f(x) = \lambda f(x)$ if $\lambda \neq 0$, and $0 \circ f(x) = 0$. As a result, $\mathcal{H}$ becomes a linear space.

b) *Action of the operators in* $\mathcal{H}^0$. We define a representation $U_{\tilde{g}}$ of G^X in $\mathcal{H}^0$ by

(2) $$(U_{\tilde{g}}\tilde{f})(x) = \lambda T(g(x)) f(x),$$

where $\lambda(\tilde{g}, \tilde{f})$ is a function of $\tilde{g}$ and $\tilde{f}$ such that $\lambda(\tilde{g}, c\tilde{f}) = \lambda(\tilde{g}, \tilde{f})$ for any $c \neq 0$, and we extend $U_{\tilde{g}}$ by additivity to all elements (1).

LEMMA 6.1. *The operators $U_{\tilde{g}}$ form a representation of G^X if and only if the function $\lambda(\tilde{g}, \tilde{f})$ satisfies the following additional condition for any $\tilde{g}_1, \tilde{g}_2 \in G^X$ and $\tilde{f} \in H^X$: $\lambda(\tilde{g}_1, \tilde{f}_1)\lambda(\tilde{g}_2, \tilde{f}) = \lambda(\tilde{g}_1\tilde{g}_2, \tilde{f})$ where $f_1(x) = T(g_2(x))f(x)$.*

If the weaker relation

$$\lambda(\tilde{g}_1, \tilde{f}_1)\lambda(\tilde{g}_2, \tilde{f}) = c(\tilde{g}_1, \tilde{g}_2)\lambda(\tilde{g}_1\tilde{g}_2, \tilde{f})$$

holds, then the $U_{\tilde{g}}$ form a projective representation.

The proof is obvious.

c) *Construction of a unitary representation.* Let $H_0 \subset H$ be the set of elements ξ such that $l(\xi) = 0$. By the invariance of l, H_0 is an invariant subspace of H. Suppose that an invariant positive definite scalar product (ξ_1, ξ_2) is defined in H.

We construct a scalar product in the space $\mathscr{H}^0$ from the scalar product (ξ_1, ξ_2) in H. To do this we first fix a vector $\xi_0 \in H$ such that $l(\xi_0) = 1$. For any pair of elements $\tilde{f}_1 = f_1(\cdot)$ and $\tilde{f}_2 = f_2(\cdot)$ of H^X we define our scalar product as follows:

$$\langle \tilde{f}_1, \tilde{f}_2 \rangle = l(\tilde{f}_1)\overline{l(\tilde{f}_2)} \exp\left(\frac{1}{l(\tilde{f}_1)\overline{l(\tilde{f}_2)}} \int (f_1'(x), f_2'(x))\, dm(x)\right), \tag{3}$$

where $f_i'(x) = f_i(x) - l(\tilde{f}_i)\xi_0$ are elements of H_0 for any $x \in X$. (We recall that $l(f(x))$ is independent of x. Instead of $l(f(x))$ we write $l(\tilde{f})$, where $\tilde{f} = f(\cdot)$.)

We extend this scalar product to the whole space $\mathscr{H}^0$ by linearity.

LEMMA 6.2. *The Hermitian form on $\mathscr{H}^0$ defined by* (3) *is positive definite.*

PROOF. We choose arbitrary elements $\tilde{f}_1, \ldots, \tilde{f}_n$ of H^X and prove that the matrix $\langle \tilde{f}_i, \tilde{f}_j \rangle$ is positive definite. We have

$$\frac{\langle \tilde{f}_i, \tilde{f}_j \rangle}{l(\tilde{f}_i)\overline{l(\tilde{f}_j)}} = \sum_{n=0}^{\infty} \frac{1}{n!} \left(\frac{1}{l(\tilde{f}_i)\overline{l(\tilde{f}_j)}} \int (f_i'(x), f_j'(x))\, dm(x)\right)^n. \tag{4}$$

Since $a_{ij} = \dfrac{1}{l(\tilde{f}_i)\overline{l(\tilde{f}_j)}} \int (f_i'(x), f_j'(x))\,dm(x)$ is positive definite, by Schur's lemma each term of (4) is positive definite, and the lemma is proved.

Thus, $U_{\tilde{g}}$ acts in a pre-Hilbert space. Let us see how the multiplier $\lambda(\tilde{g}, \tilde{f})$ can be chosen so that the representation is unitary.

For this purpose we first construct from ξ_0 a function $\beta(g)$ with values in H: $\beta(g) = T(g)\xi_0 - \xi_0$. Since l is invariant we have $l(\beta(g)) = l(T(g)\xi_0) - l(\xi_0) = 0$, that is, $\beta(g) \in H_0$ for any $g \in G$. It is not hard to see that $\beta(g)$ is a cocycle with values in H_0, in other words, it satisfies the relation $\beta(g_1) + T(g_1)\beta(g_2) = \beta(g_1 g_2)$ for any $g_1, g_2 \in G$.

We observe that $\beta(g)$ depends on the way we have fixed the vector $\xi_0 \in H$.

LEMMA 6.3. *If we set*

$$\lambda(\tilde{g}, \tilde{f}) = c(\tilde{g}) \exp\left(-\frac{1}{l(\tilde{f})} \int [(T(g(x)) f'(x), \beta(g(x))) + \tfrac{1}{2}\| \beta(g(x)) \|^2]\, dm(x)\right), \tag{5}$$

where $f'(x) = f(x) - l(\tilde{f})\xi_0$, $|c(\tilde{g})| = 1$, then the operators $U_{\tilde{g}}$ defined by (2) *are unitary and form a projective representation. Specifically, $U_{\tilde{g}_1} U_{\tilde{g}_2} = c(\tilde{g}_1, \tilde{g}_2) U_{\tilde{g}_1 \tilde{g}_2}$, where*

$$c(\tilde{g}_1, \tilde{g}_2) = \frac{c(\tilde{g}_1)\, c(\tilde{g}_2)}{c(\tilde{g}_1 \tilde{g}_2)} \times \exp\left(i \int \operatorname{Im}(T(g(x))\beta(g_2(x)), \beta(g_1(x)))\, dm(x)\right). \tag{6}$$

The proof comes from a direct verification.

REMARK. This condition on $\lambda(\tilde{g}, \tilde{f})$ is also necessary.

We now state our final result. A linear topological space H is given and also a representation $T(g)$ of G in H. A linear functional l is given in H that is invariant under the action of G, that is, $l(T(g)\xi) = l(\xi)$ for any $g \in G$ and $\xi \in H$. We define a scalar product (ξ_1, ξ_2) in the subspace H_0 of elements ξ such that $l(\xi) = 0$.

A representation of G^X is constructed as follows. We consider continuous mappings $\tilde{f} = f(\cdot)$: $X \to H$ such that $l(f(x)) = \text{const} \neq 0$. We introduce the space $\mathcal{H}^0$ whose elements are the formal sums $\tilde{f}_1 \dotplus \ldots \dotplus \tilde{f}_n$ with the relations $\lambda_1\tilde{f} + \lambda_2\tilde{f} = (\lambda_1 + \lambda_2)\tilde{f}$ if $\lambda_1 + \lambda_2 \neq 0$, $\tilde{f} + (-\tilde{f}) = 0$. We construct the scalar product:

$$\langle \tilde{f}_1, \tilde{f}_2 \rangle = l(\tilde{f}_1)\overline{l(\tilde{f}_2)} \exp\left(\frac{1}{l(\tilde{f}_1)\overline{l(\tilde{f}_2)}} \int (f_1'(x), f_2'(x))\, dm(x)\right),$$

where $f_i'(x) = f_i(x) - l(\tilde{f}_i)\xi_0$, and ξ_0 is a fixed vector in H such that $l(\xi_0) = 1$. This scalar product is then extended to the whole space $\mathcal{H}^0$. We denote by $\mathcal{H}$ the completion of $\mathcal{H}^0$ in this scalar product.

The operators $U_{\tilde{g}}$ are defined by the formula $(U_{\tilde{g}}f)(x) = \lambda(\tilde{g}, \tilde{f})T(g(x))f(x)$, where

$$\lambda(\tilde{g}, \tilde{f}) = c(\tilde{g}) \exp\left(-\frac{1}{l(\tilde{f})} \int \left[(T(g(x))f'(x), \beta(g(x))) + \frac{1}{2}\|\beta(g(x))\|^2\right] dm(x)\right),$$

$$\beta(g) = T(g)\xi_0 - \xi_0, \quad |c(\tilde{g})| = 1,$$

and are extended by additivity to sums of the form (1) and then to the completion. These operators are unitary and form a projective representation of G^X, namely, $U_{\tilde{g}_1} U_{\tilde{g}_2} = c(\tilde{g}_1\tilde{g}_2)U_{\tilde{g}_1\tilde{g}_2}$, where $c(\tilde{g}_1, \tilde{g}_2)$ is defined by (6).

2. Construction of a representation of G^X, where $G = PSL(2, \mathbf{R})$. We now apply the general construction described above to the case of the group $G = PSL(2, \mathbf{R})$, given in the second form.

We define H as the space of all continuous functions on the circle $|\xi| = 1$ in which the representation acts according to the following formula:

$$(T(g)f)(\zeta) = f\left(\frac{\alpha\zeta + \bar{\beta}}{\beta\zeta + \bar{\alpha}}\right)|\beta\zeta + \bar{\alpha}|^{-2}, \tag{7}$$

and the invariant linear functional l is

$$l(f) = \frac{1}{2\pi}\int_0^{2\pi} f(e^{it})\, dt.$$

In the subspace H_0 of functions $f(\zeta)$ for which $l(f) = 0$ we specify a scalar product as follows:

$$\|f\|^2 = \sum_{n\neq 0} \frac{1}{|n|}|a_n|^2, \quad \text{where} \quad f(e^{it}) = \sum_{n\neq 0} a_n e^{int}, \tag{8}$$

or, in integral form,

$$\|f\|^2 = c\int_0^{2\pi}\int_0^{2\pi} \ln\left|\sin\frac{t_1 - t_2}{2}\right| f(e^{it_1})\overline{f(e^{it_2})}\, dt_1\, dt_2. \tag{9}$$

It is clear from (8) that this scalar product is positive definite, and from

(9) that it is invariant under the operators $T(g)$ of the form (7). (We recall that $\int_0^{2\pi} f(e^{it})\,dt = 0.$)

We now construct $\mathcal{H}^0$. We fix in H the function $\xi_0 = f_0(\zeta) \equiv 1$. Then $\beta(g) = T(g)f_0 - f_0$; hence, $\beta(g, \zeta) = |\beta\zeta + \bar{\alpha}|^{-2} - 1$, where $g = \begin{pmatrix} \alpha & \beta \\ \bar{\beta} & \bar{\alpha} \end{pmatrix}$.

Note that $\beta(g, \zeta)$ takes only real values. We examine the set H^X of continuous functions $f(x, \zeta)$ satisfying the following condition:

$$\frac{1}{2\pi}\int_0^{2\pi} f(x, e^{it})\,dt = 1 \quad \text{for all} \quad x \in X.$$

The elements of $\mathcal{H}^0$ are all possible finite formal linear combinations of such functions: $\sum \lambda_i \circ f_i(x, \zeta)$, $\lambda_i \in \mathbf{C}$. A scalar product is defined for any pair of functions $\tilde{f}_1 = f_1(x, \zeta)$ and $\tilde{f}_2 = f_2(x, \zeta)$ in H^X by the formula

$$\langle \tilde{f}_1, \tilde{f}_2 \rangle = \exp\left(c \int \ln\left|\sin\frac{t_1 - t_2}{2}\right| f_1'(x, e^{it_1})\overline{f_2'(x, e^{it_2})}\,dt_1\,dt_2\,dm(x)\right),$$

where $f_i' = f_i - 1$, and is then extended by linearity to the whole space $\mathcal{H}^0$.

The representation operator $U_{\tilde{g}}$ is defined by the formula

$$(U_{\tilde{g}}\tilde{f})(x, \zeta) = \lambda(\tilde{g}, \tilde{f}) \circ f\left(x, \frac{\alpha(x)\zeta + \overline{\beta(x)}}{\beta(x)\zeta + \overline{\alpha(x)}}\right) |\beta(x)\zeta + \overline{\alpha(x)}|^{-2},$$

$f \in H^X$, where

$$\lambda(\tilde{g}, \tilde{f}) = \exp\left(-c \int \ln\left|\sin\frac{t_1 - t_2}{2}\right| \times \right.$$
$$\left. \times \left(T(g(x))f'(x, e^{it_1}) + \frac{1}{2}\beta(g(x), e^{it_1}),\ \beta(g(x), e^{it_2})\right) dt_1\,dt_2\,dm(x)\right),$$

$f' = f - 1$, and is then extended by linearity first to the whole space $\mathcal{H}^0$, and then to its completion $\mathcal{H}$.

We note that in the case considered here the scalar product of the vectors $T(g_1(x))\beta(g_2(x))$ and $\beta(g_1(x))$ is real. Therefore, by Lemma 6.3 (see the expression for $c(\tilde{g}_1, \tilde{g}_2)$), the operators $U_{\tilde{g}}$ form a representation of G^X.

We make here an essential remark. We can take for H the subspace of functions on the unit circle $|\zeta| = 1$ that are boundary values of analytic functions analytic (or anti-analytic) in the interior of the unit disc. Then we obtain other representations of G^X, which are projective.

3. Another construction of a representation of G^X, where $G = PSL(2, \mathbf{R})$. It is sometimes convenient to define representations of G not on functions on the circle but on functions on the line. Let G be given in the first form.

We consider the space H of all real continuous functions that satisfy the following condition: $f(t) = O(t^{-2})$ as $t \to \pm\infty$; and we define a representation of G in H by the following formula:

$$(T(g)f)(t) = f\left(\frac{\alpha t+\gamma}{\beta t+\delta}\right)|\beta t+\delta|^{-2}.$$

We further define in H a G-invariant linear functional $l(f)$:

$$l(f) = \int_{-\infty}^{+\infty} f(t)\,dt.$$

Let $H_0 \subset H$ be the subspace of functions on which $l(f) = 0$. In H_0 there is an invariant positive definite scalar product

$$(f_1, f_2) = -\int_{-\infty}^{+\infty}\int_{-\infty}^{+\infty} \ln|t_1-t_2|\, f_1(t_1)\overline{f_2(t_2)}\,dt_1\,dt_2.$$

We now fix the function $f_0(t) = \frac{\pi^{-1}}{1+t^2}$ in H, for which $l(f_0) = 1$. We also set $\beta(g, t) = (T(g)f_0)(t) - f_0(t)$.

We proceed to the construction of $\mathcal{H}^0$. We consider the set H^X of continuous functions $f(x, t)$, $x \in X$, $t \in \mathbf{R}$, satisfying the following conditions:

1) $f(x, t) = O(t^{-2})$ as $t \to \pm\infty$;

2) $\int_{-\infty}^{+\infty} f(x, t)dt = 1$ for any $x \in X$.

The elements of $\mathcal{H}^0$ are all possible formal linear combinations of such functions: $\sum \lambda_i \circ f_i(x, t)$, $\lambda_i \in \mathbf{C}$.

A scalar product is defined for any pair $f_1(x, t)$, $f_2(x, t)$ by the following formula:

$$\langle f_1(x, t), f_2(x, t)\rangle = \exp\left(-\int \ln|t_1 - t_2|\, f_1'(x, t_1)\overline{f_2'(x, t_2)}\,dt_1dt_2dm(x)\right),$$

where $f_i'(x, t) = f_i(x, t) - \frac{\pi^{-1}}{1+x^2}$.

The representation operator acts as follows:

$$U_{\tilde{g}}f(x, t) = \lambda(\tilde{g}, \tilde{f})\circ f\left(x, \frac{\alpha(x)t+\gamma(x)}{\beta(x)t+\delta(x)}\right)|\beta(x)t+\delta(x)|^{-2},$$

where

$$\lambda(\tilde{g}, \tilde{f}) = \exp\left(\int \ln|t_1-t_2| \times \right.$$
$$\left. \times\left(T(g(x))f'(x, t_1) + \frac{1}{2}\beta(g(x), t), \beta(g(x), t_2)\right)dt_1\,dt_2\,dm(x)\right).$$

REMARK. If we use instead of H only the subspace H^+ (or H^-) of functions which are boundary values of analytic functions in the upper (or lower) half-plane, then we obtain other (projective) representations of G^X.

4. Equivalence of the representation $U_{\tilde{g}}$ to the representation constructed in §5. THEOREM 6.2 *The representation $U_{\tilde{g}}$ of G^X, $G = PSL(2, \mathbf{R})$ constructed here is equivalent to the representation constructed in* §5. *Hence* *it follows that $U_{\tilde{g}}$ is irreducible.*

PROOF. For the proof it is sufficient to construct an isometric mapping of $\mathcal{H}^0$ into the representation space of §5 that commutes with the action of G^X in these spaces.

Let us examine the construction of the representation in §6.2. In it $\mathcal{H}^0$ consists of formal linear combinations of functions $f(x, \zeta)$, $|\zeta| = 1$, such that $\frac{1}{2\pi}\int_0^{2\pi} f(x, e^{it})dt = 1$ for any $x \in X$. We expand $f(x, e^{it})$ in a Fourier series in t: $f(x, e^{it}) = 1 + \sum_{n \neq 0} a_n(x)e^{int}$. We associate with $f(x, t) \in \mathcal{H}^0$ an element of the space $\mathcal{H}_V$ constructed in §5: $F = (1, u(x, n), \ldots, \ldots, u(x_1, n_1), \ldots, u(x_k, n_k), \ldots)$, where $u(x, n) = (-1)^n a_n(x)$. We extend this to a linear mapping of the whole space $\mathcal{H}^0$ onto $\mathcal{H}_V$. From the definition of the norm in these spaces it follows easily that the mapping so constructed is an isometry. Furthermore, it can be shown that it commutes with the action of G^X in these spaces.

§7. Construction with a Gaussian measure

1. To explain the construction of this section we find it convenient to make some modifications in the general constructions in §6.1.

Let G be a topological group, E a real Hilbert space, and suppose that an orthogonal representation $T(g)$ of G in E and a cocycle with values in E are given, that is, a function β: $G \to E$ satisfying the relation $\beta(g_1) + T(g_1)\beta(g_2) = \beta(g_1g_2)$.

From the part (T, β) we construct a new unitary representation of G. In what follows we change the notation for the group and write Γ in place of G, because in the examples Γ can be both G and G^X.

First we construct a new (complex) space $\mathcal{H}^0$ whose elements are formal finite linear combinations of elements $\xi_i \in E$:

$$\lambda_1 \circ \xi_1 + \ldots + \lambda_n \circ \xi_n, \quad \lambda_i \in \mathbf{C}. \tag{1}$$

In contrast to the preceding section, $\lambda \circ \xi$ and $\lambda\xi$ are now regarded as distinct.

Thus, the original space E has a natural embedding in $\mathcal{H}^0$ as a subset (not as a subspace!).

We now define a scalar product in $\mathcal{H}^0$. Namely, for elements $\xi_1, \xi_2 \in H$ we define a scalar product by the formula $\langle \xi_1, \xi_2 \rangle = \exp(\xi_1, \xi_2)$, where the round parentheses denote the scalar product in E, and then we extend this scalar product by linearity to all formal linear combinations like (1). It is easy to verify that the Hermitian form so introduced is positive definite (see the proof of Lemma 6.2).

Let $\mathcal{H}$ be the completion of $\mathcal{H}^0$ in the scalar product just introduced. We define a representation U_γ of Γ in $\mathcal{H}^0$. The action of operators U_γ,

$\gamma \in \Gamma$, on elements $\xi \in E$ is given as follows:

$$U_{\gamma^{-1}}\xi = \exp\left(-\tfrac{1}{2}\|\beta(\gamma)\|^2 - (T_\gamma\xi, \beta(\gamma))\right) \circ (T(g)\xi + \beta(g)),$$

and then we extend these operators by linearity to the whole space $\mathcal{H}^0$.

It can easily be established that the U_γ are unitary operators, hence can be extended to the whole space $\mathcal{H}$, and that these operators form a representation of Γ. Later we shall see how this construction is related to that of Araki and Streater.

2. Let us consider two examples.

a) Let $\Gamma = PSL(2, \mathbf{R})$. Let $T(g)$ be the representation of $PSL(2, \mathbf{R})$ constructed in §6.3 (or §6.2), and $\beta(g)$ the cocycle defined there. We denote by E the real subspace of H_0 also given there. From it we construct $\mathcal{H}$ and the representation U_g.

THEOREM 7.1. *The representation so constructed coincides with the canonical representation introduced in* §1.

b) Let $\Gamma = (PSL(2, \mathbf{R}))^X$. We consider the space of all mappings $\tilde{f}: X \to E$ that are measurable on X and satisfy the condition

$$\|\tilde{f}\|^2 = \int \|f(x)\|^2\, dm(x) < \infty.$$

This is a real Hilbert space. We define in it a representation $\tilde{T}$ of Γ by the formula $\tilde{T}(\tilde{g})\tilde{f} = T(g(x))f(x)$, where T is the representation of example a). Next, we introduce a cocycle $\tilde{\beta}$, setting $\tilde{\beta}(\tilde{g}) = \beta(g(x))$, where $\beta(g)$ is the cocycle of example a), and we construct from the pair $(\tilde{T}, \tilde{\beta})$ a representation $U_{\tilde{g}}$ by the procedure indicated above.

THEOREM 7.2. *The representation so constructed coincides with the representations $U_{\tilde{g}}$ constructed in* §§2–6.

3. We explain briefly another method of constructing a representation of the group, which can be specialized to yield the construction explained at the beginning of this section.

Let K be a real Hilbert space and K' its dual space. We consider the functional $\chi(\xi') = e^{-\|\xi'\|^2}$. It is normed, continuous in the norm, and positive definite on K', as on an additive group. Therefore it is the Fourier transform of a weak distribution in K: $\chi(\xi') = \int_K e^{i\langle \xi', \xi\rangle}\, d\nu(k)$. (A weak distribution is a finitely additive, normalized, non-negative measure defined on the algebra of cylinder sets in K, that is, sets of the level of Borel functions of finitely many linear functionals.)

It is known that ν can be extended to a countably additive measure in an arbitrary nuclear extension $\tilde{K}$ of K. We call this the standard Gaussian measure, and we quote two properties of this measure that we shall need.

1) The standard Gaussian measure in $\tilde{K}$ is equivalent (that is, mutually absolutely continuous) to its translations by elements of K.

2) Every orthogonal transformation of K can be uniquely extended to a linear and measurable transformation in $\tilde{K}$ that is defined almost everywhere and preserves the Gaussian measure.

We now consider the space E, introduce the pair (T, β) (See §7.1), and choose the standard Gaussian measure μ in some nuclear extension $\tilde{E}$ of E. We examine the space $L^2(\tilde{E}, \mu)$ of all square integrable complex-valued functionals on $\tilde{E}$ (more precisely, of classes of functionals that coincide almost everywhere). From the pair (T, β), we construct in $L^2(\tilde{E}, \mu)$ a representation U_g:

$$(U_{g^{-1}}F)(\varphi) = e^{-\frac{1}{2}\|\beta(g)\|^2 - \langle T(g)\beta(g), \varphi\rangle} F(T(g)\varphi + \beta(g)),$$

where $F \in L^2(\tilde{E}, \mu)$, $\varphi \in \tilde{E}$.

The following theorem can be proved:

THEOREM 7.3. *The correspondence $g \to U_g$ is a unitary representation of G in $L^2(\tilde{E}, \mu)$.*

REMARK. If there is another cocycle β' in $E' \cong E$, then we can construct the more general representation:

$$(U_{g^{-1}}^{(\beta', \beta)}F)(\varphi) = e^{i\langle\beta', \varphi\rangle} (U_{g^{-1}}F)(\varphi).$$

4. We indicate briefly the connection between the representation constructed here and that constructed in §7.1. Let $\xi \in E$. To ξ we assign the following function in $L^2(\tilde{E}, \mu)$: $\xi \longmapsto F(\varphi) = ce^{(\xi, \varphi)}$, where the number c is determined by the condition $c^2 \int e^{2(\xi, \varphi)}\, d\mu = 1, \quad c > 0.$

It is easy to verify that $c = c_0 e^{-\|\varphi\|^2}$, where c_0 is an absolute constant.

Let F_1 and F_2 correspond to the elements ξ_1 and ξ_2 of E. We compute

$$(F_1, F_2) = \int F_1(\varphi)F_2(\varphi)d\mu(\varphi).$$

It is easy to see that $(F_1, F_2) = e^{(\xi_1, \xi_2)}$, and thus is the isometric mapping of $\mathcal{H}$ into $L^2(\tilde{E}, \mu)$ given in §7.1. It can be verified that this mapping is an isomorphism. An elementary calculation shows that for a given $T(g)$ the representations in $\mathcal{H}$ and $L^2(\tilde{E}, \mu)$ are equivalent.

It is well known that the space $L^2(\tilde{E}, \mu)$, where μ is the standard Gaussian measure, can be represented in the form

$$L^2(\tilde{E}, \mu) = \exp H \cong \sum_{n=0}^{\infty} \oplus \frac{1}{n!} \underbrace{H \otimes \dots \otimes H}_{n},$$

where H is the complexification of E' and $\underbrace{H \otimes \dots \otimes H}_{n}$ is the subspace of generalized Hermite polynomials of degree n. Hence the preceding investigations show that our representation of G^X is realized in $\mathcal{H} = \exp H$ by means of the cocycle β (See §7.1). This realization coincides with the general scheme of Streater and Araki, which they have examined, however, only for certain soluble groups. In the terms used here the problem of the construction of a representation reduces to that of the discovery of the cocycle β and to the proof of the irreducibility of the representation of G^X. We have done this in the present paper for $G = PSL(2, \mathbf{R})$.

References

[1] I. M. Gel'fand and M. I. Graev, Representations of the quaternion groups over locally compact fields and function fields, Funktsional. Anal. i Prilozhen. 2 (1968) no. 1, 20–35. MR **38** # 4611.

[2] I. M. Gel'fand and I. Ya. Vilenkin, *Nekotorye primeneniya garmonicheskogo analiza. Osnashchennye gilbertory prostranstva*, Gos. Izdat. Fiz.-Mat. lit. Moscow 1961. MR **26** # 4173.
Translation: Generalized Functions, Vol. 4: Applications of harmonic analysis, Academic Press, New York and London 1964. MR **30** # 4152.

[3] I. M. Gel'fand, M. I. Graev, and I. Ya. Vilenkin, *Integral'naya geometriya i svyazannye s nei voprosy teorii predstavlenii* Gos. Izdat. Fiz.-Mat. lit. Moscow 1962. MR **28** # 3324.
Translation: Generalized Functions, Volume 5: Integràl geometry and representation theory, Academic Press, New York and London 1966. MR **34** # 7726.

[4] H. Araki, Factorizable representations of current algebra, Publ. Res. Inst. Math. Sci. **5** (1969/70), 361–422. MR **41** # 7931.

[5] I. Dixmier, Les C*-algèbres et leurs représentations, second ed., Gauthier-Villars, Paris 1969. MR **30** # 1404, **39** # 7442.

[6] A. Guichardet, Symmetric Hilbert spaces and related topics, Lecture Notes in Mathematics, Springer-Verlag, Berlin–Heidelberg–New York 1972.

[7] D. Mathon, Infinitely divisible projective representations of the Lie Algebras, Proc. Cambridge Philos. Soc. **72** (1972) 357–368.

[8] K. R. Parthasarathy, Infinitely divisible representations and positive functions on a compact group, Comm. Math. Phys. **16**, 148–156 (1970).

[9] K. R. Parthasarathy and K. Schmidt, Infinitely divisible projective representations, cocycles, and Levy-Khinchine-Araki formula on locally compact groups, Research Report 17, Manchester–Sheffield School of Probability and Statistics, 1970.

[10] K. R. Parthasarathy and K. Schmidt, Factorizable representations of current groups and the Araki-Woods embedding theorem, Acta Math. **128**, 53–71 (1972).

[11] K. R. Parthasarathy and K. Schmidt, Positive definite kernels, continuous tensor products, and central limit theorems of probability theory, Lecture Notes in Mathematics, Springer-Verlag, Berlin–Heidelberg–New York 1972.

[12] L. Pukanszky, On the Kroneker products of irreducible representations of the 2 x 2 real unimodular group, Part I, Trans. Amer. Math. Soc. **100** (1961) 116–152.

[13] R. F. Streater, Current commutation relations, continuous tensor products, and infinitely divisible group representations, Rend. Sci. Ist. Fis. E. Fermi, **11** (1969), 247–263.

[14] R. F. Streater, Continuous tensor products and current commutation relations, Nuovo Cimento **A 53** (1968), 487.

[15] R. F. Streater, Infinitely divisible representations of Lie algebras, Z. Wahrscheinlichkeitstheorie und Verw. Gebiete **19**, 1971, 67–80.

Received by the Editors,
15 June 1973

Translated by W. J. Holman

To the memory of Sergei Vasil'evich Fomin

REPRESENTATIONS OF THE GROUP OF DIFFEOMORPHISMS

A. M. Vershik, I. M. Gel'fand and M. I. Graev

This article contains a survey of results on representations of the diffeomorphism group of a non-compact manifold X associated with the space Γ_X of configurations (that is, of locally finite subsets) in X. These representations are constructed from a quasi-invariant measure μ on Γ_X. In particular, necessary and sufficient conditions are established for the representations to be irreducible. In the case of the Poisson measure μ a description is given of the corresponding representation ring.

Contents

Introduction

This article is a survey of results on the representations of the group Diff X of finite diffeomorphisms of a smooth non-compact manifold X. As for many infinite groups, it is rather difficult to see what the complete stock of irreducible unitary representations of this group might be. Therefore, it is of some interest to single out certain natural classes of representations.

We consider the space Γ_X of infinite configurations (that is, locally finite subsets) in X on which the group Diff X acts in a natural way. If μ is a quasi-invariant measure in Γ_X and ρ is a representation of the symmetric group $S_n (n = 1, 2, \dots)$, then we construct a unitary representation of Diff X from μ and ρ, which we call elementary. There is, therefore, a close connection between the theory of elementary representations of Diff X and the theories of quasi-invariant measures on Γ_X and representations of the symmetric groups. We note that quasi-invariant measures on Γ_X are studied in statistical physics (Gibbs measures and the simplest of them – the Poisson measure) (see, for example, [12]); and in the theory of point processes (see, for example, [17] and elsewhere). The space of infinite configurations Γ_X is, in its own right, a very important example of an infinite-dimensional manifold, and its study is one of the interesting problems of topology, analysis, and statistical physics.

Representations of Diff X that are of finite functional dimension, that is, representations associated with the space of finite configurations, were considered in [8] and [9]. In §1 we incidentally prove by a new method that the representations of a wide class are irreducible. However, we are basically interested in representations of infinite functional dimension associated with Γ_X; they can be regarded as limits of "partially finite" representations.

In this paper necessary and sufficient conditions are obtained for elementary representations to be irreducible. In the case when μ is the Poisson measure it is proved that the set of elementary representations is multiplicatively closed, that is, the tensor product of two elementary representations splits into the sum of elementary representations, and the structure of the corresponding representation ring is described.

An important property of Diff X, which distinguishes it from locally compact groups and which will become apparent in the situations we discuss, is that to a single orbit of Diff X in Γ_X there is no corresponding representation; however, one can construct a representation from a measure on Γ_X that is ergodic with respect to the action of Diff X. More interesting and more widely studied is the class of representations associated with the Poisson measure on Γ_X (see §4). The representation of Diff X in the space $L^2_\mu(\Gamma_X)$, where μ is the Poisson measure, arose (as an N/V limit) in [15]; however, the role of the Poisson measure was not noted here. It is remarkable that this same representation can be realized in a Fock space as $\mathrm{EXP}_\beta T$, where T is a representation of Diff X in $L^2_m(X)$ and β is a certain cocycle (see §4). This circumstance links the theory that we discuss here with [1] and [2];

Representations of Diff X associated with the Poisson measure on Γ_X are studied by another method in [22]. As far as we know, up to now, no measures, and in particular no Gibbs measures apart from the Poisson measures, have been discussed in connection with representations of Diff X.

Representations of the cross product of the additive group $C^\infty(X)$ and the group Diff X are investigated in a number of very interesting physics papers (see [15] for a list of references; see also [19] and [20]). All the representations of Diff X discussed in this article extend to representations of the cross product $C^\infty(X)\cdot$Diff X; the mathematical part of the results of [15] is contained in this paper.

§0. Basic definitions and some preliminary information

1. The group Diff X. Everywhere, X denotes a connected manifold of class C^∞. Diff X denotes the group of all diffeomorphisms $\psi: X \to X$ that are the identity outside a compact set (depending on ψ). The group Diff X is assumed to be furnished with the natural topology: a sequence ψ_n is regarded as tending to ψ if ψ and every ψ_n is the identity outside a certain compact set K and if ψ_n, together with all its derivatives, tends to ψ uniformly on K. If $Y \subset X$ is an arbitrary open subset, then Diff Y denotes the subgroup of diffeomorphisms $\psi \in$ Diff X that are the identity on $X \setminus Y$.

2. The groups S^∞ and S_∞. We denote by S^∞ the group of all permutations of the sequence of natural numbers, by $S_\infty \subset S^\infty$ the subgroup of all finite permutations, and by S^n the group of all permutations of the numbers $1, \ldots, n$ $(n = 1, 2, \ldots)$. We regard the S_n as subgroups of S_∞; thus, $S_1 \subset \ldots \subset S_n \subset \ldots$ and $S_\infty = \varinjlim S_n$. In what follows, S_0 is understood to mean the trivial group.

3. The configuration spaces Γ_X and B_X. Any locally finite subset of X is called a configuration[1] in X, that is a subset $\gamma \subset X$ such that $\gamma \cap K$ is finite for any compact set $K \subset X$. By this definition, any configuration is either a finite or a countable subset of X; if X is compact, then all configurations in X are finite.

Let us denote by Γ_X the space of all infinite and by B_X the space of all finite configurations in X. The group of diffeomorphisms Diff X acts naturally on Γ_X and B_X. The space of finite configurations B_X decomposes

---

[1] Sometimes a configuration is defined differently, allowing points $x \in X$ to be included in γ with repetitions; with such a definition a configuration is not a subset of X.

into a countable union of subsets that are transitive under Diff X: $B_X = \bigsqcup_{n \geq 0} B_X^{(n)}$, where $B_X^{(n)}$ is the collection of all n-point subsets in X. We note that $B_X^{(0)}$ consists of a single element – the empty set ϕ.

For any subset $Y \subset X$ with compact closure the space Γ_X splits into the product $\Gamma_X = B_Y \times \Gamma_{X \setminus Y}$. Consequently, since $B_Y = \bigsqcup_{n \geq 0} B_Y^{(n)}$, we have $\Gamma_X = \bigsqcup_{n \geq 0} B_Y^{(n)} \times \Gamma_{X \setminus Y}$, and all the subsets in this decomposition are invariant under Diff Y.

4. The space $\widetilde{X}^\infty$ and the topology in Γ_X. Let us consider the infinite product $X^\infty = \prod_{i=1}^{\infty} X_i$, $X_i = X$, furnished with the weak topology. The group S^∞ acts naturally on X^∞. We define the subset $\widetilde{X}^\infty \subset X^\infty$ as the set of all sequences $(x_1, \ldots, x_n, \ldots) \in X^\infty$ such that: 1) $x_i \neq x_j$ when $i \neq j$ and 2) the sequence $x_1, \ldots, x_n, \ldots$ has no accumulation points in X. The space $\widetilde{X}^\infty$ is invariant under the action of Diff X and S^∞, and the S^∞-orbit of any point of $\widetilde{X}^\infty$ is closed.

There is a natural bijection $\widetilde{X}^\infty / S^\infty \to \Gamma_X$. We introduce the corresponding quotient topology in Γ_X; this topology is Hausdorff and metrizable. Similarly, the bijections $\widetilde{X}^n / S_n \to B_X^{(n)}$, where

$$\widetilde{X}^n = \{(x_1, \ldots, x_n) \in X^n;\ x_i \neq x_j \text{ when } i \neq j\},$$

and S_n acts on X^n as the permutation group of the coordinates, give the topology on $B_X^{(n)} (n = 1, 2, \ldots)$ and hence on $B_X = \bigsqcup_{n \geq 0} B_X^{(n)}$.

It is easy to see that Γ_X is, as a topological space, the projective limit of the spaces B_K. Namely, $\Gamma_X = \varprojlim (B_K, \pi_{KK'})$, where K runs through the open submanifolds in X with compact closures, and $\pi_{KK'}: B_K \to B_{K'} (K' \subset K)$ is the restriction of the configuration $\gamma \in B_K$ to K', that is, $\pi_{KK'} \gamma = \gamma \cap K'$.

5. Quasi-invariant and ergodic measures. Let G be a group acting on a space Y. A measure μ given on some G-invariant σ-algebra in Y is said to be quasi-invariant under G if the inverse image of any measurable set of positive measure, under any transformation g: $Y \to Y$ with $g \in G$, has positive measure. If μ is quasi-invariant, then the measures μ and $g\mu$ (where $g\mu$ is defined as the image of μ, that is, $g\mu(C) = \mu(g^{-1}C)$) are equivalent; the density of $g\mu$ with respect to μ at a point $y \in Y$ is denoted by $\dfrac{d\mu(g^{-1}y)}{d\mu(y)}$. The class of σ-finite measures equivalent to μ is called the type of μ.

A quasi-invariant measure μ in Y is said to be ergodic with respect to the action of G if every measurable set $A \subset Y$ such that $\mu(gA \ \Delta \ A) = 0$ for any $g \in G$ is either a null set or a set of full measure.

We discuss measures on Γ_X, and other spaces connected with Γ_X, that are quasi-invariant under the action of Diff X, and we construct from these measures unitary representations of Diff X.

6. Measures in the configuration space Γ_X. We define,[1] as usual, the σ-algebra $\mathfrak{A}(\Gamma_X)$ of Borel sets on Γ_X. Henceforth, when we talk of measures on Γ_X, we mean[2] complete, non-negative, Borel, normalized, countably-additive measures μ. Since the structure of a complete metric space can be introduced in Γ_X, for any Borel measure μ the space (Γ_X, μ) is (after taking the completion of the σ-algebra $\mathfrak{A}(\Gamma_X)$ with respect to μ) a Lebesgue space [11], and the technique of conditional decomposition (conditional measures, and so on) can be applied. The same applies to other spaces and fibre bundles over Γ_X that occur in this paper.

Many measures on Γ_X arising for various reasons in statistical physics and probability theory turn out to be quasi-invariant and ergodic under Diff X. The following example is classical.

POISSON MEASURE. Given any positive[3] smooth measure m on a manifold X, we consider the union $\Delta_X = B_X \cup \Gamma_X$ of all configurations on X. We define the measure of each subset $\{\gamma \in \Delta_X ; |\gamma \cap U| = n\}$ by

$$\mu\{\gamma \in \Delta_X;\ |\gamma \cap U| = n\} = \frac{[\lambda m(U)]^n}{n!} e^{-\lambda m(U)},$$

where $\lambda > 0$ is fixed. By Kolmogorov's theorem there exists a unique measure on $\mathfrak{A}(\Gamma_X)$ defined by these conditions. It is called the Poisson measure with parameter λ (associated with the measure m on X).

Let us note the following important properties of the Poisson measure μ, which follow immediately from its definition.

1) When $m(X) < \infty$, the measure μ is concentrated on the set B_X of finite configurations, and when $m(X) = \infty$, it is concentrated on Γ_X.

2) Suppose that the manifold $X = X_1 \cup \ldots \cup X_n$ is split arbitrarily into finitely many disjoint measurable subsets, that $\Delta_X = \Delta_{X_1} \times \ldots \times \Delta_{X_n}$ is the corresponding decomposition of Δ_X into a direct product, and that μ_i is the projection of the Poisson measure μ onto Δ_{X_i} $(i = 1, \ldots, n)$. Then $\mu = \mu_1 \times \ldots \times \mu_n$. This property of the Poisson measure is called *infinite decomposability*.

3) The Poisson measure is quasi-invariant under Diff X and invariant under the subgroup Diff$(X, m) \subset$ Diff X of diffeomorphisms preserving m. Here,

---

[1] Note that $\mathfrak{A}(\Gamma_X)$ is σ-generated by sets of the form $C_{U,n} = \{\gamma \in \Gamma_X; |\gamma \cap U| = n\}$, where U runs over the compact sets in X $(n = 0, 1, \ldots)$.

[2] In statistical physics a measure μ on Γ_X is usually called a state (see, for example, [12]) and in probability theory and the theory of mass observation it is usually called a point random process (see, for example, [17]).

[3] By a positive smooth measure we mean a measure with positive density at all points $x \in X$.

$$(1) \qquad \frac{d\mu(\psi^{-1}\gamma)}{d\mu(\gamma)} = \prod_{x\in\gamma} \frac{dm(\psi^{-1}x)}{dm(x)}$$

(the product makes sense, because by the finiteness of ψ, almost all the factors are equal to 1).

4) If $m(X) = \infty$, then the Poisson measure μ is ergodic with respect to Diff X. Furthermore (see §4), if dim $X > 1$, then the Poisson measure is ergodic with respect to Diff (X, m).

Any measure in B_X that is quasi-invariant under Diff X is equivalent to a sum of smooth positive measures on $B_X^{(n)}$. In particular, any two quasi-invariant measures on $B_X^{(n)}$ are equivalent. Let us note that for any $Y \subset X$, where $\bar{Y}$ is compact, the projection of any quasi-invariant measure in Γ_X onto $B_Y = \bigsqcup_{n\geq 0} B_Y^{(n)}$ is non-zero for all n.

§1. The ring of representations of Diff X associated with the space of finite configurations

We discuss here the simplest class of representations of Diff X. These representations have finite functional dimension; from the point of view of orbit theory they have been discussed in detail by Kirillov [9].

1. **The representations V^ρ.** We associate with each pair (n, ρ), where ρ is a unitary representation of the symmetric group S_n in a space $W (n = 0, 1, \ldots)$, a unitary representation V^ρ of Diff X. The construction of V^ρ is similar to Weyl's construction of the irreducible finite-dimensional representations of the general linear group.

Given a positive smooth measure m on X, we define m_n in X^n to be the product measure: $m_n = m \times \ldots \times m$. We consider the space $L^2_{m_n}(X^n, W)$ of functions F on X^n with values in the representation space W of ρ such that

$$\| F \|^2 = \int \| F(x_1, \ldots, x_n) \|^2_W \, dm(x_1) \ldots dm(x_n) < \infty.$$

A unitary representation U_n of Diff X is given on $L^2_{m_n}(X^n, W)$ by the formula

$$(1) \qquad (U_n(\psi) F)(x_1, \ldots, x_n) = \prod_{k=1}^{n} J_\psi^{1/2}(x_k) F(\psi^{-1}x_1, \ldots, \psi^{-1}x_n),$$

where $J_\psi(x) = \dfrac{dm(\psi^{-1}x)}{dm(x)}$. Let us denote by $H_{n,\rho}$ the subspace of functions $F \in L^2_{m_n}(X^n, W)$ such that $F(x_{\sigma(1)}, \ldots, x_{\sigma(n)}) = \rho^{-1}(\sigma)F(x_1, \ldots, x_n)$ for any $\sigma \in S_n$. It is obvious that $H_{n,\rho}$ is invariant under Diff X.

We define the representation V^ρ of Diff X as the restriction of U_n from $L^2_{m_n}(X^n, W)$ to $H_{n,\rho}$.

In the particular case when ρ is the unit representation of S_n, then V^ρ acts by (1) on the space of scalar functions $F(x_1, \ldots, x_n)$ that are symmetric in all the arguments.

It is obvious that if m is replaced on X by any other smooth positive measure, each V^ρ is replaced by an equivalent representation.

Let us construct another realization of V^ρ, which will be useful later on. Let $\widetilde{X}^n \subset X^n$ be the submanifold of points $(x_1, \ldots, x_n) \in X^n$ with pairwise distinct coordinates. We consider the fibration p of $\widetilde{X}^n$ by the orbits of S_n, p: $\widetilde{X}^n \to \mathrm{B}_X^{(n)}$. Note that $p \circ \psi = \psi \circ p$ for any $\psi \in \mathrm{Diff}\ X$. Suppose that we are given any measurable cross section s: $\mathrm{B}_X^{(n)} \to \widetilde{X}^n$. Obviously, for any $\psi \in \mathrm{Diff}\ X$ and $\gamma \in \mathrm{B}_X^{(n)}$ the elements $s(\psi^{-1}\gamma)$ and $\psi^{-1}(s\gamma)$ lie in the same fibre of p, and we define a function σ on $\mathrm{Diff}\ X \times \mathrm{B}_X^{(n)}$ with values in S_n by the formula
$s(\psi^{-1}\gamma) = [\psi^{-1}(s\gamma)]\,\sigma(\psi, \gamma)$, where[1] $(x_1, \ldots, x_n)\sigma = (x_{\sigma(1)}, \ldots, x_{\sigma(n)})$.

Let $\mu = pm_n$ be the projection onto $\mathrm{B}_X^{(n)}$ of the measure $m_n = m \times \ldots \times m$ on X^n. We denote by $L^2_\mu(\mathrm{B}_X^{(n)}, W)$ the space of functions F on $\mathrm{B}_X^{(n)}$ with values in W such that

$$\| F \|^2 = \int \| F(\gamma) \|^2_W \, d\mu(\gamma) < \infty.$$

We define the representation V^ρ of Diff X in $L^2_\mu(\mathrm{B}_X^{(n)}, W)$ by

$$(V^\rho(\psi)F)(\gamma) = \left(\frac{d\mu(\psi^{-1}\gamma)}{d\mu(\gamma)}\right)^{1/2} \rho(\sigma(\psi, \gamma))\, F(\psi^{-1}\gamma). \tag{2}$$

It is not difficult to check that this representation is equivalent to the one constructed earlier. To see this it is sufficient to consider the map s^*: $H_{n,\sigma} \to L^2_\mu(\mathrm{B}_X^{(n)}, W)$ induced by the cross section s, $((s^*F)(\gamma) = F(s\gamma))$. It is easy to verify that s^* is an isomorphism and that the operators $V^\rho(\psi)$ in $H_{n,\rho}$ go over under s^* to operators of the form (2).

In the particular case when ρ is the unit representation of S_n, then V^ρ acts on $L^2_\mu(\mathrm{B}_X^{(n)})$ according to the formula

$$(V^\rho(\psi)F)(\gamma) = \left(\frac{d\mu(\psi^{-1}\gamma)}{d\mu(\gamma)}\right)^{1/2} F(\psi^{-1}\gamma).$$

2. **Properties of the representations V^ρ.** From the definition of V^ρ we obtain immediately the following result.

PROPOSITION 1. *For any representations ρ_1 and ρ_2 of S_n $(n = 0, 1, \ldots)$ there is an equivalence* $V^{\rho_1 \oplus \rho_2} \cong V^{\rho_1} \oplus V^{\rho_2}$.

DEFINITION (see [18]). The *exterior product* $\rho_1 \circ \rho_2$ of representations ρ_1 of S_{n_1} and ρ_2 of S_{n_2} is the representation of $S_{n_1+n_2}$ induced by the

---

[1] σ is a 1-cocycle of Diff X with values in the set of measurable maps $\mathrm{B}_X^{(n)} \to S_n$ (see Appendix 2).

representation $\rho_1 \times \rho_2$ of $S_{n_1} \times S_{n_2}$: $\rho_1 \circ \rho_2 = \mathrm{Ind}_{S_{n_1} \times S_{n_2}}^{S_{n_1+n_2}} (\rho_1 \times \rho_2)$. We are assuming that S_{n_1} and S_{n_2} are embedded in $S_{n_1+n_2}$ as the subgroups of permutations of $1, \ldots, n_1$ and of $n_1 + 1, \ldots, n_n + n_2$, respectively. Note (see [18]) that exterior multiplication is commutative and associative. The following fact parallels standard results about representations of the classical groups in the Weyl realization.

PROPOSITION 2. *For any $n_1, n_2 = 0, 1, 2, \ldots$ and any representations ρ_1 and ρ_2 of S_{n_1} and S_{n_2}, respectively, there is an equivalence* $V^{\rho_1 \circ \rho_2} \cong V^{\rho_1} \otimes V^{\rho_2}$.

COROLLARY. *The set of representations V^ρ is closed under the operation of tensor multiplication.*

THEOREM 1. 1) *If ρ is an irreducible representation of S_n, then the representation V^ρ of* Diff X *is irreducible.* 2) *Two representations V^{ρ_1} and V^{ρ_2}, where ρ_1 and ρ_2 are irreducible representations of S_{n_1} and S_{n_2}, respectively, are equivalent if and only if $n_1 = n_2$ and $\rho_1 \sim \rho_2$.*

PROOF. We consider V^{ρ_n}, where ρ_n is the regular representation of S_n ($n = 0, 1, 2, \ldots$). It is easy to see that V^{ρ_n} is equivalent to the representation in $\overset{n}{\otimes} L^2_m(X)$ given by

$$(V^{\rho_n}(\psi) F)(x_1, \ldots, x_n) = \prod_{k=1}^{n} J_\psi^{1/2}(x_k) F(\psi^{-1}x_1, \ldots, \psi^{-1}x_n).$$

Results of Kirillov ([9], Theorem 4) imply that the V^{ρ_n} are pairwise disjoint and that the number of interlacings of V^{ρ_n} is $n!$, that is, equal to the number of interlacings of ρ_n. Hence and from Proposition 1 the assertion of the theorem follows immediately.

When dim $X > 1$, a stronger assertion is true, which we prove independently of the results in [9]. Namely, let m be an arbitrary smooth positive measure on X such that $m(X) = \infty$. We denote by Diff(X, m) the subgroup of diffeomorphisms $\psi \in$ Diff X that leave m invariant.

THEOREM 2. *If* dim $X > 1$, *then the assertion of Theorem 1 is true for the restrictions of the V^ρ to* Diff(X, m).

The proof will depend on the following two assertions.

LEMMA 1. *For any natural number n and any set of distinct points $x_1, \ldots, x_n$ in X there exist neighbourhoods $O_1, \ldots, O_n$, corresponding to $x_1, \ldots, x_n$, with the following properties:*

1) *the closure $\bar{O}_i$ of O_i is C^∞-diffeomorphic to a disc, $\bar{O}_i \cap \bar{O}_j = \phi$ when $i \neq j$ and $m(O_1) = \ldots = m(O_n)$;*

2) *for any permutation $(k_1, \ldots, k_n)$ of $1, \ldots, n$ there is a diffeomorphism $\psi \in$ Diff(X, m) such that $\psi(\bar{O}_i) = \bar{O}_{k_i}$ $(i = 1, \ldots, n)$.*

PROOF. It is sufficient to consider the case when X is an open ball and m is the Lebesgue measure in X. In this case it is easy to check that for any x_i and x_j, $i \neq j$, there is a diffeomorphism $\psi_{ij} \in$ Diff(X, m) with the following properties:

1) for any sufficiently small $\varepsilon > 0$ we have
$\psi_{ij} D^{\varepsilon}_{x_i} = D^{\varepsilon}_{x_j}$, $\psi_{ij} D^{\varepsilon}_{x_j} = D^{\varepsilon}_{x_i}$, where D^{ε}_{x} is a disc of radius ε with centre at $x \in X$;

2) the diffeomorphism ψ_{ij} is the identity in neighbourhoods of x_k for which $k \neq i, j$.

Hence the assertion of the lemma follows immediately.

LEMMA 2. *For any open connected submanifold $Y \subset X$ with compact closure, the subspace $\widetilde{L}^2_m(Y) \subset L^2_m(Y)$ of functions f on Y such that $\int_Y f(y)dm(y) = 0$ is irreducible under the operators of the representation of* $\mathrm{Diff}(Y, m)$: $(U(\psi)f(y) = f(\psi^{-1}y)$.

PROOF. First we claim that for any non-trivial invariant subspace $\mathcal{L} \subset \widetilde{L}^2_m(Y)$ and any neighbourhood $O \subset Y$, where O is C^∞-diffeomorphic to a disc, there is a vector $f \in \mathcal{L}$, $f \neq 0$, such that $\operatorname{supp} f \subset O$. For let us take an arbitrary vector $f^{(1)} \in \mathcal{L}$, $f^{(1)} \neq 0$. Since $f^{(1)} \neq$ const on Y, there is a $y_0 \in Y$ such that $f^{(1)} \neq$ const in any neighbourhood O' of y_0. Consequently, there exists a diffeomorphism $\psi \in \mathrm{Diff}(Y, m)$ such that $\operatorname{supp} \psi \subset O'$ and $f^{(1)}(\psi y) \not\equiv f^{(1)}(y)$. We put $f^{(2)}y = f^{(1)}(\psi y) - f^{(1)}(y)$. Then $f^{(2)} \in \mathcal{L}$, $f^{(2)} \neq 0$ and $\operatorname{supp} f^{(2)} \subset O'$. If the neighbourhood O' is sufficiently small, then, by Lemma 1, there is a diffeomorphism $\psi_1 \in \mathrm{Diff}(Y, m)$ with $\psi_1 O' \subset O$ that carries $f^{(2)}$ into a vector f with $\operatorname{supp} f \subset O$.

Let us suppose that $\widetilde{L}^2_m(Y) = \mathcal{L}_1 \oplus \mathcal{L}_2$, where $\mathcal{L}_1$ and $\mathcal{L}_2$ are non-zero invariant subspaces. We fix neighbourhoods O and O' in Y such that $\bar{O}$ and $\bar{O}'$ are C^∞-diffeomorphic to discs, $\bar{O} \cap \bar{O}' = \phi$, and $m(O) = m(O')$. From what has been proved, there are $f_i \in \mathcal{L}_i$, $f_i \neq 0$, such that $\operatorname{supp} f_i \subset O$ $(i = 1, 2)$.

It is obvious that we can find a neighbourhood $O_1 \subset O$, where $\bar{O}_1$ is C^∞-diffeomorphic to a disc, and a diffeomorphism $\psi \in \mathrm{Diff}(O, m)$ such that $\int_{O_1} f_1(\psi y)\overline{f_2(y)}dm(y) \neq 0$; without loss of generality we may assume that $\psi = 1$. For any $\varepsilon > 0$ we can write $O = O_1 \cup O^\varepsilon \cup (O \setminus (O_1 \cup O^\varepsilon))$, where $\bar{O}^\varepsilon$ is C^∞-diffeomorphic to a disc, $\bar{O}_1 \cap \bar{O}^\varepsilon = \phi$, and $m(O \setminus (O_1 \cup O^\varepsilon)) < \varepsilon$. It is not difficult to prove that there is a diffeomorphism $\psi_\varepsilon \in \mathrm{Diff}(Y, m)$ that is the identity on O_1 and such that $m(\psi_\varepsilon O^\varepsilon \setminus O') < \varepsilon$ (see, for example, [3], Lemma 1.1). Since $O \cap O' = \phi$, we have

$$\int_O f_1(\psi_\varepsilon y)\overline{f_2(y)}\,dm(y) = \int_{O_1} f_1(y)\overline{f_2(y)}\,dm(y) +$$

$$+ \int_{O^\varepsilon \setminus \psi_\varepsilon^{-1} O'} f_1(\psi_\varepsilon y)\overline{f_2(y)}\,dm(y) + \int_{O \setminus (O_1 \cup O^\varepsilon)} f_1(\psi_\varepsilon y)\overline{f_2(y)}\,dm(y).$$

Consequently, because $\mathscr{L}_1$ and $\mathscr{L}_2$ are orthogonal,

$$\int_{O_1} f_1(y)\overline{f_2(y)}\,dm(y) + \int_{O^\varepsilon \setminus \psi_\varepsilon^{-1}O'} f_1(\psi_\varepsilon y)\overline{f_2(y)}\,dm(y) + \\ + \int_{O \setminus (O_1 \cup O^\varepsilon)} f_1(\psi_\varepsilon y)\overline{f_2(y)}\,dm(y) = 0.$$

Since the second and third terms in this equation can be made arbitrarily small, we have $\int_{O_1} f_1(y)\overline{f_2(y)}dm(y) = 0$, which is a contradiction.

PROOF OF THEOREM 2. Let us realize the representation $V^\rho = V^{n,\rho}$ of Diff X as acting on the subspace $H_{n,\rho} \subset L^2_{m_n}(X^n, W)$, where W is the space of the representation ρ of S_n (for the definition of $H_{n,\rho}$, see §1.1). In this realization the operators of the representation of Diff(X, m) have the following form:

$$(V^{n,\rho}(\psi)F)(x_1, \ldots, x_n) = F(\psi^{-1}x_1, \ldots, \psi^{-1}x_n), \quad \psi \in \mathrm{Diff}\,(X, m).$$

Let $O_1, \ldots, O_n$ be arbitrary disjoint neighbourhoods in X satisfying conditions 1 and 2 of Lemma 1. We denote by $H^{n,\rho}_{O_1,\ldots,O_n}$ the subspace of functions of $H_{n,\rho}$ that are concentrated on $\bigcup_{(k_1,\ldots,k_n)} (O_{k_1} \times \ldots \times O_{k_n}) \subset X^n$ where $(k_1, \ldots, k_n)$ runs over all permutations of $(1, \ldots, n)$; obviously there is a natural isomorphism

$$H^{n,\rho}_{O_1,\ldots,O_n} \cong L^2_m(O_1) \otimes \ldots \otimes L^2_m(O_n) \otimes W.$$

We consider the subspace

$$\widetilde{H}^{n,\rho}_{O_1,\ldots,O_n} \cong \widetilde{L}^2_m(O_1) \otimes \ldots \otimes \widetilde{L}^2_m(O_n) \otimes W,$$

where $\widetilde{L}^2_m(O_i) \subset L^2_m(O_i)$ is the orthogonal complement to the subspace of contsants. From the definition it follows that $\widetilde{H}^{n,\rho}_{O_1,\ldots,O_n}$ is invariant under under the subgroup $G_{O_1,\ldots,O_n}$ of diffeomorphisms $\psi \in \mathrm{Diff}(X, m)$ such that $\psi(O_1 \cup \ldots \cup O_n) = O_1 \cup \ldots \cup O_n$. We denote by $V^{n,\rho}_{O_1,\ldots,O_n}$ the restriction of the representation $V^{n,\rho}$ of $G_{O_1,\ldots,O_n}$ to $H^{n,\rho}_{O_1,\ldots,O_n}$.

Note that the subgroup $G^0_{O_1,\ldots,O_n} \subset G_{O_1,\ldots,O_n}$ of diffeomorphisms that are the identity on $O_1, \ldots, O_n$ acts trivially on $H^{n,\rho}_{O_1,\ldots,O_n}$ and that by Lemma 1 the factor group $G_{O_1,\ldots,O_n}/G^0_{O_1,\ldots,O_n}$ is isomorphic to the cross product of Diff$(O_1, m) \times \ldots \times$ Diff(O_n, m) with S_n. The assertion how follows easily from this and from Lemma 2.

The representations $V^{n,\rho}_{O_1,\ldots,O_n}$ of $G_{O_1,\ldots,O_n}$, where ρ runs over the inequivalent irreducible representations of S_n, are irreducible and pairwise *inequivalent.*

We now claim that *the representation* $V^{n,\rho}_{O_1,\ldots,O_n}$ *of* $G_{O_1,\ldots,O_n}$ *occurs in* $V^{n,\rho}$ *with multiplicity* 1 *and not at all in representations* $V^{n,\rho'}$, *where* $\rho' \neq \rho$, *nor in representations* $V^{n',\rho'}$, *where* $n' < n$.

For let H' be the orthogonal complement to $\widetilde{H}^{n,\rho}_{O_1,\ldots,O_n}$ in $H_{n,\rho}$. We split H' into the sum of subspaces that are primary with respect to $\mathrm{Diff}(O_1, m) \times \ldots \times \mathrm{Diff}(O_n, m)$. It is not difficult to see that in each of these subspaces at least one of the subgroups $\mathrm{Diff}(O_i, m)$ $(i = 1, \ldots, n)$ acts trivially. But the representation of each subgroup $\mathrm{Diff}(O_i, m)$ in $\widetilde{H}^{n,\rho}_{O_1,\ldots,O_n}$ is a multiple of a non-trivial irreducible representation. Consequently, the representations $V^{n,\rho'}_{O_1,\ldots,O_n}$ are not contained in H', nor for the same reason in $H_{n',\rho}$, $n' < n$.

From the properties of $V^{n,\rho}_{O_1,\ldots,O_n}$ we have just established it follows immediately that the representations V^ρ of $\mathrm{Diff}(X, m)$ are pairwise inequivalent. We claim that they are irreducible.

Let $\mathcal{L} \subset H_{n,\rho}$ be a subspace invariant under $\mathrm{Diff}(X, m)$, $\mathcal{L} \neq 0$. Then for any collection $O_1, \ldots, O_n$ of disjoint neighbourhoods satisfying conditions 1 and 2 of Lemma 1 either $\widetilde{H}^{n,\rho}_{O_1,\ldots,O_n} \subset \mathcal{L}$, or $\widetilde{H}^{n,\rho}_{O_1,\ldots,O_n} \cap \mathcal{L} = 0$. It is not difficult to see that the spaces $\widetilde{H}^{n,\rho}_{O_1,\ldots,O_n}$ generate $H_{n,\rho}$, therefore, $\widetilde{H}^{n,\rho}_{O_1,\ldots,O_n} \subset \mathcal{L}$ for some collection $O_1, \ldots, O_n$. But then, by Lemma 1, $\mathcal{L}$ contains the whole of $\widetilde{H}^{n,\rho}_{O_1,\ldots,O_n}$ and hence coincides with $H_{n,\rho}$. The theorem is now proved.

REMARK. Let us denote by $\mathfrak{A}$ the group of all (classes of coinciding mod 0) invertible measurable transformations of X that preserve the measure m (the dimension of X is arbitrary); we furnish $\mathfrak{A}$ with the weak topology. The representation V^ρ of $\mathrm{Diff}(X, m) \subset \mathfrak{A}$ extends naturally to a representation of $\mathfrak{A}$ and the resulting representation V^ρ of $\mathfrak{A}$ is continuous in the weak topology. It is easy to show that in the weak topology $\mathrm{Diff}(X, m)$, for $\dim X > 1$, is everywhere dense in $\mathfrak{A}$. This makes it possible to prove Theorem 2 anew, reducing its proof to those of the analogous assertions for $\mathfrak{A}$, which are easily verified. On the other hand, this path enables us to establish Theorem 1 for any weakly dense subgroup of $\mathfrak{A}$, that is, to prove the following proposition.

THEOREM 3. *The assertions of Theorem* 1 *are true for the restrictions of the representations* V^ρ *of* $\mathfrak{A}$ *to any subgroup* $G \subset \mathfrak{A}$ *that is weakly dense in* $\mathfrak{A}$.

3. The representation ring $\mathcal{R}$. We consider the free module $\mathcal{R}$ over $\mathbf{Z}$ on the set of all pairwise inequivalent irreducible representations V^ρ of $\mathrm{Diff}\, X$ as basis. By the propositions in § 1.2, the tensor product $V^{\rho_1} \otimes V^{\rho_2}$ of irreducible representations V^{ρ_1} and V^{ρ_2} decomposes into a sum of irreducible representations V^ρ and therefore is an element of $\mathcal{R}$.

In this way a ring structure is defined in $\mathcal{R}$, where multiplication is the tensor product.

Let us introduce another ring $R(S)$ associated with the representations of the symmetric groups S_n (see [18]). We denote by $R(S_n)$ the free module over **Z** on the set of pairwise inequivalent irreducible representations of S_n $(n = 0, 1, 2, \ldots)$ as basis (where $R(S_0) = \mathbf{Z}$). We consider the **Z**-module $R(S) = \bigoplus_{n=0}^{\infty} R(S_n)$ and give a ring structure to $R(S)$ by defining multiplication as the exterior product. From the propositions in §1.2 we obtain immediately the following result.

THEOREM 4. *The ring $\mathcal{R}$ generated by the representations V^ρ of* Diff X *is isomorphic to* $R(S)$.

For the map $\rho \to V^\rho$, where ρ runs over the representations of S_n $(n = 0, 1, \ldots)$ extends to a ring isomorphism $R(S) \to \mathcal{R}$.

REMARK. There exists a natural ring isomorphism

$$\theta: R(S) \to \mathbf{Z}[a_1, a_2, \ldots],$$

where a_n is the n-th elementary symmetric function in an infinite number of unknowns, $n = 1, 2, \ldots$; for the definition of θ see, for example, [18]. By the theorem we have proved, there is a ring isomorphism $\mathcal{R} \to \mathbf{Z}[a_1, a_2, \ldots]$, where to each representation V^ρ there corresponds the symmetric function $\theta(\rho)$.

These symmetric functions in an infinite number of unknowns have the usual properties of characters: each representation V^ρ is uniquely determined by its symmetric function, on adding two representations their corresponding symmetric functions are added, and on taking the tensor product they are multiplied.

§2. Quasi-invariant measures in the space of infinite configurations

Before turning to the discussion of representations of Diff X associated with the space of infinite configurations Γ_X, we ought first of all to study in detail measures in Γ_X, and in fibrations over it, that are quasi-invariant under Diff X. We have already recalled that there are many such measures with various properties (see §0.6); these measures arise (in another connection) in statistical physics, probability theory, and elsewhere.

The ergodic theory for infinite dimensional groups differs in many ways from the theory for locally compact groups (see, for example, [6]). In particular, the action of Diff X in Γ_X is such that in Γ_X there is no quasi-invariant measure that is concentrated on a single orbit.[1] In addition, care is needed because an infinite-dimensional group can act transitively, but not ergodically, on an infinite-dimensional space [13]. This explains the

[1] For locally compact groups such a measure exists and is equivalent to the transform of the Haar measure on the group.

somewhat lengthy proof of the lemma in §2.1, which at first glance would appear obvious.

1. **Lemma on quasi-invariant measures on $B_Y^{(n)} \times \Gamma_{X-Y}$.** LEMMA 1. *Let $Y \subset X$ be a connected open submanifold with compact closure, let μ_n be a measure on $B_Y^{(n)} \times \Gamma_{X-Y}$ that is quasi-invariant under the subgroup* Diff Y, *and let μ_n' and μ_n'' be the projections of μ_n onto $B_Y^{(n)}$ and Γ_{X-Y}, respectively. Then μ_n is equivalent to $\mu_n' \times \mu_n''$ ($n = 0, 1, 2, \dots$).*

REMARK. If μ_n is the restriction to $B_Y^{(n)} \times \Gamma_{X-Y}$ of a fixed quasi-invariant measure μ on Γ_X, then the measures μ_n'' on Γ_{X-Y} are, generally speaking, not equivalent. It is easy to show that the equivalence of the measures μ_n'' on Γ_{X-Y} ($n = 0, 1, 2, \dots$) corresponds precisely to the equivalence of the measures μ and $\mu' \times \mu''$ on Γ_X, where μ' and μ'' are the projections of μ onto B_Y and Γ_{X-Y}.[1]

First we prove the following geometrically obvious proposition.

PROPOSITION 1. *In* Diff Y *there is a countable set of one-parameter subgroups G_l such that the group $G \subset$* Diff Y *generated by them acts transitively in $B_Y^{(n)}$ ($n = 1, 2, \dots$).*

PROOF. We suppose first that dim $Y = 1$. We specify in Y a countable basis of neighbourhoods U_r, $\bar{U}_r \subset Y$, that are diffeomorphic to $\mathbf{R}^1$. We fix for each r a diffeomorphism φ_r: $\mathbf{R}^1 \to U_r$. Under φ_r the group of translations on $\mathbf{R}^1$ goes over into a one-parameter group of diffeomorphisms $x \to f_t(x)$ on U_r ($-\infty < t < \infty$), which acts transitively on U_r. The map φ_r can always be chosen so that the diffeomorphisms $x \to f_t(x)$ on U_r extend trivially to a diffeomorphism on the whole of Y. It is not difficult to check that the sequence of groups $\{G_l\}$ constructed in this way satisfies the required condition.

Now let dim $Y = p$, where $p > 1$. We specify in $\mathbf{R}^p$ a countable set of one-parameter subgroups $H_l \subset$ Diff $\mathbf{R}^p$ such that the group generated by them acts transitively in $\mathbf{R}^p$; the construction of such a family presents no difficulty.

Now we take a countable basis of neighbourhoods U_r in Y, diffeomorphic to $\mathbf{R}^p$, and fix diffeomorphisms φ_r: $\mathbf{R}^p \to U_r$. Let us denote by G_{lr} the image of H_l under φ_r. The elements of G_{lr} can be extended trivially to diffeomorphisms over the whole of Y, and so G_{lr} can be regarded as a one-parameter subgroup of Diff Y.

For a fixed r, the subgroups G_{lr} generate a group which acts transitively in U_r and leaves the points of $Y \setminus U_r$ fixed. Hence it is obvious that the group $G \subset$ Diff Y generated by all the G_{lr} acts n-transitively in Y ($n = 1, 2, \dots$), and Proposition 1 is proved.

The following proposition is concerned with the theory of measurable currents of a quasi-invariant measure.

---

[1] When μ is the Poisson measure, the equivalence $\mu \sim \mu' \times \mu''$ is a direct consequence of the property of being infinitely decomposable. The assertion of the lemma in this case is trivial.

PROPOSITION 2. *Suppose that* $\mathbf{R}^1$ *acts measurably*[1] *on the Lebesgue space* (X, μ) *with quasi-invariant measure* μ *and that* ζ *is a measurable partitioning of* (X, μ) *that is fixed* mod 0 *under* $\mathbf{R}^1$. *Then for almost all* $C \in \zeta$ *the conditional measures* μ^C *on* C *are quasi-invariant under* $\mathbf{R}^1$.

PROOF. For any $t \in \mathbf{R}$ and $C \in \zeta$ we put

$$q(t, C) = \inf_A \mu^C(T^t A),$$

where the inf is taken over all subsets $A \subset C$, with $\mu^C(A) = 1$. We also use the notation $q_0(t, C) = q(T, C)q(-t, C)$. Obviously, the condition $\mu^C \sim T^t\mu^C$ is equivalent to $q_0(t, C) = 1$.

Since the action of $\mathbf{R}^1$ on (X, μ) is measurable and ζ is a measurable partitioning, $q(t, C)$, and hence also $q_0(t, C)$, are measurable as functions on $\mathbf{R}^1 \times X_\zeta$ $(X_\zeta = X/\zeta)$.

Since μ is quasi-invariant, for any fixed $t \in \mathbf{R}^1$ we have $\mu^C \sim T^t\mu^C$ for almost all $C \in \zeta$ with respect to the measure μ_ζ on X_ζ (μ_ζ is the projection of μ); hence $q_0(t, C) = 1$ almost everywhere with respect to μ_ζ on X_ζ. Hence, by Fubini's theorem for $(\mathbf{R}^1 \times X_\zeta, m \times \mu_\zeta)$, where m is the Lebesgue measure on $\mathbf{R}^1$, for almost all $C \in \zeta$ with respect to μ_ζ we have: $m\{t \in \mathbf{R}^1; q_0(t, C) \neq 1\} = 0$.

On the other hand, the set of $t \in \mathbf{R}^1$ for which $\mu^C \sim T^t\mu^C$ (for a fixed C) forms a group. Thus, for almost all $C \in \zeta$ the set $\{t \in \mathbf{R}^1; q_0(t, C) = 1\}$ is a subgroup of $\mathbf{R}^1$ of full Lebesgue measure. But every subgroup of a locally compact group with full Haar measure coincides with the whole group [4]. Consequently, for almost all $C \in \zeta$ $\{t \in \mathbf{R}^1; q_0(t, C) = 1\} = \mathbf{R}^1$, that is, for almost all $C \in \zeta$ the measure μ^C is quasi-invariant under the action of $\mathbf{R}^1$, and Proposition 2 is proved.

PROOF OF LEMMA 1. Let $\{G_l\}$ be a countable set of one-parameter subgroups of Diff Y such that the group G generated by them acts transitively in $\mathrm{B}_Y^{(n)}$; such a set exists by Proposition 1.

Since $G_l \cong \mathbf{R}^1$, it follows from Proposition 2 that for almost all, (in the sense of μ_n''), configurations $\gamma \in \Gamma_{X-Y}$ the conditional measure μ_n^γ on $\mathrm{B}_Y^{(n)}$ is quasi-invariant under each G_l $(l = 1, 2, \ldots)$, hence also under the whole group G generated by them.

On the other hand, the measures on $\mathrm{B}_Y^{(n)}$ that are quasi-invariant under G are all equivalent to each other, and consequently to μ_n'. For on $B_Y^{(n)}$, as on every smooth manifold, there is, up to equivalence, a unique measure that is quasi-invariant under a group of diffeomorphisms acting transitively, namely, the smooth measure with everywhere positive density.

---

[1] That is the map $\mathbf{R}^1 \times X \to X$ $((g, x) \to gx)$ is measurable as a map between spaces with measures $m \times \mu$ and μ respectively, where m is the Lebesgue measure in $\mathbf{R}^1$.

Thus, for almost all $\gamma \in \Gamma_{X-Y}$ (in the sense of μ_n'') the conditional measure μ_n^γ on $B_Y^{(n)}$ is equivalent to μ_n', and the lemma is proved.

2. **Measurable indexings in Γ_X.** We say that i is an indexing in Γ_X if for each configuration $\gamma \in \Gamma_X$ there is a bijective map $i(\gamma, \cdot)$: $\gamma \to N$, $N = \{1, 2, \ldots\}$.

We denote by $\Gamma_{X,1}$ the subset of elements $(\gamma, x) \in \Gamma_X \times X$ such that $x \in \gamma$, and we associate with each indexing i a bijective map $\Gamma_{X,1} \to \Gamma_X \times N$, defined by $(\gamma, x) \to (\gamma, i(\gamma, x))$. If this map is measurable in both directions (with respect to Borel σ-algebras on $\Gamma_{X,1}$ and $\Gamma_X \times N$), then the indexing i is called *measurable*.[1]

Let i be a measurable indexing. We introduce a sequence of measurable maps a_k: $\Gamma_X \to X$ $(k = 1, 2, \ldots)$ defined by the conditions: $a_k(\gamma) \in \gamma$, $i(\gamma, a_k(\gamma)) = k$ (that is, $a_k(\gamma)$ is the k-th element of the configuration γ). We associate with i a cross section s: $\Gamma_X \to \widetilde{X}^\infty$, defined by $s(\gamma) = (a_1(\gamma), \ldots, a_n(\gamma), \ldots)$. It is not difficult to verify that the set $s\,\Gamma_X$ is measurable and that the bijective map $\Gamma_X \to s\Gamma_X$ is measurable in both directions.

For any $\psi \in \mathrm{Diff}\, X$ and $\gamma \in \Gamma_X$, the elements $s(\psi^{-1}\gamma) \in \widetilde{X}^\infty$ and $\psi^{-1}(s\gamma) \in \widetilde{X}^\infty$ belong to the same S^∞-orbit in $\widetilde{X}^\infty$. We define a map σ: $\mathrm{Diff}\, X \times \Gamma_X \to S^\infty$ by $s(\psi^{-1}\gamma) = [\psi^{-1}(s\gamma)]\,\sigma(\psi, \gamma)$; the notation here means $(x_1, \ldots, x_n, \ldots)\,\sigma = (x_{\sigma(1)}, \ldots, x_{\sigma(n)}, \ldots)$.

Let us now introduce the idea of an *admissible indexing.* We are given an increasing sequence $X_1 \subset \ldots \subset X_k \subset \ldots$ of connected open subsets with compact closures such that $X = \bigcup_k X_k$.

DEFINITION. We say that a measurable indexing i is admissible (with respect to the given sequence $X_1 \subset \ldots \subset X_n \subset \ldots$) if the map σ: $\mathrm{Diff}\, X \times \Gamma_X \to S^\infty$ defined by it satisfies the following condition: if $\mathrm{supp}\, \psi \subset X_k$ and $|\gamma \cap X_k| = n$, then $\sigma(\psi, \gamma) \in S_n$ $(k = 1, 2, \ldots;$ $n = 0, 1, \ldots)$.

In particular, $\sigma(\psi, \gamma) \in S_\infty$ for any $\psi \in \mathrm{Diff}\, X$ and $\gamma \in \Gamma_X$.

It is not difficult to construct examples of admissible indexings. For example, the following indexing, which was proved to be measurable in [17], is admissible.

Let a continuous metric be given on X. With each positive integer k we associate a covering $(X_{kl})_{l=1,2,\ldots}$ of X by disjoint measurable subsets with diameters not exceeding $1/k$, satisfying the following two conditions.

1) the partitioning $X = X_{k1} \cup \ldots \cup X_{kl} \cup \ldots$ is a refinement of $X = X_1 \cup (X_2 \setminus X_1) \cup \ldots \cup (X_n \setminus X_{n-1}) \cup \ldots$;

2) Each set X_n is covered by finitely many of the sets X_{kl}.

---

[1] If a measure μ is given in Γ_X, then indexings need be given only on subsets of full measure in Γ_X, and we make no distinction between indexings that coincide mod 0.

It is obvious that such a covering exists. We number its elements so that if $X_{ki} \subset X_n$ and $X_{kj} \subset X \setminus X_n$, then $i < j$ $(n = 1, 2, \ldots)$. For any $x \in X$ and $k \in N$ we put $f_k(x) = l$, if $x \in X_{kl}$. The correspondence $x \to (f_1(x), \ldots, f_k(x), \ldots)$ is a morphism from X to the set of all sequences of positive integers. We define an ordering in X by putting $x' \prec x''$ if $(f_1(x'), \ldots f_k(x'), \ldots) < (f_1(x''), \ldots, f_k(x''), \ldots)$ in the lexicographic ordering.

For any $\gamma \in \Gamma_X$ and $x \in \gamma$, the set $\{x' \in \gamma;\, x' \prec x\}$ is finite, because it is contained in the compact set $\bar{X}_{11} \cup \ldots \cup \bar{X}_{1 f_1(x)}$. Consequently, for any $\gamma \in \Gamma_X$, the set of elements $x \in \gamma$ is a sequence with respect to the ordering introduced in X; we denote by $i(\gamma, x)$ the number of elements $x' \prec \gamma$ in this sequence. The map $(\gamma, x) \to i(\gamma, x)$ so constructed is a measurable indexing (see [17]). It is not difficult to prove that it is also admissible.

3. Convolution of measures. DEFINITION. The convolution $\mu_1 * \mu_2$ (see, for example, [17]) of two measures μ_1 and μ_2 on the space of all configurations $\Delta_X = \Gamma_X \cup B_X$ is defined as the image of the product measure $\mu_1 \times \mu_2$ on $\Delta_X \times \Delta_X$ under the map $(\gamma_1, \gamma_2) \to \gamma_1 \cup \gamma_2$.

REMARK. This definition agrees with the usual definition for the convolution of two measures in the space $\mathcal{F}(X)$ of generalized functions on $X(\Delta_X$ is embedded in $\mathcal{F}(X)$ by $\gamma \to \sum\limits_{x \in \gamma} \delta_x)$, because the union of (disjoint) configurations corresponds to the sum of their images in $\mathcal{F}(X)$.

It is obvious that $\psi(\mu_1 * \mu_2) = \psi\mu_1 * \psi\mu_2$ for any $\psi \in \operatorname{Diff} X$, where $\psi\mu$ is the image of μ under the diffeomorphism ψ. Hence *the convolution of quasi-invariant* (*under* Diff X) *measures is itself quasi-invariant.*

Note that *if μ_1 and μ_2 are Poisson measures with parameters λ_1 and λ_2, respectively, then their convolution $\mu_1 * \mu_2$ is the Poisson measure with parameter $\lambda_1 + \lambda_2$.* (This fact follows easily from the definition of the Poisson measure).

Later on we shall be interested in the case when one of the factors is a quasi-invariant measure concentrated on Γ_X, and the second is a smooth positive measure m_n concentrated on $B_X^{(n)} (n = 1, 2, \ldots)$. Since all smooth positive measures m_n on $B_X^{(n)}$ are equivalent, the type of the measure $\mu * m_n$ depends only on the type of μ and on n.

Let us agree to call the type of $\mu * m_n$ on Γ_X the *n-point augmentation* of μ and to denote it by $n \circ \mu$. Thus, with each measure μ on Γ_X there is associated a sequence of measures $0 \circ \mu \sim \mu, 1 \circ \mu, \ldots, n \circ \mu, \ldots$ defined up to equivalence. Note that $n_1 \circ (n_2 \circ \mu) \sim (n_1 + n_2) \circ \mu$ for any n_1 and n_2.

Here we establish the following properties of the operation $\circ$.

1) *For any quasi-invariant measure μ on Γ_X there exists a quasi-invariant measure μ' such that $1 \circ \mu' \sim \mu$.*

2) *If a measure μ on Γ_X is ergodic, then $1 \circ \mu$ is also ergodic.*

To prove this we give an admissible indexing i on Γ_X and let $s\colon \Gamma_X \to \widetilde{X}^\infty$ be the cross section defined by this indexing (see §2.2). We denote by Y_s the image of Γ_X under s and by Δ_s the minimal S_∞-invariant subset of $\widetilde{X}^\infty$ containing Y_s; obviously, Δ_s is the disjoint union $\Delta_s = \bigcup_{\sigma \in S_\infty} Y_s\sigma$. Since $Y_s \subset \widetilde{X}^\infty$ is measurable and S_∞ countable, Δ_s is a measurable subset of $\widetilde{X}^\infty$. Since the indexing i is admissible, it follows that Δ_s is invariant under Diff X.

Let μ be a measure on Γ_X that is quasi-invariant under Diff X. Let c be an arbitrary positive function on S_∞ such that $\sum_{\sigma \in S_\infty} c(\sigma) = 1$; we introduce a measure $\widetilde{\mu}$ on $\widetilde{X}^\infty$ by the formula:

$$\widetilde{\mu} = \sum_{\sigma \in S_\infty} c(\sigma)(s\mu)\sigma,$$

where $(s\mu)\sigma$ is the image of μ under the map $\gamma \to (s\gamma)\sigma$. In other words, for any measurable subset $A \subset \widetilde{X}^\infty$

$$\widetilde{\mu}(A) = \sum_{\sigma \in S_\infty} c(\sigma)\mu[p(A \cap Y_s\sigma)] \tag{1}$$

where p is the projection $\widetilde{X}^\infty \to \Gamma_X$.

Obviously, $\widetilde{\mu}(\Delta_s) = 1$. Note that the choice of the positive function c on S_∞ does not play a role in defining $\widetilde{\mu}$, because the measures on $\widetilde{X}^\infty$ constructed from two such functions are equivalent. From the definition of $\widetilde{\mu}$ it follows easily that:

a) $p\widetilde{\mu} = \mu$ where $p\widetilde{\mu}$ is the projection of $\widetilde{\mu}$ onto Γ_X.

b) the measure $\widetilde{\mu}$ on $\widetilde{X}^\infty$ is quasi-invariant under both Diff X and S_∞.

We cite without proof two further simple assertions.

PROPOSITION 3. *If a normalized measure* μ_1 *on* $\widetilde{X}^\infty$ *is quasi-invariant under* Diff X *and if* $p\mu_1 = \mu$ *and* $\mu_1(\Delta_s) = 1$, *then* $\mu_1 \sim \widetilde{\mu}$.

PROPOSITION 4. *If a measure* μ *in* Γ_X *is ergodic, then* $\widetilde{\mu}$ *is also ergodic with respect to* Diff X.

Let us decompose the space $X^\infty = \prod_{i=1}^{\infty} X_i$, where $X_i = X$, into the direct product $X^\infty = X \times \prod_{i=2}^{\infty} X_i$ and consider the induced map $h = X \times \widetilde{X}^\infty \to \widetilde{X}^\infty$ (that is, $h(x; \{x_k\}_{k=1}^{\infty}) = (\{x'_k\}_{k=1}^{\infty}; x'_1 = x, x'_k = x_{k-1}$ when $k > 1$)).

PROPOSITION 5. $h(m \times \mu) \sim \widetilde{m * \mu}$, *where* m *is an arbitrary smooth positive measure on* X.

PROOF. Consider the diagram

$$\begin{array}{ccc} X \times \widetilde{X}^\infty & \xrightarrow{h} & \widetilde{X}^\infty \\ {\scriptstyle p_1}\downarrow & & \downarrow{\scriptstyle p} \\ X \times \Gamma_X & \xrightarrow{\widetilde{h}} & \Gamma_X \end{array},$$

where $p_1 = \mathrm{Id} \times p$, $\widetilde{h}(x, \gamma) = \gamma \cup \{x\}$. Obviously, this is commutative, and $(p \circ h)(m \times \widetilde{\mu}) = (\widetilde{h} \circ p_1)(m \times \widetilde{\mu}) = m * \mu$. Further, the measure $h(m \times \widetilde{\mu})$ is quasi-invariant and concentrated on Δ_s (since the indexing i is admissible, the point $h(x, s\gamma)$ belongs to the same S_∞-orbit as $s(\gamma \cup \{x\})$). Consequently, by Proposition 3, $h(m \times \widetilde{\mu}) \sim \widetilde{m * \mu}$.

The proof of the following assertion is similar to that of Lemma 1 in §2.1.

PROPOSITION 6. *Every measure μ in $\widetilde{X}^\infty$ that is quasi-invariant under* Diff X *is equivalent to the product $m_n \times \mu_n$ of its projections in the factorization $\widetilde{X}^\infty = \widetilde{X}^n \times \widetilde{X}^\infty_{n+1}$; moreover, m_n is equivalent to a positive smooth measure on X^n, and μ_n is quasi-invariant under* Diff X.

COROLLARY. *A quasi-invariant measure $\widetilde{\mu}$ in $\widetilde{X}^\infty$ is ergodic if and only if it is regular (that is, satisfies the* 0 – 1 *law).*

PROOF OF PROPERTY 1). Let μ be a quasi-invariant measure on Γ_X. By Proposition 6, $\mu \sim h(m \times \mu_1)$, where m is a smooth positive measure on X, μ_1 is a quasi-invariant measure on $\widetilde{X}^\infty$, and h: $X \times \widetilde{X}^\infty \to \widetilde{X}^\infty$ is the map induced by the direct product (see above). Since i is admissible, μ_1, like μ, is concentrated on Δ_s; consequently, by Proposition 3, $\mu_1 \sim \widetilde{\mu'}$ is a quasi-invariant measure in Γ_X. By Proposition 5, $\widetilde{\mu} \sim h(m \times \widetilde{\mu}') \sim \widetilde{m * \mu'}$; consequently, $\mu \sim m * \mu'$, as required.

PROOF OF PROPERTY 2). If the measure μ in Γ_X is ergodic, then by Proposition 4, the measure $\widetilde{\mu}$ in $\widetilde{X}^\infty$ is ergodic; consequently, by the corollary to Proposition 6, $\widetilde{\mu}$ is regular. Obviously, $m \times \widetilde{\mu}$ is then also regular and therefore ergodic. Consequently, the measure $\widetilde{m * \mu} \sim h(m \times \widetilde{\mu})$ is also ergodic and hence, so is its projection $m * \mu$.

DEFINITION. We say that a quasi-invariant measure μ is *saturated* if $1 \circ \mu \sim \mu$ (and consequently, $n \circ \mu \sim \mu$ for any n).

It is not difficult to verify that *the Poisson measure is saturated* (this follows from the property of being infinitely decomposable).

We now give a criterion for a measure μ to be saturated. The map T: $X^\infty \to X^\infty$, defined by $(Tx)_i = x_{i+1}$ $(i = 1, 2, \ldots)$ is called *left translation* in X^∞. Obviously, the subset $\widetilde{X}^\infty$ is T-invariant.

PROPOSITION 7. *For a quasi-invariant measure on Γ_X to be saturated it is necessary and sufficient that the measure $\widetilde{\mu}$ on $\widetilde{X}^\infty$ corresponding to it (defined by means of a fixed admissible indexing) is quasi-invariant under the left translation T.*

PROOF. From the definition of the left translation T it follows that $\widetilde{\mu} \sim h(m \times T\widetilde{\mu})$. On the other hand, by Proposition 3, $\widetilde{1 \circ \mu} \sim h(m \times \widetilde{\mu})$. Hence it is obvious that the condition $1 \circ \mu \sim \mu$ is equivalent to $\widetilde{\mu} \sim T\widetilde{\mu}$.

An example of a non-saturated measure μ will be given in Appendix 1.

4. The space $\Gamma_{X,n}$ and Campbell's measure on $\Gamma_{X,n}$. We consider the Cartesian product $\Gamma_X \times X^n$ $(n = 1, 2, \ldots)$ and denote by $\Gamma_{X,n}$ the set of elements $(\gamma; x_1, \ldots, x_n) \in \Gamma_X \times X^n$, where $\gamma \in \Gamma_X$, $x_i \in X$, such that $x_i \in \gamma$ $(i = 1, \ldots, n)$ and $x_i \neq x_j$ when $i \neq j$. Further, we put $\Gamma_{X,0} = \Gamma_X$.

Obviously, $\Gamma_{X,n}$ is closed in $\Gamma_X \times X^n$.

Now $\Gamma_{X,n}$ can be regarded as a fibre space, $\pi: \Gamma_{X,n} \to \Gamma_X$, whose fibre over a point $\gamma \in \Gamma_X$ is the collection of all ordered n-point subsets in γ.

Let us denote by $\mathfrak{A}_n$ the σ-algebra of all Borel sets in $\Gamma_{X,n}$. We associate with each subset $C \in \mathfrak{A}_n$ a function on Γ_X: $\nu_C(\gamma) = \{$the number of points $(x_1, \ldots, x_n) \in X^n$ such that $(\gamma: x_1, \ldots, x_n) \in C\}$.

From the continuity of π it follows that ν_C is a Borel function.

DEFINITION. Let μ be a measure on Γ_X. The *Campbell measure* on $\Gamma_{X,n}$ associated with μ is the measure $\tilde{\mu}$ on $\mathfrak{A}_n$ defined by

$$\tilde{\mu}(C) = \int_{\Gamma_X} \nu_C(\gamma)\, d\mu(\gamma), \qquad C \in \mathfrak{A}_n.$$

A Campbell measure $\tilde{\mu}$ induces on the fibres of the fibration $\pi: \Gamma_{X,n} \to \Gamma_X$ a uniform measure, which is 1 at each point of the fibre.

We define in $\Gamma_{X,n}$ the actions of the groups Diff X and S_n:

$$\psi: (\gamma; x_1, \ldots, x_n) \mapsto (\psi^{-1}\gamma; \psi^{-1}x_1, \ldots, \psi^{-1}x_n),$$
$$\sigma: (\gamma; x_1, \ldots, x_n) \mapsto (\gamma; x_{\sigma(1)}, \ldots, x_{\sigma(n)}).$$

Obviously, ψ and σ are continuous and $\psi \circ \sigma = \sigma \circ \psi$ for any $\psi \in$ Diff X and $\sigma \in S_n$. The next result is easy to establish.

LEMMA 2. *The Campbell measure $\tilde{\mu}$ on $\Gamma_{X,n}$ corresponding to a measure μ on Γ_X is invariant under S_n. If the measure μ on Γ_X is invariant under* Diff X, *then the Campbell measure $\tilde{\mu}$ is also quasi-invariant under* Diff X, *and*

$$\left.\frac{d\tilde{\mu}(\psi^{-1}c)}{d\tilde{\mu}(c)}\right|_{c=(\gamma;\, x_1, \ldots, x_n)} = \frac{d\mu(\psi^{-1}\gamma)}{d\mu(\gamma)}.$$

Now let i be a measurable indexing in Γ_X. We denote by $\tilde{N}^n$ the set of all n-tuples of natural numbers $(i_1, \ldots, i_n)$, where $i_p \neq i_q$ when $p \neq q$ $(p, q = 1, \ldots, n)$. We define a map

$$\Gamma_{X,n} \to \Gamma_X \times \tilde{N}^n \tag{2}$$

by

$$(\gamma; x_1, \ldots, x_n) \mapsto (\gamma; i(\gamma, x_1), \ldots, i(\gamma, x_n)).$$

This map is bijective, measurable in both directions, and carries the Campbell measure $\tilde{\mu}$ on $\Gamma_{X,n}$ into the measure $\mu \times \nu$ on $\Gamma_X \times \tilde{N}^n$, where ν is the measure on $\tilde{N}^n$, that is equal to 1 at each point on $\tilde{N}^n$. Thus, the space $(\Gamma_{X,n}, \tilde{\mu})$ can be identified with $(\Gamma_X \times N^n, \mu \times \nu)$.

Under this identification the actions of S_n and Diff X go over from $\Gamma_{X,n}$ to $\Gamma_X \times \tilde{N}^n$. It is not difficult to verify that the action of these groups on $\Gamma_X \times \tilde{N}^n$ are given by:

$$\sigma: (\gamma, a) \mapsto (\gamma, a\sigma),$$
$$\psi: (\gamma, a) \mapsto (\psi^{-1}\gamma, \sigma(\psi, \gamma)a),$$

where $a = (i_1, \ldots, i_n)$; $a\sigma = (i_{\sigma(1)}, \ldots, i_{\sigma(n)})$ $\sigma \in S_n$;
$\sigma a = (\sigma(i_1), \ldots, \sigma(i_n))$ $\sigma \in S^\infty$; and $\sigma(\psi, \gamma)$ is the function on Diff $X \times \Gamma_X$ with values in S^∞ defined by i (see §2.2).

5. The map $\Gamma_X \times X^n \to \Gamma_{X,n}$. Let us consider the spaces $\Gamma_X \times X^n$ and $\Gamma_{X,n}$ together with their σ-algebras of Borel subsets (see §2.4). Let μ be a quasi-invariant measure on Γ_X, and m_n a smooth, positive measure on X^n. In $\Gamma_X \times X^n$ we specify the measure $\mu \times m_n$ and in $\Gamma_{X,n}$ the Campbell measure $\widetilde{n \circ \mu}$ corresponding to $n \circ \mu \sim \mu * m_n$ on Γ_X.

A map α: $\Gamma_X \times X^n \to \Gamma_{X,n}$ is given by the following formula:

$$\alpha(\gamma; x_1, \ldots, x_n) = (\gamma \cup \{x_1, \ldots, x_n\}; x_1, \ldots, x_n);$$

α is taken to be defined on the subset of elements $(\gamma: x_1, \ldots, x_n) \in \Gamma_X \times X^n$ for which $\gamma \cap \{x_1, \ldots, x_n\} = \phi$; it is not difficult to verify that this subset and its image in $\Gamma_{X,n}$ are sets of full measure. Nor is it difficult to check that α is measurable in both directions and commutes with the action of Diff X on both $\Gamma_X \times X^n$ and $\Gamma_{X,n}$.

THEOREM. *The image $\alpha(\mu \times m_n)$ of the measure $\mu \times m_n$ on $\Gamma_X \times X^n$ under α is equivalent to the Campbell measure $\widetilde{n \circ \mu}$.*

PROOF. We carry out the proof for the case $n = 1$; the arguments for arbitrary n are similar.

We define maps $\alpha_1: \Gamma_X \times X \to \Gamma_X$ and $\alpha_2: \Gamma_{X,1} \to \Gamma_X$ by $\alpha_1(\gamma, x) = \gamma \cup \{x\}$, $\alpha_2(\gamma, x) = \gamma$. It is obvious that the following diagram commutes:

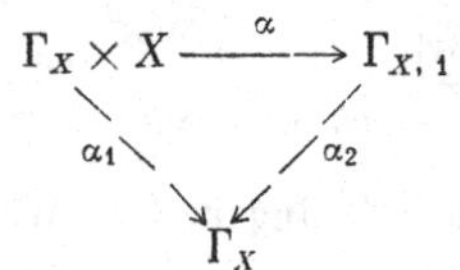

The image of the measure $\mu \times m$ on $\Gamma_X \times X$ under $\alpha_1 = \alpha_2\alpha$ is $1 \circ \mu$. Hence it follows that $\alpha(\mu \times m) \prec \widetilde{1 \circ \mu}$. It remains to prove that $\alpha(\mu \times m) \succ \widetilde{1 \circ \mu}$.

First we construct a measurable indexing in Γ_X in the following way. We fix a continuous metric ρ in X and a point $x_0 \in X$. We consider the subset of Γ_X

$$(3) \qquad \{\gamma \in \Gamma_X: \rho(x_0, x) \neq \rho(x_0, x') \text{ for any } x \neq x' \text{ in } \gamma\}$$

and the preimage of (3) under α_2 in $\Gamma_{X,1}$. As is easy to see, these subsets are of full measure in Γ_X and $\Gamma_{X,1}$, respectively, and it is to be understood in what follows that it is these subsets which are meant by Γ_X and $\Gamma_{X,1}$.

We prescribe an ordering on each configuration $\gamma \in \Gamma_X$, putting $x \prec x'$ for any $x, x' \in \gamma$, if $\rho(x_0, x) < \rho(x_0, x')$. For any $(\gamma, x) \in \Gamma_{X,1}$ we denote by $i(\gamma, x)$ the number of the element $x \in \gamma$, as given by the ordering on γ. It is not difficult to see that i is a measurable indexing. We now introduce a sequence of measurable maps $a_k : \Gamma_X \to X$ $(i = 1, 2, \ldots)$, where $a_k(\gamma)$ is the k-th element in the configuration γ.

Now let $C \subset \Gamma_{X,1}$ be an arbitrary measurable set of positive Campbell measure: $\widetilde{1 \circ \mu}(C) > 0$; we have to prove that then $(\mu \times m)(\alpha^{-1} C) > 0$. Since $\Gamma_{X,1}$ splits into the countable union of subsets $\{(\gamma, a_k(\gamma));\ \gamma \in \Gamma_X\}$ $(k = 1, 2, \ldots)$, we may assume without loss of generality that $C \subset \{(\gamma, a_k(\gamma))\}$ for some k. We introduce the notation $C_n = \{(\gamma', x) \in \Gamma_X \times X;\ \gamma' \cup \{x\} \in \alpha_2 C,\ a_n(\gamma \cup \{x\})) = x\}$ $(n = 1, 2, \ldots)$.

Note that the condition $\widetilde{1 \circ \mu}(C) > 0$ is equivalent to

$$(4) \qquad (\mu \times m)\{(\gamma', x) \in \Gamma_X \times X;\ \gamma' \cup \{x\} \in \alpha_2 C\} > 0.$$

In its turn (4) is equivalent to the existence of a natural number l, for which $(\mu \times m)\ \{(\gamma', x) \in \Gamma_X \times X,\ \gamma' \cup \{x\} \in \alpha_2 C,\ a_l(\gamma' \cup \{x\}) = x\} > 0$, that is, $(\mu \times m)\ (\widetilde{C}_l) > 0$.

On the other hand, since $C \subset \{(\gamma, a_k(\gamma))\}$, it follows that

$$\alpha^{-1} C = \{(\gamma', x) \in \Gamma_X \times X;\ \gamma' \cup \{x\} \in \alpha_2 C,\ a_k(\gamma' \cup \{x\}) = x\},$$

that is, $\alpha^{-1} C = \widetilde{C}_k$. Thus, the proof of the lemma reduces to proving the following assertion: *for any natural numbers k and l the conditions $(\mu \times m)\ (\widetilde{C}_k) > 0$ and $(\mu \times m)\ (\widetilde{C}_l) > 0$ are equivalent.*

Let us prove this assertion. We write $X_r = \{x \in X;\ \rho(x_0, x) < r\}$, where $r > 0$ is an arbitrary rational number $(X_0 = \phi)$. We fix a positive integer $n \geqslant \max(k, l)$ and introduce the following subsets in $\Gamma_X \times X$:

$$U^p_{r_1, \ldots, r_n} = \{(\gamma', x) \in \Gamma_X \times X;\ |\gamma' \cap (X_{r_i} \setminus X_{r_{i-1}})| = 1 \text{ when } i \neq p,$$
$$|\gamma' \cap (X_{r_p} \setminus X_{r_{p-1}})| = 0,\ x \in X_{r_p} \setminus X_{r_{p-1}}\},$$

where $r_1, \ldots, r_n$ are rational numbers such that $0 = r_0 < r_1 < \ldots < r_n$ $(p = 1, \ldots, n)$. It is obvious that the sets $U^p_{r_1, \ldots, r_n}$ cover $\widetilde{C}_p$; consequently, the condition $(\mu \times m)\ (\widetilde{C}_l) > 0$ amounts to the existence of some $U^l_{r_1, \ldots, r_n}$ such that

$$(5) \quad (\mu \times m)\{(\gamma', x) \in U^l_{r_1, \ldots, r_n};\ \gamma' \cup \{x\} \in \alpha_2 C,\ a_l(\gamma' \cup \{x\}) = x\} > 0$$

We note that $U^l_{r_1, \ldots, r_n} \subset (B^{(n-1)}_{X_{r_n}} \times \Gamma_{X \setminus X_n}) \times X$ and make use of the fact that by Lemma 1 of §2.1 the restriction μ_{n-1} of μ to $B^{(n-1)}_{X_{r_n}} \times \Gamma_{X \setminus X_{r_n}}$ is equivalent to the product $\underbrace{m \times \ldots \times m}_{n-1} \times \mu''_{n-1}$, where μ''_{n-1} is the projection of μ_{n-1} onto $\Gamma_{X \setminus X_{r_n}}$. So we obtain that (5) is equivalent to the following condition:

$$(6) \quad (m \times \ldots \times m \times \mu''_{n-1})\{(x_1, \ldots, x_n; \gamma') \in X^n \times \Gamma_{X \setminus X_{r_n}}; \\ x_i \in X_{r_i} \setminus X_{r_{i-1}},\ i=1, \ldots, n;\ \gamma' \cup \{x_1, \ldots, x_n\} \in \alpha_2 C\} > 0.$$

Thus, $(\mu \times m)(\widetilde{C}_l) > 0$ amounts to the condition that (6) is satisfied for some collection of rational numbers $0 = r_0 < r_1 < \ldots < r_n$. But from the same arguments it follows that the condition $(\mu \times m)(\widetilde{C}_k) > 0$ also is equivalent to (6). Consequently, the conditions $(\mu \times m)(\widetilde{C}_l) > 0$ and $(\mu \times m)(\widetilde{C}_k) > 0$ are equivalent, as required.

§3. Representations of Diff X defined by quasi-invariant measures in the space of infinite configurations (elementary representations)

1. Definition of elementary representations. Let μ be a quasi-invariant measure in the space of infinite configurations Γ_X. We introduce a series of unitary representations of Diff X associated with μ. First we consider the space $L^2_\mu(\Gamma_X)$. In it a unitary representation U_μ of Diff X is defined by[1]

$$(1) \qquad (U_\mu(\psi) f)(\gamma) = \left(\frac{d\mu(\psi^{-1}\gamma)}{d\mu(\gamma)}\right)^{1/2} f(\psi^{-1}\gamma)$$

We do not study the properties of U_μ separately, but examine straightaway a wider class – the elementary representations. For the Poisson measure μ these representations are additive generators in the representation ring determined by U_μ (see §4).

Although the proof of the irreducibility and other properties of U_μ are simpler than in the general case, we prefer to study all the elementary representations simultaneously.

DEFINITION. A representation of Diff X is called elementary if it is of the form $U_\mu \otimes V^\rho$, where U_μ is the representation in $L^2_\mu(\Gamma_X)$ given by (1), and V^ρ is the representation defined in §1.

Thus, each elementary representation is given by a quasi-invariant measure μ on Γ_X and a representation ρ of the symmetric group S_n ($n = 0, 1, 2, \ldots$).

THEOREM 1. *If μ is an ergodic measure on Γ_X and ρ is an irreducible representation of S_n, then the elementary representation $U_\mu \otimes V^\rho$ of Diff X is irreducible.*

REMARK 1. The converse assertion is obvious.

REMARK 2. Another convenient formulation of Theorem 1 is: When μ is ergodic, then U_μ is *absolutely irreducible*, that is, remains irreducible after taking the tensor product with any irreducible representation V^ρ.

Essentially, the whole of §3 is devoted to a proof of Theorem 1. But first we construct some other useful realizations of elementary representations.

---

[1] If μ is concentrated not on Γ_X, but on $B_X^{(n)}$, then (1) gives the representation $V^{\rho_n^0}$ (see §1), where ρ_n^0 is the unit representation of S_n.

2. The representations U_μ^ρ. Let ρ be a unitary representation of S_n in a space W ($n = 0, 1, \ldots$). We consider the space $L^2_{\tilde\mu}(\Gamma_{X,n}, W)$ of functions F on $\Gamma_{X,n}$ with values in W such that

$$\| F \| = \int_{\Gamma_{X,n}} \| F(c) \|_W^2 \, d\tilde\mu(c) < \infty;$$

$\tilde\mu$ is the Campbell measure on $\Gamma_{X,n}$ corresponding to the measure μ on Γ_X (see §2.4). A unitary representation U of Diff X is given in $L^2_{\tilde\mu}(\Gamma_{X,n}, W)$ by

$$(U(\psi)F)(\gamma; x_1, \ldots, x_n) = \left(\frac{d\mu(\psi^{-1}\gamma)}{d\mu(\gamma)}\right)^{1/2} F(\psi^{-1}\gamma; \psi^{-1}x_1, \ldots, \psi^{-1}x_n).$$

We denote by $H_{\mu,n,\rho}$ the subspace of functions $F \in L^2_{\tilde\mu}(\Gamma_{X,n}, W)$ such that $F(\gamma, x_{\sigma(1)}, \ldots, x_{\sigma(n)}) = \rho^{-1}(\sigma)F(\gamma, x_1, \ldots, x_n)$ for any $\sigma \in S_n$. Obviously, $H_{\mu,n,\rho}$ is invariant under Diff X.

DEFINITION. The restriction of the representation U of Diff X from $L^2_\mu(\Gamma_{X,n}, W)$ to $H_{\mu,n,\rho}$ is denoted by U_μ^ρ.

In the particular case when ρ is the unit representation of S_0, $U_\mu^\rho \cong U_\mu$, where U_μ is the representation in $L^2_\mu(\Gamma_X)$ defined by (1).

REMARK. If in this construction of U_μ^ρ the space Γ_X of infinite configurations is replaced by the space $B_X^{(k)}$ of k-point configurations ($k \geqslant n$), then we obtain instead of U_μ^ρ the representation $V^{\rho\circ\overset{0}{\rho}_{k-n}}$ defined in §1, where ρ^0_{k-n} is the unit representation of S_{k-n}.

THEOREM 2. $U_\mu \otimes V^\rho \cong U^\rho_{n\circ\mu}$, where ρ is a representation of S_n. (For the definition of $n \circ \mu$, see §2.3).

PROOF. In §2.5 we have established an isomorphism between spaces with measures

$$(\Gamma_X \times X^n, \mu \times m_n) \to (\Gamma_{X,n}, \widetilde{n \circ \mu}) \tag{2}$$

($\widetilde{n \circ \mu}$ is the Campbell measure on $\Gamma_{X,n}$ corresponding to $n \circ \mu$), which commutes with the action of Diff X. Let us consider the isomorphism of Hilbert spaces $L^2_{\widetilde{n\circ\mu}}(\Gamma_{X,n}, W) \to L^2_\mu(\Gamma_X) \otimes L^2_{m_n}(X^n, W)$ induced by (2), where W is the space of the representation ρ of S_n. It is easy to verify that the image of $H_{\mu,n,\rho} \subset L^2_{\widetilde{n\circ\mu}}(\Gamma_{X,n}, W)$ is $L^2_\mu(\Gamma_X) \otimes H_{n,\rho}$, where $H_{n,\rho} \subset L^2_{m_n}(X^n, W)$ is the subspace of the representation V^ρ (see §1) and that the operators $U_{n\circ\mu}(\psi)$ in $H_{\mu,n,\rho}$ go over to $U_\mu(\psi) \otimes V^\rho(\psi)$.

COROLLARY 1. *The class of the representations U_μ^ρ is the same as that of the elementary representations $U_\mu \otimes V^\rho$.*

For on the one hand, $U_\mu \otimes V^\rho \cong U^\rho_{n\circ\mu}$; and on the other hand, for any quasi-invariant measure μ on Γ_X there is another quasi-invariant measure μ' such that $\mu \sim n \circ \mu'$ (see §2.3) and, consequently,

$U^\rho_\mu \cong U_{\mu'} \otimes V^\rho$.

COROLLARY 2. *If μ is a saturated measure (that is, $1 \circ \mu \sim \mu$) and, in particular, if μ is the Poisson measure, then $U^\rho_\mu \cong U_\mu \otimes V^\rho$.*

THEOREM 3. *If μ is an ergodic measure on Γ_X and ρ an irreducible representation of S_n, then the representation U^ρ_μ of* Diff X *is irreducible.*

We note that Theorem 1 follows immediately from Theorems 2 and 3. For let ρ be an irreducible representation of S_n and μ an ergodic measure on Γ_X. Since $n \circ \mu$ is also ergodic (see §2.3), $U_\mu \otimes V^\rho \cong U^\rho_{n\circ\mu}$ is irreducible by Theorem 3.

3. Another realization of U^ρ_μ. Let ρ be a unitary representation of S_n in W. We consider the set $\widetilde{N}^n$ of all n-tuples $a = (i_1, \ldots, i_n)$ of natural numbers, where $i_p \neq i_q$, when $p \neq q$. We define an action of S^∞ on $\widetilde{N}^n$ by $a \to \sigma a = (\sigma(i_1), \ldots, \sigma(i_n))$, $\sigma \in S^\infty$. We denote by $l^2(\widetilde{N}^n, W)$ the space of functions φ on $\widetilde{N}^n$ with values in W such that

$$\| \varphi \|^2 = \sum_{a\in\widetilde{N}^n} \| \varphi(a) \|^2_W < \infty.$$

We consider in $l^2(\widetilde{N}^n, W)$ the subspace H^ρ of all functions $\varphi \in l^2(\widetilde{N}^n, W)$ such that $\varphi(i_{\sigma(1)}, \ldots, i_{\sigma(n)}) = \rho^{-1}(\sigma)\varphi(i_1, \ldots, i_n)$ for any $\sigma \in S_n$. Obviously, H^ρ is invariant under the action of S^∞. Now let i be an arbitrary measurable indexing in Γ_X and σ the map Diff $X \times \Gamma_X \to S^\infty$ defined by it.

LEMMA 1. *The representation U^ρ_μ of* Diff X *is equivalent to the representation in $L^2_\mu(\Gamma_X) \cdot \otimes H^\rho$, defined by*

$$(U^\rho_\mu(\psi) f)(\gamma, a) = \left(\frac{d\mu(\psi^{-1}\gamma)}{d\mu(\gamma)} \right)^{1/2} f(\psi^{-1}\gamma, \sigma(\psi, \gamma) a). \tag{3}$$

PROOF. In §2.4 an isomorphism was established between spaces with measures:

$$(\Gamma_X \times \widetilde{N}^n, \mu \times \nu) \to (\Gamma_{X,n}, \widetilde{\mu}) \tag{4}$$

(ν is the measure on $\widetilde{N}^n$ that is 1 at each point). We consider the isomorphism of Hilbert spaces $L^2_{\widetilde{\mu}}(\Gamma_{X,n}, W) \to L^2_\mu(\Gamma_X) \otimes l^2(\widetilde{N}^n, W)$ induced by (4). It is easy to verify that the image of the subspace $H_{\mu,n,\rho} \subset L^2_{\widetilde{\mu}}(\Gamma_{X,n}, W)$ on which U^ρ_μ acts is $L^2_\mu(\Gamma_X) \otimes H^\rho$ and that the operators $U^\rho_\mu(\psi)$ in $H_{\mu,n,\rho}$ go over to operators of the form (3).

4. The decomposition of the space of the representation U^ρ_μ of Diff X into a sum of subspaces that are primary with respect to the subgroup Diff X_k. Let

$$X_1 \subset \ldots \subset X_k \subset \ldots \tag{5}$$

be an increasing sequence of open connected subsets with compact closures such that $X = \bigcup_k X_k$.

We fix an admissible indexing i (with respect to (5)); let $\sigma(\psi, \gamma)$ be the map Diff $X \times \Gamma_X \to S_\infty$ defined by it. By Lemma 1, the representation U^ρ_μ may be realized in $L^2_\mu(\Gamma_X) \otimes H$; here $H = H^\rho$ is the space of functions φ on $\widetilde{N}^n$ with values in the space W of the representation ρ of S_n such that $\|\varphi\|^2 = \sum_{a \in \widetilde{N}^n} \|\varphi(a)\|^2_W < \infty$ and $\varphi(a\sigma) = \rho^{-1}(\sigma)\varphi(a)$ for any $\sigma \in S_n$. The representation operators are given by (3).

We decompose $L^2_\mu(\Gamma_X) \otimes H$ into a direct sum of subspaces that are primary with respect to the subgroups Diff $X_k \subset$ Diff X (that is, those that are the identity on $X \setminus X_k$) ($k = 1, 2, \ldots$).

First we decompose Γ_X into a countable union of spaces that are invariant under Diff X_k:

$$\Gamma_X = \coprod_{r=0}^{\infty} \mathrm{B}^{(r)}_{X_k} \times \Gamma_{X \setminus X_k}, \tag{6}$$

where $\mathrm{B}^{(r)}_{X_k}$ is the space of r-point subsets in X_k. It follows from (6) that

$$L^2_\mu(\Gamma_X) \otimes H = \bigoplus_{r=0}^{\infty} (L^2_{\mu_r}(\mathrm{B}^{(r)}_{X_k} \times \Gamma_{X \setminus X_k}) \otimes H),$$

where μ_r is the restriction of μ to the subset $\mathrm{B}^{(r)}_{X_k} \times \Gamma_{X \setminus X_k} \subset \Gamma_X$. It remains to decompose each term in this sum into a direct sum of invariant subspaces that are primary with respect to Diff X_k.

Next we split H into the direct sum of subspaces that are primary with respect to the symmetric group $S_r \subset S_\infty$. This decomposition can be presented in the following way: $H = \bigoplus_i (W^i_r \otimes C^i_r)$, where W^i_r are the spaces in which the irreducible and pairwise inequivalent representations ρ^i_r of S_r act; C^i_r is the space on which S_r acts trivially.

As a result, we obtain a decomposition into the direct sum:

$$L^2_{\mu_r}(\mathrm{B}^{(r)}_{X_k} \times \Gamma_{X \setminus X_k}) \otimes H = \bigoplus_i (L^2_{\mu_r}(\mathrm{B}^{(r)}_{X_k} \times \Gamma_{X \setminus X_k}) \otimes W^i_r \otimes C^i_r).$$

All the terms of this decomposition are invariant under Diff X. For since the indexing is admissible, it follows from $\psi \in$ Diff X_k and $\gamma \in \mathrm{B}^{(r)}_{X_k} \times \Gamma_{X \setminus X_k}$ that $\sigma(\psi, \gamma) \in S_r$. We claim that these subspaces are primary and disjoint.

We denote by μ'_r and μ''_r the projections of μ_r onto $\mathrm{B}^{(r)}_{X_k}$ and $\Gamma_{X \setminus X_k}$, respectively. By Lemma 1 of §2, the measure μ_r on $\mathrm{B}^{(r)}_{X_k} \times \Gamma_{X \setminus X_k}$ is equivalent to the product $\mu'_r \times \mu''_r$ of μ'_r and μ''_r. Consequently, there is an isomorphism

$$\tau_r\colon L^2_{\mu_r}(\mathrm{B}^{(r)}_{X_k} \times \Gamma_{X \setminus X_k}) \to L^2_{\mu'_r}(\mathrm{B}^{(r)}_{X_k}) \otimes L^2_{\mu''_r}(\Gamma_{X \setminus X_k}),$$

defined by

$$\tau_r F = \left(\frac{d\mu_r}{d\mu'_r\, d\mu''_r}\right)^{1/2} F.$$

We denote by the same letter the trivial extension of τ_r to an isomorphism

$$\tau_r\colon L^2_{\mu_r}(\mathrm{B}^{(r)}_{X_k}\times\Gamma_{X\setminus X_k})\otimes W^i_r\otimes C^i_r\to(L^2_{\mu'_r}(\mathrm{B}^{(r)}_{X_k})\otimes W^i_r)\otimes(L^2_{\mu''_r}(\Gamma_{X\setminus X_k})\otimes C^i_r).$$

We denote the elements of $\mathrm{B}^{(r)}_{X_k}$ by $\gamma^{(r)}$ and define a map σ_r: Diff $X_k\times\mathrm{B}^{(r)}_{X_k}\to S_r$ by $\sigma_r(\psi,\gamma^{(r)})=\sigma(\psi,\gamma)$, where $\gamma\in\mathrm{B}^{(r)}_{X_k}\times\Gamma_{X\setminus X_k}$, $\gamma\cap X_k=\gamma^{(r)}$. This is well defined, because if $\gamma\cap X_k=\gamma'\cap X_k$, then $\sigma(\psi,\gamma)=\sigma(\psi,\gamma')$. Immediately from the definition of τ_r we derive the next result.

LEMMA 2. *Under the isomorphism τ_r the operators $U(\psi)=U^\rho_\mu(\psi)$, $\psi\in$ Diff X_k go over to operators $\tau_rU(\psi)\tau_r^{-1}$ of the following form: $\tau_rU(\psi)\tau_r^{-1}=U^i_r(\psi)\otimes I$, where I is the unit operator in $L^2_{\mu''_r}(\Gamma_{X\setminus X_k})\otimes C^i_r$ and $U^i_r(\psi)$ is an operator in $L^2_{\mu'_r}(\mathrm{B}^{(r)}_{X_k})\otimes W^i_r$, that is, in the space of functions on $\mathrm{B}^{(r)}_{X_k}$ with values in W^i_r defined by*

$$(U^i_r(\psi)F)(\gamma^{(r)})=\left(\frac{d\mu'_r(\psi^{-1}\gamma^{(r)})}{d\mu'_r(\gamma^{(r)})}\right)^{1/2}\rho^i_r(\sigma_r(\psi,\gamma^{(r)}))\,F(\psi^{-1}\gamma^{(r)}).$$

The representation U^i_r of Diff X_k is equivalent to $V^{\rho^i_r}$ defined in §1.1. By Proposition 3 of §1.2, all the representations $V^{\rho^i_r}$ of Diff X_k are irreducible and mutually pairwise inequivalent. Therefore Lemma 2 has the following corollaries.

COROLLARY 1. *The representations of* Diff X_k *in the subspaces $L^2_{\mu_r}(\mathrm{B}^{(r)}_{X_k}\times\Gamma_{X\setminus X_k})\otimes W^i_r\otimes C^i_r$ $(r=0,1,2,\ldots;\ i=1,2,\ldots)$ are primary and disjoint.*

COROLLARY 2. *Any invariant subspace under* Diff X_k

$$\mathscr{L}^i_{k,r}\subset(L^2_{\mu'_r}(\mathrm{B}^{(r)}_{X_k})\otimes W^i_r)\otimes(L^2_{\mu''_r}(\Gamma_{X\setminus X_k})\otimes C^i_r)$$

is of the form $\mathscr{L}^i_{k,r}=(L^2_{\mu'_r}(\mathrm{B}^{(r)}_{X_k})\otimes W^i_r)\otimes D^i_r$, where $D^i_r\subset L^2_{\mu''_r}(\Gamma_{X\setminus X_k})\otimes C^i_r$.

COROLLARY 3. *Any subspace $\mathscr{L}\subset L^2_\mu(\Gamma_X)\otimes H$ that is invariant under* Diff X_k *splits into the direct sum $\mathscr{L}=\bigoplus_{r,i}\mathscr{L}^i_{k,r}$, where*

$$\mathscr{L}^i_{k,r}=\mathscr{L}\cap(L^2_{\mu_r}(\mathrm{B}^{(r)}_{X_k}\times\Gamma_{X\setminus X_k})\otimes W^i_r\otimes C^i_r). \tag{7}$$

5. Proof of Theorem 3. We use the notation and results of the preceding subsection. We denote by $L^\infty_\mu(\Gamma_X)$ the space of essentially bounded functions on Γ_X with respect to μ. Now $L^\infty_\mu(\Gamma_X)$ is a ring with the usual multiplication. Further, if $f\in L^2_\mu(\Gamma_X)\otimes H$ and $\varphi\in L^\infty_\mu(\Gamma_X)$, then $\varphi f\in L^2_\mu(\Gamma_X)\otimes H$.

LEMMA 3. *If $\mathscr{L}\subset L^2_\mu(\Gamma_X)\otimes H$ is invariant under* Diff X, *then $\mathscr{L}$ is invariant under multiplication by elements of $L^\infty_\mu(\Gamma_X)$.*

PROOF. We denote by μ_{X_k} the projection of μ onto B_{X_k}, and by $L^\infty_{\mu_{X_k}}(\mathrm{B}_{X_k})$ the space of essentially bounded functions of B_{X_k} with respect to the measure μ_{X_k} $(k = 1, 2, \ldots)$. We identify each space $L^\infty_{\mu_{X_k}}(\mathrm{B}_{X_k})$ with its image under the natural map $L^\infty_{\mu_{X_k}}(\mathrm{B}_{X_k}) \to L^\infty_\mu(\Gamma_X)$ (that is, the space of essentially bounded functions on Γ_X that are constant on the fibres of the fibration $\Gamma_X \to \mathrm{B}_{X_k}$). We consider the union $\bigcup_k L^\infty_{\mu_{X_k}}(\mathrm{B}_{X_k})$.

It is obvious that for any $f \in L^2_\mu(\Gamma_X) \otimes H$ and $\varphi \in L^\infty_\mu(\Gamma_X)$ the product $\varphi f \in L^2_\mu(\Gamma_X) \otimes H$ is approximated in $L^2_\mu(\Gamma_X) \otimes H$ by elements $\varphi' f$, where $\varphi' \in \bigcup_k L^\infty_{\mu_{X_k}}(\mathrm{B}_{X_k})$. Therefore, to prove the lemma it is sufficient to check that $\mathscr{L}$ is invariant under multiplication by elements of $L^\infty_{\mu_{X_k}}(\mathrm{B}_{X_k})$ $(k = 1, 2, \ldots)$.

We fix k and denote by $L^\infty_{\mu'_n}(\mathrm{B}^{(n)}_{X_k})$ the subspace of functions in $L^\infty_{\mu_{X_k}}(\mathrm{B}_{X_k})$ that are concentrated on $\mathrm{B}^{(n)}_{X_k} \times \Gamma_{X-X_k}$, (here $\mathrm{B}^{(n)}_{X_k}$ is the subspace of n-point subsets in X_k) $(n = 0, 1, \ldots)$. Obviously, for any $f \in L^2_\mu(\Gamma_X) \otimes H$ and $\varphi \in L^\infty_{\mu_{X_k}}(\mathrm{B}_{X_k})$ the product φf is approximated in $L^2_\mu(\Gamma_X) \otimes H$ by finite sums of elements $\varphi_n f$, where $\varphi_n \in L^\infty_{\mu'_n}(\mathrm{B}^{(n)}_{X_k})$. Thus, the proof of the lemma reduces to the following assertion:

If $\mathscr{L} \subset L^2_\mu(\Gamma_X) \otimes H$ is invariant under Diff X_k, *then $\mathscr{L}$ is invariant under multiplication by elements of $L^\infty_{\mu'_n}(\mathrm{B}^{(n)}_{X_k})$ $(n = 0, 1, 2, \ldots)$.*

Let us prove this assertion. We use the decomposition $\mathscr{L} = \bigoplus_{r,i} \mathscr{L}^i_{k,r}$, where

$$\mathscr{L}^i_{k,r} = \mathscr{L} \cap (L^2_{\mu_r}(\mathrm{B}^{(r)}_{X_k} \times \Gamma_{X\setminus X_k}) \otimes W^i_r \otimes C^i_r). \tag{8}$$

It is sufficient to prove the assertion for each subspace $\mathscr{L}^i_{k,r}$ separately. Note that when $r \neq n$, the supports of the functions in $L^r_{\mu_r}(\mathrm{B}^{(r)}_{X_k} \times \Gamma_{X\setminus X_k})$ and in $L^\infty_{\mu'_n}(\mathrm{B}^{(n)}_{X_k})$ do not intersect. Therefore, it is only necessary to consider the case $r = n$.

Let $\widetilde{\mathscr{L}}^i_{k,n}$ be the image of $\mathscr{L}^i_{k,n}$ under τ_n. By Corollary 2 to Lemma 2, $\widetilde{\mathscr{L}}^i_{k,n}$ has the form $\widetilde{\mathscr{L}}^i_{k,n} = L^2_{\mu'_n}(\mathrm{B}^{(n)}_{X_k}) \otimes W^i_r \otimes D^i_r$.

Hence it is clear that $\widetilde{\mathscr{L}}^i_{k,n}$ is invariant under multiplication by functions in $L_{\mu'_n}(\mathrm{B}^{(n)}_{X_k})$. Since the corresponding elements in $\mathscr{L}^i_{k,n}$ and $\widetilde{\mathscr{L}}^i_{k,n}$ differ only by the factor $\left(\frac{d\mu'_n \, d\mu''_n}{d\mu_n}\right)^{1/2}$, the space $\mathscr{L}^i_{k,n}$ is also invariant under multiplication by elements of $L^\infty_{\mu'_n}(\mathrm{B}^{(n)}_{X_k})$, and the lemma is proved.

We consider the space $H = H^\rho$, a factor in the tensor product $L^2_\mu(\Gamma_X) \otimes H$. A representation R of S_∞ is defined in H by

$(R(\sigma)\varphi)(a) = \varphi(\sigma^{-1}a)$.

Note that $R \cong \mathrm{Ind}_{S_n \times S_\infty^n}^{S_\infty}(\rho \times I)$, where S_∞^n is the subgroup of finite permutations leaving $1, \ldots, n$ fixed and I is the unit representation of S_∞^n. Hence the next result follows easily.

LEMMA 4. *The representation R of S_∞ is irreducible.*

LEMMA 5. *Every subspace $\mathcal{L} \subset L_\mu^2(\Gamma_X) \otimes H$ that is invariant under* Diff X *is also invariant under the operators $I \otimes R(\sigma)$, $\sigma \in S_\infty$, where I is the unit operator in $L_\mu^2(\Gamma_X)$.*

PROOF. Since $S_\infty = \varinjlim S_p$, it is sufficient to prove that $\mathcal{L}$ is invariant under the operators $I \otimes R(\sigma)$, $\sigma \in S_p$ $(p = 1, 2, \ldots)$.

We use the notation and results of §3.4. Let p be any fixed positive integer. We consider the subspace

$$\mathcal{L}_{k,p} = \sum_{r=p}^{\infty} \sum_i \oplus \mathcal{L}_{k,r}^i \quad (k = 1, 2, \ldots),$$

where $\mathcal{L}_{k,r}^i$ is defined by (8).

Clearly the union $\bigcup_{k=1}^{\infty} \mathcal{L}_{k,p}$ is everywhere dense in $\mathcal{L}$. Therefore, it is sufficient to prove that each space $\mathcal{L}_{k,r}^i$, $r \geqslant p$, is invariant under $I \otimes R(\sigma)$.

We consider the image $\widetilde{\mathcal{L}}_{k,r}^i$ of $\mathcal{L}_{k,r}^i$ under τ_r. By Corollary 2 to Lemma 2 $\widetilde{\mathcal{L}}_{k,r}^i$ and hence its preimage $\mathcal{L}_{k,r}^i$ is invariant under $I \otimes R(\sigma)$, $\sigma \in S_r$. Since $p \leqslant r$, we have $S_p \subset S_r$, and so $\mathcal{L}_{k,r}^i$ is also invariant under the operators $I \otimes R(\sigma)$, $\sigma \in S_p$. The lemma is now proved.

LEMMA 6. *Every subspace $\mathcal{L} \subset L_\mu^2(\Gamma_X) \otimes H$ invariant under* Diff X *is of the form $\mathcal{L} = L_\mu^2(A) \otimes H$, where $A \subset \Gamma_X$ is a measurable subset.*

PROOF. It follows from Lemmas 4 and 5 that

$$\mathcal{L} = E \otimes H,$$

where $E \subset L_\mu^2(\Gamma_X)$, and from Lemma 3 that E is invariant under multiplication by functions from $L_\mu^\infty(\Gamma_X)$. Consequently, $E = L_\mu^2(A)$, where A is a measurable subset in Γ_X, and the lemma is proved.

The assertion of Theorem 3 follows immediately from Lemma 6. For let $\mathcal{L} \subset L_\mu^2(\Gamma_X) \otimes H$ be an invariant subspace with respect to Diff X. Then from Lemma 6 we have $\mathcal{L} = L_\mu^2(A) \otimes H$, where $A \subset \Gamma_X$ is a measurable subset. Consequently, since μ is ergodic, either $\mathcal{L} = 0$ or $\mathcal{L} = L_\mu^2(\Gamma_X) \otimes H$. This proves Theorem 3, and with it Theorem 1.

REMARK 1. The assertion of the theorem remains true for any subgroup $G \subset$ Diff X satisfying the following requirements:

1) G acts ergodically on Γ_X;

2) for any open connected submanifold $Y \subset X$ with compact closure, Theorem 1 of §1 about the representations V^ρ of Diff Y remains true for

the restriction of the V^ρ to $G \cap \mathrm{Diff}\ Y$.

REMARK 2. The proof of Theorem 3 can be simplified considerably in the case $n = 0$, that is, for the representation U_μ in $L^2_\mu(\Gamma_X)$. In this case it reduces to proving Lemma 1 of §2.1 and establishing a functional version of the $0-1$ law. In the simplest case when $\mu \sim \mu'_{X_k}$ $(k = 1, 2, \ldots)$ $(\mu'_{X_k}, \mu''_{X_k}$ are the projections of μ onto B_{X_k} and Γ_{X-X_k}, respectively), this law consists of the following. Let $\mathscr{L}$ be a subspace of $L^2_\mu(\Gamma_X)$. If in terms of the decomposition $L^2_\mu(\Gamma_X) \cong L^2_{\mu'_{X_k}}(B_{X_k}) \otimes L^2_{\mu''_{X_k}}(\Gamma_{X-X_k})$ the subspace $\mathscr{L} \subset L^2_\mu(\Gamma_X)$ for any k has the form

$$\mathscr{L} \cong L^2_{\mu'_{X_k}}(B_{X_k}) \otimes C_k, \qquad C_k \subset L^2_{\mu''_{X_k}}(\Gamma_{X-X_k}),$$

then[1] either $\mathscr{L} = 0$ or $\mathscr{L} = L^2_\mu(\Gamma_X)$. An extra difficulty comes from the fact that there is no equivalence $\mu \sim \mu'_{X_k} \times \mu''_{X_k}$, generally speaking, and only the weaker relation $\mu \sim \sum_{r=0}^{\infty} \mu'_r \times \mu''_r$ is true (for the definition of μ'_r and μ''_r see p. 25).

§4. Representations of Diff X generated by the Poisson measure

1. Properties of the Poisson measure. Let X be a non-compact manifold with a smooth positive measure m, $m(X) = \infty$, and let $\mu = \mu_\lambda$ be the Poisson measure on Γ_X with parameter λ corresponding to the measure m on X (for the definition of the Poisson measure, see §0.2).

Some basic properties of Poisson measure were stated in §0.6.

LEMMA 1. *If* $\dim X > 1$, *then for any two μ-measurable sets $A_1, A_2 \subset \Gamma_X$ with positive measure there exists a diffeomorphism $\psi \in \mathrm{Diff}(X, m)$ such that*[2] $\mu(A_1 \cap \psi A_2) > \frac{1}{2}\mu(A_1)\mu(A_2)$. ($\mathrm{Diff}(X, m)$ *is the subgroup of diffeomorphisms preserving* m.)

PROOF. First we recall some definitions and facts. By a cyclindrical set in Γ_X we mean a set $A \subset \Gamma_X$, of positive measure, of the form $A = \pi_Y^{-1} A'$, where Y is a compact set in X, $A' \subset B_Y$ is a measurable subset and π_Y is the natural map $\Gamma_X \to B_Y$ $(\pi_Y \gamma = \gamma \cap Y)$; Y is called the *carrier*[3] of A. Since the Poisson measure μ is infinitely decomposable, it follows that if the carriers of two cylindrical sets A_1 and A_2 intersect in a set of measure 0, then $\mu(A_1 \cap A_2) = \mu(A_1)\mu(A_2)$.

We recall that any μ-measurable set $C \subset \Gamma_X$ can be approximated by cylindrical sets (that is, for any $\varepsilon > 0$ there is a cylindrical set A such that $\mu(C \Delta A) < \varepsilon$).

---

[1] We recall that in these terms the usual $0-1$ law would be formulated as follows. Let $f \in L^2_\mu(\Gamma_X)$. If $f = 1_k \otimes f_k$ for any k, where 1_k is the constant in $L^2_{\mu'_{X_k}}(B_{X_k})$ and $f_k \in L^2_{\mu''_{X_k}}(\Gamma_{X-X_k})$, then $f = \mathrm{const}$.

[2] For $X = \mathbf{R}^1$ the lemma is false, because in this case $\mathrm{Diff}(X, m)$ is trivial.

[3] Of course, the carrier of a cylindrical set is not uniquely defined.

Hence it is clear that it is sufficient to prove the assertion of the lemma for cylindrical sets A_1 and A_2. Without loss of generality we may suppose further that $X = \mathbf{R}^n$, $n > 1$.

Let us establish the following property of Diff(X, m): if Y_1 and Y_2 are two compact sets in X, then there is a diffeomorphism $\psi \in$ Diff(X, m) such that $Y_1 \cap \psi Y_2 = \phi$.

For if Y_1, Y_2 are compact in $X = \mathbf{R}^n$ and m is the Lebesgue measure in $\mathbf{R}^n$, then there exists a disc containing Y_1 and Y_2 and a rotation ψ of the disc (this preserves m) such that $Y_1 \cap \psi Y_2 = \phi$; this rotation ψ can be extended beyond the boundary of the disc to a finite diffeomorphism of $\mathbf{R}^n$ preserving m. If now m is an arbitrary smooth positive measure in $\mathbf{R}^n$, then it is sufficient to use a lemma (see [21]), which states that any open ball in $\mathbf{R}^n$, $n > 1$, with a smooth measure m can be mapped diffeomorphically onto itself so that m goes over to the Lebesgue measure.

Let Y_1 and Y_2 be the carriers of A_1 and A_2. By what has just been proved, we can find a diffeomorphism $\psi \in$ Diff(X, m) such that $Y_1 \cap \psi Y_2 = \phi$. But then $\mu(A_1 \cap \psi A_2) = \mu(A_1)\mu(\psi A_2) = \mu(A_1)\mu(A_2)$, and hence $\mu(A_1 \cap \psi A_2) > \frac{1}{2}\mu(A_1)\mu(A_2)$. The lemma is now proved.

THEOREM 1. *If* dim $X > 1$, *then the Poisson measure* μ *in* Γ_X *is ergodic with respect to* Diff(X, m).

PROOF. Let $A \subset \Gamma_X$, $\mu(A) > 0$, be a subset that is invariant mod 0 under Diff(X, m); we must prove that $\mu(A) = 1$. Suppose the contrary: that $\mu(\Gamma_X \setminus A) > 0$. Then by Lemma 1 there is a $\psi \in$ Diff(X, m) such that $\mu(\Gamma_X \setminus A) \cap \psi A) > 0$; hence, since A is invariant, $\mu((\Gamma_X \setminus A) \cap A) > 0$, which is false. This proves the theorem.

2. The representation of Diff X generated by the Poisson measure. Let $\mu = \mu_\lambda$ be the Poisson measure on Γ_X with parameter $\lambda > 0$. In §3 we have associated with each quasi-invariant measure μ on Γ_X a unitary representation U_μ of Diff X in $L^2_\mu(\Gamma_X)$ defined by

$$(U_\mu(\psi) f)(\gamma) = \left(\frac{d\mu(\psi^{-1}\gamma)}{d\mu(\gamma)}\right)^{1/2} f(\psi^{-1}\gamma),$$

and also a set of elementary representations U^ρ_μ. For the Poisson measure μ_λ the theory of such representations can be advanced considerably further than in the general case. In particular, it is possible to describe the corresponding representation ring. Furthermore, for μ_λ the representations U_{μ_λ} can be realized in the form $\mathrm{EXP}_\beta T$ (see [1] and [2]).

In what follows we write U_λ *(instead of* U_{μ_λ}*) for representation generated by the Poisson measure* μ_λ.

Since μ_λ is ergodic, by Theorem 1 of §3, U_λ is irreducible.

3. The spherical function of the representation U_λ. Let us assume that dim $X > 1$. We consider the subgroup Diff$(X, m) \subset$ Diff X of diffeomorphisms preserving the measure m; for us this subgroup will play a role similar to that of maximal compact subgroups in the theory of representations of semisimple Lie groups.

Since μ is invariant under Diff(X, m) and U_λ is infinitely decomposable, the restriction of U_λ to Diff(X, m) is given by

$$(U_\lambda(\psi)f)(\gamma) = f(\psi^{-1}\gamma), \quad \psi \in \text{Diff}\,(X, m). \tag{1}$$

In view of Theorem 1 there is in $L^2_{\mu_\lambda}(\Gamma_X)$ one, and up to a multiplicative factor, only one vector that is invariant under Diff(X, m), namely, $f_0 \equiv 1$.

DEFINITION. The following function on Diff X is called the *spherical function* of U_λ:

$$u_\lambda(\psi) = \langle U_\lambda(\psi)f_0, f_0\rangle,$$

where the brackets denote the inner product in $L^2_{\mu_\lambda}(\Gamma_X)$.

Let us find an explicit form for the spherical function. Let supp $\psi \subset Y$, $m(Y) < \infty$. We denote by $\tilde{\mu}_\lambda$ the projection of μ_λ onto B_Y and by $\tilde{\mu}^n_\lambda$ the restriction of $\tilde{\mu}_\lambda$ to $B^{(n)}_Y$. Then

$$u_\lambda(\psi) = \int_{\Gamma_X} \Big(\prod_{x\in\gamma} \mathcal{J}^{1/2}_\psi(x)\Big)\, d\mu_\lambda(\gamma) = \int_{B_Y} \Big(\prod_{x\in\gamma} \mathcal{J}^{1/2}_\psi(x)\Big)\, d\tilde{\mu}_\lambda(\gamma) =$$

$$= \sum_{n=0}^{\infty} \int_{B^{(n)}_Y} \Big(\prod_{x\in\gamma} \mathcal{J}^{1/2}_\psi(x)\Big)\, d\tilde{\mu}^n_\lambda(\gamma) =$$

$$= \sum_{n=0}^{\infty} \frac{e^{-\lambda m(Y)}[\lambda m(Y)]^n}{n!} \Big(\frac{1}{m(Y)} \int_Y \mathcal{J}^{1/2}_\psi(x)\, dm(x)\Big)^n =$$

$$= e^{-\lambda m(Y)} e^{\lambda \int_Y \mathcal{J}^{1/2}_\psi(x)\,dm(x)} = e^{\lambda \int_Y (\mathcal{J}^{1/2}_\psi(x) - 1)\, dm(x)}$$

So we obtain

$$u_\lambda(\psi) = \exp\Big(\lambda \int_X (\mathcal{J}^{1/2}_\psi(x) - 1)\, dm(x)\Big), \tag{2}$$

where $\mathcal{J}_\psi(x) = \dfrac{dm(\psi^{-1}x)}{dm(x)}$. Since f_0 is defined invariantly in the representation space $L^2_{\mu_\lambda}(\Gamma_X)$ and since, by (2), $u_{\lambda_1} \neq u_{\lambda_2}$ when $\lambda_1 \neq \lambda_2$, we obtain the following theorem.

THEOREM 2. *The representations U_{λ_1} and U_{λ_2} of* Diff X (dim $X > 1$) *are not equivalent when $\lambda_1 \neq \lambda_2$.*

4. The Gaussian form of the representation U_λ of Diff X. Let us consider the *real* Hilbert space $H = L^2_m(X)$, where m is a smooth positive measure on X. A unitary representation T of Diff X is given in H by $(T(\psi)f)(x) = \mathcal{J}^{1/2}_\psi(x) f(\psi^{-1}x)$, where $\mathcal{J}_\psi(X) = \dfrac{dm(\psi^{-1}x)}{dm(x)}$. A 1-cocycle

β: Diff $X \to H$ is given by $[\beta(\psi)](x) = \mathcal{J}_\psi^{1/2}(x) - 1$.

In accordance with [1] and [2] this is a way of constructing a new representation $\widetilde{U} = \mathrm{EXP}_\beta T$ of Diff X.

We denote by $\widetilde{\mu}$ a measure in the space $\mathcal{F}(X)$ of generalized functions on X given by its characteristic functional:

$$\int_{\mathcal{F}(X)} e^{i\langle F,\, f\rangle}\, d\widetilde{\mu}(F) = e^{-\|f\|^2/2}, \tag{3}$$

where $\|\cdot\|$ is the norm in $L^2_m(X)$. We call $\widetilde{\mu}$ the *standard Gaussian measure* in $\mathcal{F}(X)$.

The representation $\widetilde{U} = \mathrm{EXP}_\beta T$ is given in $L^2_{\widetilde{\mu}}(\mathcal{F}(X))$ by

$$(\widetilde{U}(\psi)\Phi)(F) = e^{i\langle F,\, \beta(\psi)\rangle}\Phi(T^*(\psi)F),$$

where the operator $T^*(\psi)$ is defined by $\langle T^*(\psi)F, f\rangle = \langle F, T(\psi)f\rangle$.

LEMMA 2. *The vector $\Phi_0 \equiv 1$ in $L^2_{\widetilde{\mu}}(\mathcal{F}(X))$ is cyclic with respect to* Diff X.

PROOF. Since $(\widetilde{U}(\psi)\Phi_0)(F) = e^{i\langle F,\beta(\psi)\rangle}$, the assertion of the lemma is just that the set of functionals $e^{i\langle F,\beta(\psi)\rangle}$, $\psi \in$ Diff X, is total in $L^2_{\widetilde{\mu}}(\mathcal{F}(X))$, that is, the minimal linear subspace $\mathcal{L} \subset L^2_{\widetilde{\mu}}(\mathcal{F}(X))$ containing them is $L^2_{\widetilde{\mu}}(\mathcal{F}(X))$ itself. It is known [1] that the functionals of the form

$$\Phi(F) = \prod_{i=1}^{n} \langle F, f_i\rangle, \tag{4}$$

where $f_1, \ldots, f_n$ are smooth finite functions on X ($n = 0, 1, 2, \ldots$) form a total set in $L^2_{\widetilde{\mu}}(\mathcal{F}(X))$; therefore, it is sufficient to prove that $\mathcal{L}$ contains all functionals of the form (4).

Let f be any smooth finite function on X satisfying $\int f(x)dm(x) = 0$, let τ be any real number such that $1 - \tau f(x) > 0$ for all $x \in X$. A measure m_τ in X is given by $dm_\tau(x) = (1 - \tau f(x))dm(x)$. The measures m and m_τ coincide outside a compact set $Y \supset$ supp f, and $m(Y) = m_\tau(Y)$. Therefore, by a theorem of Moser [21], there exists a diffeomorphism $\psi \in$ Diff X carrying m to m_τ, that is, $\mathcal{J}_\psi(x) = 1 - \tau f(x)$. But then $\beta(\psi) = \sqrt{(1 - \tau f(x))} - 1$, and hence the functional $e^{i\langle F, \sqrt{(1-\tau f(x))}-1\rangle}$ belongs to $\mathcal{L}$ for any sufficiently small τ. Hence all the terms in the expansion of $e^{i\langle F, \sqrt{(1-\tau f(x))}-1\rangle}$ as a power series in τ also belong to L.

The coefficient of τ^n in this expansion is $c_n\langle F, f\rangle^n$, $c_n \neq 0$, apart from terms of the form $\prod_{i=1}^{k}\langle F, f_i\rangle$, $k < n$. Therefore, by induction on n, we can verify that $\mathcal{L}$ contains all functionals of the form $\langle F, f\rangle^n$, where $\int f(x)dm(x) = 0$, and hence those of the form $\langle F, f\rangle^n$, where f is an arbitrary smooth finite function. Since the functionals (4) can be presented as linear combinations of the $\langle F, f\rangle^n$, they also belong to $\mathcal{L}$, and the lemma is proved.

If in the definition of $\widetilde{U}$ we replace the 1-cocycle β by $s\beta$, where s is any real number, we obtain a one-parameter family of unitary representations of Diff X: $\widetilde{U}_s = \mathrm{EXP}_{s\beta} T$.

It is obvious that the assertion of lemma 2 remains true for all representations $\widetilde{U}_s$, $s \neq 0$.

THEOREM 3. *If* $s \neq 0$, *then* $\widetilde{U}_s \cong U_{s^2}$, *where* U_{s^2} *is the representation of* Diff X *generated by the Poisson measure with parameter* s^2 (see ≠ 4.2).

PROOF. We compute the matrix element $\langle \widetilde{U}_s(\psi)\Phi_0, \Phi_0\rangle$, where $\Phi_0 \equiv 1$. From formula (3) for the characteristic functional of $\widetilde{\mu}$ we immediately obtain

$$\langle \widetilde{U}_s(\psi)\Phi_0, \Phi_0\rangle = \exp\left(s^2 \int_X (\mathcal{J}_\psi^{1/2}(x) - 1)\, dm(x)\right) = u_{s^2}(\psi),$$

where u_{s^2} is the spherical function of U_{s^2}. The assertion of the theorem follows from this and the fact that Φ_0 is cyclic.

Let us now consider the special case $s = 0$. Then (see [1]) $\widetilde{U}_0$ splits into the direct sum: $\widetilde{U}_0 = T^0 \oplus T^1 \oplus \ldots \oplus T^n \oplus \ldots$, where T^0 is the unit representation and T^n $(n \geqslant 1)$ is the n-th symmetrized power of T introduced at the beginning of §3.4. In the notation of §1 we have $T^n = V^{\rho_n^0}$, where ρ_n^0 is the unit representation of S_n. Thus, $\widetilde{U}_0 = \bigoplus_{n=0}^{\infty} V^{\rho_n^0}$.

5. Elementary representations of Diff X associated with the Poisson measure. According to §3, the tensor product $U_\lambda^\rho = U_\lambda \otimes V^\rho$ of representations U_λ and V^ρ of Diff X is called an *elementary representation*. By Theorem 1 of §3, an *elementary representation* U_λ^ρ *is irreducible if and only if* ρ *is an irreducible representation of* S_n $(n = 0, 1, 2, \ldots)$.

THEOREM 4. *Two irreducible elementary representations* $U_{\lambda_1}^{\rho_1}$ *and* $U_{\lambda_2}^{\rho_2}$ *of* Diff X, $\dim X > 1$, *are equivalent if and only if* $\lambda_1 = \lambda_2$ *and* ρ_1 *and* ρ_2 *are equivalent representations of* S_n.

PROOF. We restrict U_λ^ρ to Diff(X, m). From §4.4 it follows that $\overline{U}_\lambda \cong \bigoplus_{n=0}^{\infty} V^{\rho_n}$, where ρ_n^0 is the unit representation of S_n (the bar denotes the restriction to Diff(X, m)). Consequently,

$$\overline{U}_\lambda^\rho \cong \overline{V}^\rho \oplus \sum_{n=1}^{\infty} \overline{V}^{\rho \circ \rho_n^0}. \tag{5}$$

From the decomposition (5) it follows easily on the basis of Theorem 2 in §2 that $\overline{U}_{\lambda_1}^{\rho_1} \nsim \overline{U}_{\lambda_2}^{\rho_2}$ when $\rho_1 \nsim \rho_2$. Hence, a fortiori, $U_{\lambda_1}^{\rho_1} \nsim U_{\lambda_2}^{\rho_2}$ when $\rho_1 \nsim \rho_2$. It only remains to discuss the representations $U_{\lambda_1}^\rho$ and $U_{\lambda_2}^\rho$.

Let $U_{\lambda_1}^\rho \nsim U_{\lambda_2}^\rho$. We denote by f_{λ_i} the vacuum vector in the representation space of U_{λ_i} $(i = 1, 2)$ and by F an arbitrary vector in the representation space of V^ρ. From (5) it follows easily that under an isomorphism of the representation spaces of $U^\rho_{\lambda_1}$ and $U^\rho_{\lambda_2}$ the vector $f_{\lambda_1} \otimes F$ goes over into the vector $f_{\lambda_2} \otimes F$; therefore,

$$\langle U^\rho_{\lambda_1}(\psi) f_{\lambda_1} \otimes F,\ f_{\lambda_1} \otimes F \rangle = \langle U^\rho_{\lambda_2}(\psi) f_{\lambda_2} \otimes F,\ f_{\lambda_2} \otimes F \rangle$$

for any $\psi \in \mathrm{Diff}\, X$. Hence $u_{\lambda_1} = u_{\lambda_2}$, where u_λ is the spherical function for U_λ (see §4.3), and therefore, $\lambda_1 = \lambda_2$.

§5. The ring of elementary representations generated by the Poisson measure

1. The decomposition of the tensor product $U_{\lambda_1} \otimes U_{\lambda_2}$ of representations of Diff X into irreducible representations. First we prove a general theorem about representations $\mathrm{EXP}_\beta T$. Let G be an arbitrary group and T a unitary representation of G in a real space H; let β: $G \to H$ be a 1-cocycle. Then a new unitary representation $U_s = \mathrm{EXP}_{s\beta} T$ can be defined as in [1] and [2], where s is an arbitrary real number. Let H' be the dual space to H and $\widetilde{\mu}$ a measure in any nuclear completion $\widetilde{H}$ of H', defined by the characteristic functional:

$$\int e^{i\langle F,\, f\rangle}\, d\mu(F) = e^{-\|f\|^2/2}.$$

The operators of the representation $\widetilde{U}_s = \mathrm{EXP}_{s\beta} T$ act in the complex Hilbert space $L^2_{\widetilde{\mu}}(\widetilde{H})$ according to the formula:

$$(\widetilde{U}_s(g)\Phi)(F) = e^{is\langle F,\, \beta(g)\rangle} \Phi(T^*(g) F).$$

THEOREM 1. *If $s_1^2 + s_2^2 = s_1'^2 + s_2'^2$, then $\widetilde{U}_{s_1} \otimes \widetilde{U}_{s_2} \cong \widetilde{U}_{s_1'} \otimes \widetilde{U}_{s_2'}$.*

PROOF. We define operators A_t, $t \in \mathbf{R}$, in $L^2_{\widetilde{\mu}}(\widetilde{H}) \otimes L^2_{\widetilde{\mu}}(\widetilde{H})$ by the formula:

$$(A_t\Phi)(F_1,\ F_2) = \Phi(\cos t\, F_1 + \sin t\, F_2,\ \ -\sin t\, F_1 + \cos t\, F_2).$$

From the definition of the Gaussian measure $\widetilde{\mu}$ it follows that A_t for any $t \in \mathbf{R}$ is a unitary operator. Further, from the definition of $\widetilde{U}_s$ it follows easily that

$$A_t^{-1}(\widetilde{U}_{s_1}(g) \otimes \widetilde{U}_{s_2}(g)) A_t = \widetilde{U}_{s_1 \cos t + s_2 \sin t}(g) \otimes \widetilde{U}_{-s_1 \sin t + s_2 \cos t}(g).$$

Consequently, $\widetilde{U}_{s_1} \otimes \widetilde{U}_{s_2} \cong \widetilde{U}_{s_1 \cos t + s_2 \sin t} \otimes \widetilde{U}_{-s_1 \sin t + s_2 \cos t}$ for any $t \in \mathbf{R}$; hence the assertion of the theorem follows immediately.

COROLLARY. $\widetilde{U}_{s_1} \otimes \widetilde{U}_{s_2} \cong \widetilde{U}_{\sqrt{(s_1^2 + s_2^2)}} \otimes \widetilde{U}_0$.

Now let U_λ be the representation of Diff X generated by the Poisson measure with parameter λ (see §4), let V^ρ be the representation of Diff X defined in §1 (ρ runs over the representations of S_n).

THEOREM 2.

$$U_{\lambda_1} \otimes U_{\lambda_2} \cong \bigoplus_{n=0}^{\infty} (U_\lambda \otimes V^{\rho_n^0}), \tag{1}$$

where $\lambda = \lambda_1 + \lambda_2$, *and* ρ_n^0 *is the unit representation of* S_n. *Every term in* (1) *is an irreducible representation of* Diff X.

PROOF. Let T be a representation of Diff X in $L_m^2(X)$ defined by $(T(\psi)f)(x) = \mathcal{J}_\psi^{1/2}(x)f(\psi^{-1}x)$. A 1-cocycle β: Diff $X \to \widetilde{L_m^2}(X)$ is given by $[\beta(\psi)](x) = \mathcal{J}_\psi^{1/2}(x) - 1$. We consider the representation $\widetilde{U}_s = \mathrm{EXP}_{s\beta}T$ of Diff X. By Theorem 1 we have, for any $\lambda_1 > 0$, $\lambda_2 > 0$

$$\widetilde{U}_{\sqrt{\lambda_1}} \otimes \widetilde{U}_{\sqrt{\lambda_2}} \cong \widetilde{U}_{\sqrt{\lambda_1+\lambda_2}} \otimes \widetilde{U}_0. \tag{2}$$

On the other hand, it was proved in §4.4 that $\widetilde{U}_{\sqrt{\lambda}} \cong U_\lambda$ for any $\lambda > 0$, and $\widetilde{U}_0 = \bigoplus_{n=0}^{\infty} V^{\rho_n^0}$. Consequently, (2) implies that

$U_{\lambda_1} \otimes U_{\lambda_2} \cong \bigoplus_{n=0}^{\infty} (U_\lambda \otimes V^{\rho_n^0})$, where $\lambda = \lambda_1 + \lambda_2$.

The irreducibility of $U_\lambda \otimes V^{\rho_n^0}$ follows from the main theorem of §3.

2. The decomposition of the tensor product of two elementary representations of Diff X associated with the Poisson measure.

THEOREM 3. $U_{\lambda_1}^{\rho_1} \otimes U_{\lambda_2}^{\rho_2} \cong \bigoplus_{n=0}^{\infty} U_{\lambda_1+\lambda_2}^{\rho_1 \circ \rho_2 \circ \rho_n^0}$.

(For the definition of the operation $\rho_1 \circ \rho_2$, see §1.)

PROOF. By definition, $U_{\lambda_1}^{\rho_1} = U_{\lambda_1} \otimes V^{\rho_1}$, $U_{\lambda_2}^{\rho_2} = U_{\lambda_2} \otimes V^{\rho_2}$. Consequently, $U_{\lambda_1}^{\rho_1} \otimes U_{\lambda_2}^{\rho_2} \cong (U_{\lambda_1} \otimes U_{\lambda_2}) \otimes (V^{\rho_1} \otimes V^{\rho_2})$. By Theorem 2, $U_{\lambda_1} \otimes U_{\lambda_2} = \bigoplus_{n=0}^{\infty} (U_{\lambda_1+\lambda_2} \otimes V^{\rho_n^0})$. Further, $V^{\rho_1} \otimes V^{\rho_2} \cong V^{\rho_1 \circ \rho_2}$ for any ρ_1, ρ_2 (see §1). Hence the required result is obtained straightaway.

COROLLARY. *The set of representations of* Diff X *that split into the direct sum of irreducible representations of the form* $U_\lambda \otimes V^\rho$ *is closed under the operation of taking the tensor product.*

§6. Representations of Diff X associated with infinitely divisible measures

The group Diff X acts naturally in the space $\mathcal{F}(X)$ of generalized functions on X. Therefore, representations of Diff X can be constructed for any quasi-invariant measure $\widetilde{\mu}$ on $\mathcal{F}(X)$. We have already noted earlier that the configuration space Γ_X has a natural embedding in $\mathcal{F}(X)$, and, therefore, the representations considered earlier are part of a considerably wider class of representations. Here we consider a special class of measures in $\mathscr{F}(X)$ (infinitely decomposable measures), which generate the same stock of representations as the representations $U_\lambda \otimes V^\rho$ studied in §§4 and 5.

1. The measure $\widetilde{\mu}_\tau$ in the space of generalized functions on X. Let $\mathscr{F}(X)$ be the space of generalized functions on X; we define an action of Diff X on $\mathscr{F}(X)$ by $\langle \psi^* F, f\rangle = \langle F, f \circ \psi\rangle$.

Let us consider a positive definite function χ_τ on $\mathbf{R}$ of the form

$$\chi_\tau(t) = \exp\left(\int_{-\infty}^{+\infty} (e^{i\alpha t} - 1)\, d\tau(\alpha)\right),$$

where τ is a non-negative finite measure on $\mathbf{R}$ (not necessarily normalized). ($\chi_\tau(t)$ is the Fourier transform of a certain infinitely divisible measure on $\mathbf{R}$.) Let m be a fixed smooth positive measure on X, $m(X) = \infty$. A measure $\widetilde{\mu} = \widetilde{\mu}_\tau$ is given on $\mathscr{F}(X)$ by the characteristic functional

$$(1) \quad L_\tau(f) = \exp\left(\int_X \ln \chi_\tau(f(x))\, dm(x)\right) = \exp\left(\int_{\mathbf{R}}\int_X (e^{i\alpha f(x)} - 1)\, dm(x)\, d\tau(\alpha)\right).$$

We list some basic properties of $\widetilde{\mu}$, which follow easily from this definition.

1) If $X = X_1 \cup \cdots \cup X_n$ is a finite partitioning of X and $\mathscr{F}(X) = \mathscr{F}(X_1) \oplus \ldots \oplus \mathscr{F}(X_n)$ the corresponding decomposition of $\mathscr{F}(X)$ into a direct sum, then $\widetilde{\mu} = \widetilde{\mu}_1 \times \ldots \times \widetilde{\mu}_n$, where $\widetilde{\mu}_i$ is the projection of μ onto the subspace $\mathscr{F}(X_i)$, $i = 1, \ldots, n$ (infinite decomposability).

2) The measure $\widetilde{\mu} = \widetilde{\mu}_\tau$ is concentrated on the set $\mathscr{F}_0(X)$ of generalized functions of the form $\sum_{k=1}^{\infty} \alpha_k \delta_{x_k}$, $\alpha_k \neq 0$, where $\alpha_k \in \operatorname{supp} \tau$, and $\{x_k\}$ is a set without accumulation points in X (that is, a configuration in X); δ_x denotes the delta-function on X concentrated at $x \in X$.[1]

3) The measure μ is quasi-invariant under Diff X, and

$$\left.\frac{d\widetilde{\mu}(\psi^* F)}{d\widetilde{\mu}(F)}\right|_{F = \sum_{k=1}^{\infty} \alpha_k \delta_{x_k}} = \prod_{k=1}^{\infty} \mathscr{J}_\psi(x_k),$$

where $\mathscr{J}_\psi(x) = \dfrac{dm(\psi^{-1}x)}{dm(x)}$. (Since ψ is finite, only finitely many factors $\mathscr{J}_\psi(x_k)$ are distinct from 1.)

Let us note the particular case when τ is concentrated at one point, $\alpha = 1$. Then $\mathscr{F}_0(X) = \{\sum \delta_{x_i};\ \{x_i\} \in \Gamma_X\}$, where Γ_X is the configuration $\mathscr{F}_0(X) \to \Gamma_X$ in which each generalized function $\Sigma\delta_{x_i}$ is associated with a configuration $\{x_i\} \in \Gamma_X$. It is not difficult to verify that the image of μ under this map is the Poisson measure on Γ_X with parameter $\lambda = \tau(\mathbf{R})$.

[1] The converse is also true: any infinitely decomposable measure in $\mathscr{F}(X)$ concentrated on a set of the type indicated is a measure μ_τ for a certain τ; when $X = \mathbf{R}^1$, this fact is very well known (see, for example, J.L. Doob, Stochastic processes, Wiley & Sons, New York 1953.

2. The representation of Diff X associated with $\widetilde{\mu}_\tau$. A unitary representation of Diff X is given in $L^2_{\widetilde{\mu}}(\mathscr{F}(X))$ by the formula

$$(V_\tau(\psi)\Phi)(F) = \left(\frac{d\widetilde{\mu}_\tau(\psi^*F)}{d\widetilde{\mu}_\tau(F)}\right)^{1/2}\Phi(\psi^*F). \tag{2}$$

In the particular case when the measure τ on $\mathbf{R}$ is concentrated at a single point and is equal to λ at this point, the representation V_τ so constructed is equivalent to the representation U_λ corresponding to the Poisson measure with parameter λ (see the remark above). Here we shall obtain the decomposition of V_τ into elementary representations.

We denote by $\mathscr{F}_0(X)$ the set of all generalized functions of the form $\Sigma\,\alpha_k\delta_{x_k}$, $\alpha_k \neq 0$, where $\{x_k\} \in \Gamma_X$; it was mentioned above that $\mathscr{F}_0(X)$ is a subset of full measure in $\mathscr{F}(X)$. We introduce the space $\mathbf{R}^\infty = \prod_{i=1}^{\infty}\mathbf{R}_i$, $\mathbf{R}_i = \mathbf{R}$ with measure $\nu = \tau_0 \times \ldots \times \tau_0 \times \ldots$, where τ_0 is the normalized measure on $\mathbf{R}$: $\tau_0 = \frac{\tau}{\lambda}$, $\lambda = \tau(\mathbf{R})$.

Next, let $i(\gamma, x)$ be an admissible indexing (for the definition see §2) in X and consider the sequence of maps $a_k\colon \Gamma_X \to X$ $(k = 1, 2, \ldots)$ defined by $a_k(\gamma) \in \gamma$, $i(\gamma, a_k(\gamma)) = k$. We define a map

$$\pi\colon \mathscr{F}_0(X) \to \Gamma_X \times \mathbf{R}^\infty$$

by

$$\pi\Big(\sum_{x\in\gamma}\alpha_x\delta_x\Big) = (\gamma;\ \alpha_{a_1(\gamma)}, \ldots, \alpha_{a_k(\gamma)}, \ldots).$$

Standard arguments establish the following result.

LEMMA 1. *The map π is measurable in both directions; the image of $\mathscr{F}_0(X)$ is a subset of full measure in $(\Gamma_X \times \mathbf{R}^\infty; \mu \times \nu)$; the image of $\widetilde{\mu}_\tau$ under π is the product measure $\mu \times \nu$, where μ is the Poisson measure on Γ_X with parameter $\lambda = \tau(\mathbf{R})$.*

By means of π the action of Diff X can be carried over from $\mathscr{F}_0(X)$ to $\Gamma_X \times \mathbf{R}^\infty$. It is not difficult to see that the action of Diff X on $\Gamma_X \times \mathbf{R}^\infty$ is given by

$$\psi\colon (\gamma, \alpha) \mapsto (\psi^{-1}\gamma,\ \alpha\sigma(\psi, \gamma)), \tag{3}$$

where $\alpha = (\alpha_1, \ldots, \alpha_n, \ldots)$, and $\alpha\sigma = (\alpha_{\sigma(1)}, \ldots, \alpha_{\sigma(n)}, \ldots)$; here $\sigma(\psi, \gamma)$ is the map Diff $X \times \Gamma_X \to S_\infty$ defined by i (see §2).

Lemma 1 and (3) imply the next result.

LEMMA 2. *The representation V_τ of* Diff X *is equivalent to the representation acting on $L^2_\mu(\Gamma_X) \otimes L^2_\nu(\mathbf{R}^\infty)$ according to the formula*

$$(V_\tau(\psi)f)(\gamma, \alpha) = \left(\frac{d\mu(\psi^{-1}\gamma)}{d\mu(\gamma)}\right)^{1/2} f(\psi^{-1}\gamma,\ \alpha\sigma(\psi, \gamma)). \tag{4}$$

Here μ is the Poisson measure with parameter $\lambda = \tau(\mathbf{R})$, and $\nu = \tau_0 \times \ldots \times \tau_0 \times \ldots$

Let us now consider the space $L_\nu^2(\mathbf{R}^\infty) = L_{\tau_0}^2(\mathbf{R}) \otimes \ldots \otimes L_{\tau_0}^2(\mathbf{R}) \otimes \ldots$ with a given unitary representation T of S_∞: $(T(\sigma)\varphi)(\alpha) = \varphi(\alpha\sigma)$. We split $L_\nu^2(\mathbf{R}^\infty)$ into invariant subspaces.

We fix an orthonormal basis $e_0, e_1, \ldots$ in $L_{\tau_0}^2(\mathbf{R})$, where $e_0 \equiv 1$ (that is, e_0 is the function on $\mathbf{R}$ everywhere equal to 1). It is known that the vectors $e_{i_1} \otimes \ldots \otimes e_{i_k} \otimes \ldots$, where $\Sigma i_k < \infty$ (so that all the indices i_k apart from finitely many are zero) form a basis in $L_\nu^2(\mathbf{R})$. Now S_∞ permutes these vectors. We divide the set A of basis vectors into orbits under S_∞.

Let us consider all possible collections of natural numbers of the form

$$(n_1, \ldots, n_k; i_1, \ldots, i_k), \tag{5}$$

where $n_1 \geqslant \ldots \geqslant n_k$, $i_p \neq i_q$ when $p \neq q$, and $i_p > i_{p+1}$ if $n_p = n_{p+1}$; $k = 0, 1, \ldots$ ($k = 0$ corresponds to the empty set). With each collection (5) we associate a basis vector in A:

$$\underbrace{e_{i_1} \otimes \ldots \otimes e_{i_1}}_{n_1} \otimes \ldots \otimes \underbrace{e_{i_k} \otimes \ldots \otimes e_{i_k}}_{n_k} \otimes e_0 \otimes e_0 \otimes \ldots. \tag{6}$$

We denote by $A_{n_1, \ldots, n_k}^{i_1, \ldots, i_k}$ the orbit of S_∞ in A generated by (6) and by $H_{n_1, \ldots, n_k}^{i_1, \ldots, i_k}$ the subspace of $L_\nu^2(\mathbf{R}^\infty)$ spanned by the vectors of $A_{n_1, \ldots, n_k}^{i_1, \ldots, i_k}$.

It is easy to establish that the orbits $A_{n_1, \ldots, n_k}^{i_1, \ldots, i_k}$ are pairwise distinct and that their union is the whole of A. Hence it follows that $L_\nu^2(\mathbf{R}^\infty)$ splits into a direct sum of invariant subspaces

$$L_\nu^2(\mathbf{R}^\infty) = \oplus H_{n_1, \ldots, n_k}^{i_1, \ldots, i_k}; \tag{7}$$

the sum is taken over all collections $(n_1, \ldots, n_k; i_1, \ldots, i_k)$ of the type (5). Note that representations of S_∞ in $H_{n_1, \ldots, n_k}^{i_1, \ldots, i_k}$ and in $H_{n_1, \ldots, n_k}^{i'_1, \ldots, i'_k}$ are equivalent.

The decomposition (7) we have just obtained leads to the following result.

LEMMA 3. *The space $L_\mu^2(\Gamma_X) \times L_\nu^2(\mathbf{R}^\infty)$ decomposes into the direct sum of V_τ-invariant subspaces:*

$$L_\mu^2(\Gamma_X) \otimes L_\nu^2(\mathbf{R}^\infty) = \sum \oplus L_\mu^2(\Gamma_X) \otimes H_{n_1, \ldots, n_k}^{i_1, \ldots, i_k}. \tag{8}$$

The representations of Diff X *in $L_\mu^2(\Gamma_X) \otimes H_{n_1, \ldots, n_k}^{i_1, \ldots, i_k}$ and in $L_\mu^2(\Gamma_X) \otimes H_{n_1, \ldots, n_k}^{i'_1, \ldots, i'_k}$ are equivalent.*

We denote by $V_\tau^{n_1, \ldots, n_k}$ the restriction of V_τ to $L_\mu^2(\Gamma_X) \otimes H_{n_1, \ldots, n_k}^{i_1, \ldots, i_k}$.

LEMMA 4. $V_\tau^{n_1, \ldots, n_k} \cong U_\lambda^{\rho_{n_1}^0 \circ \cdots \circ \rho_{n_k}^0}$ *where $\lambda = \tau(\mathbf{R})$, and ρ_n^0 is the unit representation of S_n.*

The assertion of the lemma is easily established if we use the realization of elementary representations introduced in §3.3.

Lemmas 3 and 4 give the next result.

THEOREM 1. *The representation* V_τ *of* Diff X *in* $L^2_{\mu_\tau}(\mathcal{F}(X))$ *defined by* (2) *splits into a discrete direct sum of irreducible elementary representations of the form* U^p_λ, *where* $\lambda = \tau(\mathbf{R})$.

REMARK 1. If τ is concentrated at two points on $\mathbf{R}$, and if the measures of these points are λ_1 and λ_2, respectively, then, as is easy to show, $V_{\tilde{\tau}} \cong U_{\lambda_1} \otimes U_{\lambda_2}$. In this way we obtain from Lemmas 3 and 4, in particular, the decomposition of the tensor product $U_{\lambda_1} \otimes U_{\lambda_2}$ into irreducible representations. This was obtained by another method in §5.

REMARK 2. The representation V_τ can be treated as a continual tensor product of Poisson representations U_λ; Theorem 1 then gives a decomposition of the continual tensor product of U_λ into irreducible representations.

REMARK 3. The representation so constructed is a cyclic subrepresentation in $\mathrm{EXP}_\beta T$, where T is the representation of Diff X in the real space $L^2(X \times \mathbf{R}, m \times \tau)$ and β is the 1-cocycle: $[\beta(\psi)](x, \alpha) = \mathcal{J}_\psi^{1/2}(x) - 1$, see §7. If $\tau = \delta_{x_0}$, $x_0 \neq 0$, then we obtain the Gaussian form for the Poisson representation U_λ (see §4.4).

3. Criteria for representations V_τ **of Diff** X **to be equivalent.** By Lemma 3, the multiplicity with which $V_\tau^{n_1, \ldots, n_k}$ occurs in V_τ depends only on the numbers $n_1, \ldots, n_k$ and on the dimension of $L^2_{\tau_0}(\mathbf{R})$. Therefore, Lemmas 3 and 4 also imply the following result.

THEOREM 2. *Let* τ' *and* τ'' *be two non-negative finite measures on* $\mathbf{R}$ *such that*

1) $\tau'(\mathbf{R}) = \tau''(\mathbf{R})$;

2) *the supports of* τ' *and* τ'' *are either both infinite or consistent of the same finite number of points.*

Then the representations $V_{\tau'}$ *and* $V_{\tau''}$ *of* Diff X *are equivalent.*

By Theorem 2, each representation V_τ is given, up to equivalence, by a pair of numbers: the parameter $\lambda = \tau(\mathbf{R})$ of the Poisson measure ($0 < \lambda < \infty$) and the index h, which is equal to n if τ is concentrated on n points, and is ∞ if supp τ is infinite. It is convenient, therefore, to denote these representations by $V_{\lambda,h}$ (instead of the previous notation V_τ).

4. The tensor product of representations $V_\tau = V_{\lambda,h}$. THEOREM 3. $V_{\lambda_1,h_1} \otimes V_{\lambda_2,h_2} \cong V_{\lambda_1+\lambda_2, h_1+h_2}$; *thus, the set of representations* $V_\tau = V_{\lambda,h}$ *is closed under the operation of tensor multiplication.*

PROOF. We have $V_{\lambda_i,h_i} \cong V_{\tau_i}$, where τ_i is any non-negative finite measure on $\mathbf{R}$ scuh that $\tau_i(\mathbf{R}) = \lambda_i$ and $|\mathrm{supp}\, \tau_i| = h_i$ if $h_i < \infty$, and supp τ_i is any infinite set if $h_i = \infty$ ($i = 1, 2$). The measures τ_1 and τ_2 can always be chosen so that supp $\tau_1 \cap$ supp $\tau_2 = \phi$. Let us consider $\tau = \tau_1 + \tau_2$ on $\mathbf{R}$. Obviously, $V_\tau = V_{\lambda_1+\lambda_2, h_1+h_2}$. Therefore, it is sufficient for us to check that $V_\tau \cong V_{\tau_1} \otimes V_{\tau_2}$.

We denote by $\tilde{\mu}_1$, $\tilde{\mu}_2$, and $\tilde{\mu}$ the measures on $\mathcal{F}(X)$ corresponding to τ_1, τ_2, and τ on $\mathbf{R}$, respectively, and by $L_{\tau_1}(f)$, $L_{\tau_2}(f)$, and $L_\tau(f)$ their characteristic functionals. Since $\operatorname{supp} \tau_1 \cap \operatorname{supp} \tau_2 = \phi$, we have $L_\tau(f) = L_{\tau_1}(f)\, L_{\tau_2}(f)$. Hence

$$L^2_{\tilde\mu}(\mathcal{F}(X)) \cong L^2_{\tilde\mu_1}(\mathcal{F}(X)) \otimes L^2_{\tilde\mu_2}(\mathcal{F}(X)),$$

and so $V_\tau \cong V_{\tau_1} \otimes V_{\tau_2}$, as required.

§7. Representations of the cross product $\mathcal{G} = C^\infty(X)\cdot\operatorname{Diff} X$

1. Definition of $\mathcal{G}$ and the construction of representations. Let us introduce the (additive) group $C^\infty(X)$ of all real finite functions f on X of class C^∞. Now Diff X acts on $C^\infty(X)$ as a group of automorphisms $f \to f \circ \psi^{-1}$. In this way we can define the cross product $\mathcal{G} = C^\infty(X)\cdot\operatorname{Diff} X$ of $C^\infty(X)$ with the multiplication: $(f_1, \psi_1)(f_2, \psi_2) = (f_1 + f_2 \circ \psi^{-1}, \psi_1\psi_2)$.

Let $\tilde\mu$ be an arbitrary quasi-invariant (under Diff X) measure in the space $\mathcal{F}(X)$ of generalized functions on X. We consider $L^2_{\tilde\mu}(\mathcal{F}(X))$ and associate with each element (f, ψ) $\mathcal{G}$ the following operator $V(f, \psi)$ in $L^2_{\tilde\mu}(\mathcal{F}(X))$:

$$(V(f, \psi)\Phi)(F) = e^{i\langle F, f\rangle}\left(\frac{d\mu(\psi^*F)}{d\mu(F)}\right)^{1/2}\Phi(\psi^*F). \tag{1}$$

It is easy to check that the $V(f, \psi)$ are unitary and form a representation of $\mathcal{G}$. This representation of $\mathcal{G}$ is cyclic with respect to $C^\infty(X)$ (the constant is a cyclic vector). It is irreducible if and only if the measure $\tilde\mu$ on $\mathcal{F}(X)$ is ergodic with respect to Diff X.

2. Representations associated with infinitely decomposable measures. From now on we restrict ourselves to the measure $\tilde\mu$ on $\mathcal{F}(X)$ introduced in §6, that is, measures with characteristic functionals of the form

$$L_\tau(f) = \exp\left(\iint_{RX} (e^{i\alpha f(x)} - 1)\, dm(x)\, d\tau(\alpha)\right),$$

where m is a smooth positive measure on X, $m(X) = \infty$, and τ is a non-negative finite measure on **R**. The representation of $\mathcal{G}$ corresponding to this measure is now denoted by V_τ (the measure m on X is assumed to be fixed).

It is not difficult to prove that these measures are ergodic with respect to Diff X; consequently, *the representations V_τ of $\mathcal{G} = C^\infty(X)\cdot\operatorname{Diff} X$ are irreducible.*

LEMMA 1. *If* dim $X > 1$, *then $\tilde\mu$ is ergodic with respect to the subgroup* Diff $(X, m) \subset$ Diff X *of diffeomorphisms preserving m.*

The proof goes as for Poisson measures (see §4).

COROLLARY 1. *The restriction of V_τ to the subgroup* $C^\infty(X)\cdot\operatorname{Diff}(X, m)$ *is irreducible.*

COROLLARY 2. *The only vectors in $L^2_{\tilde{\mu}}(\mathcal{F}(X))$ that are invariant under* Diff (X, m) *are the constants.*

Let Φ_0 be the function in $L^2_{\tilde{\mu}}(\mathcal{F}(X))$ that is identically equal to 1. The following function on $\mathcal{G}$ is called the *spherical function* of V_τ:

$$u_\tau(f, \psi) = \langle V_\tau(f, \psi)\Phi_0, \Phi_0 \rangle,$$

where the brackets denote the scalar product in $L^2_{\tilde{\mu}}(\mathcal{F}(X))$. Since Φ_0 is a cyclic vector in $L^2_{\tilde{\mu}}(\mathcal{F}(X))$, the representation V_τ is uniquely determined by $u_\tau(f, \psi)$.

By a simple calculation we obtain

$$u_\tau(f, \psi) = \exp\Big(\int_{\mathbf{R}}\int_{X} (\mathcal{J}_\psi^{1/2}(x)\, e^{i\alpha f(x)} - 1)\, dm(x)\, d\tau(\alpha)\Big), \tag{2}$$

where $\mathcal{J}_\psi(x) = \dfrac{dm(\psi^{-1}x)}{dm(x)}$.

Obviously, if $\tau_1 = \tau_2$, then $u_{\tau_1} \neq u_{\tau_2}$. Since Φ_0 is defined invariantly in $L^2_{\tilde{\mu}}(\mathcal{F}(X))$, we have the following result.

LEMMA 2. *If $\tau_1 \neq \tau_2$, the representations V_{τ_1} and V_{τ_2} of* $\mathcal{G} = C^\infty(X)\cdot \mathrm{Diff}(X)$ $(\dim X > 1)$ *are inequivalent.*

3. The Gaussian form of the representations V_τ. Let us consider the complex Hilbert space H of functions on $X \times \mathbf{R}$ with the norm

$$\|\varphi\|^2 = \int\int |\varphi(x, \alpha)|^2\, dm(x)\, d\tau(\alpha).$$

A unitary representation T of $\mathcal{G}$ is given in H by

$$(T(f, \psi)\varphi)(x, \alpha) = e^{i\alpha f(x)}\mathcal{J}_\psi^{1/2}(x)\, \varphi(\psi^{-1}x, \alpha).$$

We define map β: $\mathcal{G} \to H$ by

$$[\beta(f, \psi)](x, \alpha) = e^{i\alpha f(x)}\mathcal{J}_\psi^{1/2}(x) - 1.$$

It is easy to verify that for any $g_1, g_2 \in \mathcal{G}$ we have: $\beta(g_1 g_2) = \beta(g_1) + T(g_1)\beta(g_2)$, so that β is a 1-cocycle.

Let us construct from T and the 1-cocycle β a new representation $\widetilde{V}_\tau = \mathrm{EXP}_\beta T$ of $\mathcal{G}$ (see [1], [2]). We denote the dual space of H by H'. The standard Gaussian measure on the completion $\widetilde{H}'$ of H' is the measure μ with the characteristic functional

$$\int e^{i\,\mathrm{Re}\,\langle F, f\rangle}\, d\mu(F) = e^{-\|f\|^2/2}, \qquad f \in H.$$

The representation $\widetilde{V}_\tau$ of $\mathcal{G}$ is given on $L^2_\mu(\widetilde{H}')$ by

$$(\widetilde{V}_\tau(g)\Phi)(F) = e^{i\,\mathrm{Re}\,\langle F, \beta(g)\rangle}\Phi(T^*(g)F),$$

where the operator $T^*(g)$ is given by $\langle T^*(g)F, f\rangle = \langle F, T(g)f\rangle$.

THEOREM. *Let τ be a measure on* **R** *such that $\int e^{i\alpha t} d\tau(\alpha)$ is a real function of t. Then the restriction of the representation $\widetilde{V}_\tau$ of $\mathcal{G} = C^\infty(X)\cdot$ Diff X to the cyclic subspace $\mathcal{H} \subset L^2_\mu(H')$ generated by the vector $\Omega \equiv 1$ is equivalent to the representation of V_τ defined in* §7.2.

PROOF. From the definition of $\widetilde{V}_\tau$ it follows that

$$\langle \widetilde{V}_\tau(f, \psi)\Omega, \Omega\rangle = \exp\left(\int\int (\cos(\alpha f(x))\, \mathcal{J}_\psi^{1/2}(x) - 1)\, dm(x)\, d\tau(\alpha)\right)$$

(here the diamond brackets denote the inner product in $L^2_\mu(H')$). Consequently, from the hypothesis of the theorem,

$$\langle V_\tau(f, \psi)\Omega, \Omega\rangle = \exp\left(\int\int (e^{i\alpha f(x)} \mathcal{J}_\psi^{1/2}(x) - 1)\, dm(x)\, d\tau(\alpha)\right) = u_\tau(f, \psi),$$

where u_τ is the spherical function of V_τ (see §7.2). Hence the assertion of the theorem follows immediately.

REMARK 1. A representation of $C^\infty(X)\cdot$ Diff X was constructed in [15] by means of the N/V limit. This representation coincides with that constructed here for the Poisson measure (the connection with the Poisson measure was apparently not noticed), and the transition to a Fock model in [15] is equivalent to the realization of this representation as $\mathrm{EXP}_\beta T$ (see above).

We emphasize that a representation of the cross product can be constructed for any measure in $\mathcal{F}(X)$ that is quasi-invariant under Diff X. However, only those that are constructed from an infinitely decomposable measure have the structure $\mathrm{EXP}_\beta T$, because it is only in this case that there is a vacuum vector.

REMARK 2. Instead of $C^\infty(X)$ we can consider an arbitrary group of smooth functions $C^\infty(X, G) = G^X$ on X with values in a Lie group G and the cross product $C^\infty(X, G)\cdot$ Diff X.

If a unitary representation π of G is given on a space H, then the representation T of this cross product acts naturally in the space

$\mathcal{H} = \int \oplus H_x\, dm(x)$, $H_x = H$. This is irreducible if π is irreducible. If β: $C^\infty(X, G)\cdot$ Diff $X \to \mathcal{H}$ is a non-trivial cocycle (see [1]), then in EXP $\mathcal{H}$ we get a representation $\mathrm{EXP}_\beta T$ of $C^\infty(X, G)\cdot$ Diff X.

APPENDIX 1

On the methods of defining measures on the configuration space Γ_X

1. Let $X_1 \subset \ldots \subset X_n \subset \ldots$ be a sequence of open submanifolds in X with compact closures such that $X = \bigcup_n X_n$. The projections $p_k\colon \Gamma_X \to B_{X_k}$

are given by putting $p_k\,\gamma = \gamma \cap X_k$. Let μ be a measure on Γ_X and $\mu_k = p_k \mu$ its projection on $B_{X_k}(k = 1, 2, \ldots)$. Then the measures μ_k are mutually compatible, that is, for any $k > 1$ we have $p_{lk}\mu_k = \mu_l$, where $p_{lk}\colon B_{X_k} \to B_{X_l}$ is the natural projection. A well known theorem of Kolmogorov about the extension of measures enables us to establish the converse: if $\{\mu_k\}$ is any compatible sequence of measures on $\{B_{X_k}\}$, then there is a unique measure on Γ_X such that $p_k\mu = \mu_k (k = 1, 2, \ldots)$.

We can abandon the compatibility conditions and consider sequences of measures μ_k on B_{X_k} for which $\lim_{k\to\infty} p_{lk}\mu_k = \mu^{(l)}$ exists (in the weak sense) for all l. In this case the measures $\mu^{(l)}$ on B_{X_l} are compatible and define a measure μ on Γ_X.

We also recall that Γ_X is naturally embedded in the space of generalized functions $(\gamma \to \sum_{x\in\gamma} \delta_x)$, therefore, the methods for defining measures in linear spaces are applicable here (by means of the characteristic functional and so on); see, for example, [5].

2. A fundamentally different method describing measure on Γ_X has received attention in statistical physics [7]. It generalizes the method of specifying Markov measures (by transition probabilities). It consists in giving conditional measures on B_Y (or their densities with respect to Poisson measure) as functions on $\Gamma_{X\setminus Y}$ for all compact domains $Y \subset X$ by means of a single function (the potential) on B_X. The question of existence and uniqueness of the measure on Γ_X with a given system of conditional measures is, as a rule, very difficult. Curiously enough, in this case the condition for a measure μ on Γ_X to be quasi-invariant under Diff X can be formulated very simply: all the conditional measures on B_Y, where Y is any open set with compact closure, must be equivalent to a quasi-invariant (under Diff Y) measure on B_Y. By now there are many such measures known in statistical physics that are not equivalent to the Poisson measure (Gibbs measures).

A measure on Γ_X can also be given with the help of so-called correlation functions on B_X; a correlation function defines uniquely an initial measure on Γ_X (see, for example, [12]).

3. Let us introduce yet another method of defining measures on Γ_X. We say that a normalized Borel measure μ on X^∞ (see §0.3) is admissible if:

1) $\mu(\widetilde{X}^\infty) = 1$, that is, $\widetilde{X}^\infty$ is a subset of full measure in X^∞;

2) μ is quasi-invariant under Diff X.

If μ is an admissible measure on X^∞, then its projection $\widetilde{\mu} = p\mu$ on Γ_X (that is, $\widetilde{\mu}(C) = \mu(p^{-1}C)$ for any measurable set $C \subset \Gamma_X$) is a quasi-invariant measure on Γ_X. This method of defining a measure on Γ_X is of limited interest, however, it is convenient for constructing various examples.

It is easy to show that an admissible measure μ on X^∞ is ergodic if and only if it is regular (regularity means that it satisfies the $0 - 1$ law). In particular, any admissible product measure $\mu = m_1 \times \ldots \times m_n \times \ldots$ is ergodic. The following lemma is analogous to the Borel-Cantelli lemma.

LEMMA. *The product measure $\mu = m_1 \times \ldots \times m_n \times \ldots$ on X^∞ is admissible if and only if $\sum_{i=1}^{\infty} m_i(Y) < \infty$ for any compact set $Y \subset X$.*

EXAMPLE OF AN ADMISSIBLE MEASURE. Let $X = \mathbf{R}^n$. We denote by m_a the Gaussian measure with centre at $a \in \mathbf{R}$ and with unit correlation matrix: $dm_a(x) = (2\pi)^{-n/2} e^{-\|x-a\|^2/2}\, dx$, where dx is the Lebesgue measure on $\mathbf{R}^n$. By a direct computation it is not difficult to check that $\mu = m_{a_1} \times \ldots \times m_{a_n} \times \ldots$ is admissible if, for example, $\|a_n\| \geqslant c \log n$, $n = 1, 2, \ldots$.

If $\|a_n\| \to \infty$ sufficiently quickly, then it is easy to verify that μ is concentrated on the set $\Delta_s = \prod_{\sigma \in S_\infty} (s\Gamma_X)\sigma$, where $s\colon \Gamma_X \to \widetilde{X}^\infty$ is the cross-section corresponding to a certain admissible indexing i in Γ_X. However, μ is not invariant under left translations in $\widetilde{X}^\infty$: $(Tx)_i = x_{i+1}$. Hence its projection $\widetilde{\mu} = p\mu$ is a non-saturated measure in Γ_X (see Proposition 7 in §2.3).

Another example refers to the group Diff X, where X is a compact manifold. In this case, let $\widetilde{X}^\infty$ denote the set of all sequences in X that converge in X. It is easy to verify that $\widetilde{X}^\infty/S^\infty \equiv \Gamma_X$ is the union of all countable subsets of X with a unique limit point (one for each subset). Let $x_0 \in X$, let ρ be a continuous metric in X, and let $\{m_n\}$ be a sequence of smooth measures in X such that $\lim_{n\to\infty} \int \rho(x, x_0) dm_n(x) = 0$. It is clear that the product measure $m^{x_0} = \prod_n m_n$ is concentrated on $\widetilde{X}^\infty$. We introduce the measure $\widetilde{m} = \int m^{x_0}\, dx_0$ – the mixing of the measures m^{x_0} in $\widetilde{X}^\infty$. This $\widetilde{m}$ projects onto a measure m on Γ_X that is quasi-invariant and ergodic under Diff X. Another more complicated example of a measure on the countable subsets of a compact manifold that is quasi-invariant and ergodic under Diff X was constructed earlier in [8].

APPENDIX 2

S_∞-cocyles and Fermi representations

According to the standard definition, a 1-cocycle on Diff X with values in the group $S_\infty(\Gamma_X)$ of measurable maps $\Gamma_X \to S_\infty$ is a map $\sigma\colon$ Diff $X \to S_\infty(\Gamma_X)$ satisfying the following condition:

$$\sigma(\psi_1, \gamma)\sigma(\psi_2, \psi_1^{-1}\gamma) = \sigma(\psi_1\psi_2, \gamma). \tag{1}$$

Two 1-cocycles σ_1 and σ_2 are said to be cohomologous if there exists a measurable map $\sigma_0\colon \Gamma_X \to S_\infty$ such that

$$(2) \qquad \sigma_2(\psi, \gamma) = \sigma_0(\gamma)\sigma_1(\psi, \gamma)\sigma_0^{-1}(\psi^{-1}\gamma).$$

Let i be a measurable indexing in Γ_X, $a_k = \Gamma_X \to X$ $(k = 1, 2, \ldots)$ the sequence of measurable maps defined by i (see §2.2); and $s\colon \Gamma_X \to \widetilde{X}^\infty$ the cross-section of the fibration[1] $\widetilde{X}^\infty \to \Gamma_X$ defined by i:

$$s(\gamma) = (a_1(\gamma), \ldots, a_k(\gamma), \ldots).$$

Further, let $\Delta_s = \prod_{\sigma \in S_\infty} (s\Gamma_X)\sigma$.

We say that a measurable indexing i is *correct* if for any $\gamma, \gamma' \in \Gamma_X$ the conditions $|\gamma \cap K| = |\gamma' \cap K|$ and $\gamma \cap (X \setminus K) = \gamma' \cap (X \setminus K)$ for a certain compact set $K \subset X$ imply that $a_k(\gamma) = a_k(\gamma')$ for all indices k except finitely many.[2] A cross-section $s\colon \Gamma_X \to \widetilde{X}^\infty$ defined for a correct indexing i is also called *correct.* If i is correct, then the set Δ_s is invariant under Diff X.

To each correct cross-section $s\colon \Gamma_X \to \widetilde{X}^\infty$ there corresponds a 1-cocycle σ_s defined by the following relation (see §2.2):

$$s(\psi^{-1}\gamma) = [\psi^{-1}(s\gamma)]\sigma_s(\psi, \gamma).$$

REMARK. There are examples of cocycles that are not generated by correct cross-sections.

Cocycles σ_s generated by correct cross-sections s are also called *correct.*

We give, without proof, some properties of correct cocycles σ_s.

1) The cross-section s is uniquely determined by the correct cocycle σ_s corresponding to it.

2) Any two correct cocycles σ_{s_1} and σ_{s_2} are cohomologous as cocycles with values in $S^\infty(\Gamma_X)$, that is,

$$(3) \qquad \sigma_{s_2}(\psi, \gamma) = \sigma_0(\gamma)\sigma_{s_1}(\psi, \gamma)\sigma_0^{-1}(\psi^{-1}\gamma),$$

where σ_0 is a measurable map $\Gamma_X \to S^\infty$.

3) No correct 1-cocycle is cohomologous to the trivial cocycle.

4) Let σ_s be a correct cocycle, $\sigma_0\colon \Gamma_X \to S^\infty$ a measurable map, and σ a 1-cocycle defined by

$$\sigma(\psi, \gamma) = \sigma_0(\gamma)\sigma_s(\psi, \gamma)\,\sigma_0^{-1}(\psi^{-1}\gamma).$$

For σ to be correct it is necessary and sufficient that $\sigma_0(\gamma)\sigma_0^{-1}(\psi^{-1}\gamma) \in S_\infty$ for any $\psi \in$ Diff X and $\gamma \in \Gamma_X$.

5) Two correct cocycles σ_{s_1} and σ_{s_2} are cohomologous as cocycles with values in $S_\infty(\Gamma_X)$ if and only if the cross-sections s_1 and s_2 are cofinal, that is, $s_2 = \sigma_0 \circ s_1$ where $\sigma_0 \in S_\infty(\Gamma_X)$.

[1] Note that, generally speaking, the fibration $\widetilde{X}^\infty \to \Gamma_X$ has no continuous cross-sections. It can be shown that this fibration has no continuous quotient fibrations with fibre $\mathbf{Z}_2$.

[2] The condition of correctness is weaker than the condition of admissibility introduced in §2.2

For each 1-cocycle σ: Diff $X \to S_\infty(\Gamma_X)$ we define a corresponding 1-cocycle α_σ: Diff $X \to \mathbf{Z}_2(\Gamma_X)$ by

$$\alpha_\sigma(\psi, \gamma) = \text{sign}\, \sigma(\psi, \gamma) \tag{4}$$

(sign σ is the parity of $\sigma \in S_\infty$).

It is obvious that when σ_1 and σ_2 are cohomologous, then the correspond- cocycles α_{σ_1} and α_{σ_2} are cohomologous. However, there exist cocycles σ_1 and σ_2 (even correct ones) that are not cohomologous, although the corresponding cocycles α_{σ_1} and α_{σ_2} are.

EXAMPLE. The correct cocycles $\sigma_1(\psi, \gamma)$ and $\sigma_2(\psi, \gamma) = \sigma_0\sigma(\psi, \gamma)\sigma_0^{-1}$, where $\sigma_0 \notin S_\infty$. Since sign $(\sigma_0\sigma\sigma_0^{-1})$ = sign σ, we have $\alpha_{\sigma_1} = \alpha_{\sigma_2}$. However, the cocycles σ_1 and σ_2 are themselves not cohomologous (see property 4).

A SUFFICIENT CONDITION FOR COCYCLES TO BE COHOMOLOGOUS. Let $\sigma_2(\psi, \gamma) = \sigma_0(\gamma)\sigma_1(\psi, \gamma)\sigma_0^{-1}(\psi^{-1}\gamma)$. If $\sigma_0(\gamma) = \sigma_0\tilde{\sigma}_0(\gamma)$, where $\tilde{\sigma}_0 \in S_\infty(\Gamma_X)$ and σ_0 is an arbitrary element of S^∞, then $\alpha_{\sigma_1} \sim \alpha_{\sigma_2}$.

Note that each $\mathbf{Z}_2$-cocycle α defines a $\mathbf{Z}_2$-covering of Γ_X with a given action of Diff X on it. The elements of this covering are pairs (ε, γ), $\gamma \in \Gamma_X$, $\varepsilon = \pm 1$; and Diff X acts by $\psi(\varepsilon, \gamma) = (\varepsilon\,\alpha(\psi, \gamma), \psi^{-1}\gamma)$.

$\mathbf{Z}_2$-cocycles of the form (4), where σ is a correct cocycle, are called *correct $\mathbf{Z}_2$-cocycles,* and the $\mathbf{Z}_2$-coverings of Γ_X defined by them are also called correct.

LEMMA. *Correct $\mathbf{Z}_2$-cocycles are non-trivial.*

Let i be a correct indexing in Γ_X, s: $\Gamma_X \to \widetilde{X}^\infty$ the cross-section defined by i, μ a quasi-invariant measure in Γ_X, and T a unitary representation of S_∞ acting on a Hilbert space H. We associate with the triple (i, μ, T) a unitary representation V of Diff X in the space $\mathcal{H} \subset L^2_\mu(\Delta_s, H)$ of functions f: $\Delta_s \to H$ such that

$$\|f\|^2 = \int_{\Gamma_X} \|f(s\gamma)\|_H^2\, d\mu(\gamma) < \infty; \quad f(x\sigma) = T^{-1}(\sigma) f(x)$$

for every $\sigma \in S_\infty$; the representation operators are defined by

$$(V(\psi) f)(x) = \left(\frac{d\mu(\psi^{-1}px)}{d\mu(px)}\right)^{1/2} f(\psi^{-1}x), \tag{5}$$

where p is the projection $\Delta_s \to \Gamma_X$.

ALTERNATIVE DEFINITION: V is given in the space $L^2_\mu(\Gamma_X, H)$ of functions f: $\Gamma_X \to H$ for which $\|f\|^2 = \int \|f(\gamma)\|_H^2\, d\mu(\gamma) < \infty$. the operators $V(\psi)$ have the form

$$(V(\psi) f)(\gamma) = \left(\frac{d\mu(\psi^{-1}\gamma)}{d\mu(\gamma)}\right)^{1/2} T(\sigma_s(\psi, \gamma)) f(\psi^{-1}\gamma), \tag{6}$$

where σ_s is the 1-cocycle generated by a correct cross-section s: $\Gamma_X \to \widetilde{X}^\infty$.

It is obvious that the representations of Diff X so defined are equivalent.

Note that $V = \text{Ind}_{G_\gamma}^{\text{Diff}\, X}(T \circ \pi)$, where G_γ is the subgroup of all diffeomorphisms $\psi \in \text{Diff } X$ for which $\psi\gamma = \gamma$, and π is the projection $G_\gamma \to G_\gamma \backslash G_\gamma^0 \cong S_\infty$. ($G_\gamma^0$ is the subgroup leaving every point $x \in \gamma$ fixed).

Two 1-cocycles σ and σ' are said to be equivalent with respect to a measure μ if $\sigma(\psi, \gamma) = \sigma'(\psi, \gamma)$ mod 0 with respect to μ for any $\psi \in \text{Diff } X$. Note that the right-hand side of (6) does not change if the 1-cocycle is replaced by any equivalent cocycle. Therefore, the 1-cocycles σ need only be defined up to an equivalence.

The properties 1) – 5) of 1-cocycles can be reformulated without difficulty for equivalence classes of 1-cocycles. Moreover, 2) can be made more precise as follows: the map σ_0 in (3) is uniquely determined mod 0 if μ is ergodic.

Let us consider the particular case when $T = \text{Ind}_{S_n \times S_\infty^n}^{S_\infty} (\rho \times I)$, where ρ is a unitary representation of S_n and I is the unit representation of S_∞^n (S_∞^n is the group of finite permutations of $n + 1, n + 2, \ldots$). Comparing the definition of V with that of the elementary representations U_μ^ρ (see §3) it is easy to check that $V \cong U_\mu^\rho$. Hence, in particular, in this example all the correct cocycles lead to equivalent representations.[1]

Quite a different example is the Fermi representation. Let us consider a $\mathbf{Z}_2$-cocycle $\alpha(\psi, \gamma)$ (see above) and define a representation of Diff X in $L_\mu^2(\Gamma_X)$ by

$$(V(\psi) f)(\gamma) = \alpha(\psi, \gamma) \left(\frac{d\mu(\psi^{-1}\gamma)}{d\mu(\gamma)} \right)^{1/2} f(\psi^{-1}\gamma).$$

This is called a *Fermi representation* of Diff X.

When the $\mathbf{Z}_2$-cocycle is generated by a correct cross-section $s: \Gamma_X \to \widetilde{X}^\infty$, that is, when $\alpha(\psi, \gamma) = \text{sign } \sigma_s(\psi, \gamma)$, where σ_s: Diff $X \to S_\infty(\Gamma_X)$ is a correct cocycle, then there is another convenient realization of V: the representation is given in $\mathcal{H}$ by a function $f(x)$ on $\Delta_s \subset X^\infty$ such that

1) $f(x\sigma) = \text{sign } \sigma f(x)$ for any $\sigma \in S_\infty$ (an "odd" function).

2) $\|f\|^2 = \int_{\Gamma_X} |f(s\gamma)|^2 \, d\mu(\gamma) < \infty$.

The representation operators are defined by (5). In this case it can be shown by the same arguments as in §3 that *if μ is ergodic and the cross-section $s: \Gamma_X \to \widetilde{X}^\infty$ is generated by an admissible indexing i* (in the sense of §2.2), *then the Fermi representation $V(\psi)$ is irreducible.*

REMARK 1. Apparently, there exist non-equivalent Fermi representations of Diff X constructed from the same measure μ on Γ_X (for the construction of a Fermi representation by means of an N/V limit, see [19] and [20]).

REMARK 2. The group Diff X has factor representations of type II – it is sufficient to take a representation of S_∞ of type II in H (for example,

[1] There is a more general fact: if a representation T of S_∞ can be extended to a representation of S^∞, then the representation V of Diff X corresponding to T is uniquely determined, up to equivalence, by T and a measure μ on Γ_X.

the regular representation) and to construct a representation of Diff X in $L^2_\mu(\Gamma_X, H)$ from it.

APPENDIX 3

Representations of Diff X associated with measures in the tangent bundle of the space of infinite configurations

The representations of Diff X discussed above are of "zero order", that is, they do not depend on the differentials of the diffeomorphisms. In this context let us note that these representations can be extended to representations of the group of measurable finite transformations of a manifold X with a quasi-invariant measure.

However, one can construct representations of Diff X of positive order (by a representation of order k we mean one depending essentially on the k-jet of the diffeomorphisms). A number of representations of Diff X in spaces of finite functional dimension were defined in [9]; in the terminology of this paper these representations are connected with the space B_X of finite configurations. But there are also representations of positive order connected with the space Γ_X of infinite configurations. Let us take as an example a representation of order 1 connected with this space.

Let μ be a measure in Γ_X that is quasi-invariant under the action of Diff X. We consider the "tangent bundle" $T\Gamma_X$ over Γ_X, that is a fibre bundle over Γ_X, where the fibre over $\gamma = \{x_i\}$ is the direct product $\prod_{i=1}^{\infty} T_{x_i} X$ of the tangent spaces at the points $x_i \in \gamma$.

The space $T\Gamma_X$ can be regarded as the factor space $\widetilde{TX^\infty}/S^\infty$, where $\widetilde{TX^\infty}$ is the subset of $(TX)^\infty = \prod_{i=1}^{\infty} TX_i$, $X_i = X$, consisting of points $\{(x_i, v_i);\ v_i \in T_{x_i}X_i,\ \{x_i\} \in \widetilde{X}^\infty\}$. The topology in $T\Gamma_X$ and the σ-algebra of Borel sets are induced from $\widetilde{TX^\infty}$.

Let λ_i^γ be a normalized measure in $T_{x_i} X$ that is equivalent to the Lebesgue measure, and let $\lambda^\gamma = \prod_{i=1}^{\infty} \lambda_i^\gamma$ be the product measure in $\prod_{i=1}^{\infty} T_{x_i} X$. In this way a measure λ is introduced in $T\Gamma_X$ such that its projection onto Γ_X is μ and that the conditional measures in $T^\gamma\Gamma_X = \prod_{i=1}^{\infty} T_{x_i} X$ are λ^γ. The action of Diff X on $(T\Gamma_X, \lambda)$ is defined by $\psi(\gamma, a) = (\psi\gamma, d\psi a)$, where $a = \prod_{x\in\gamma}^{\infty} a_x$, $a_x \in T_x X$ and $d\psi$ is the natural action on $T\Gamma_X$. The measure λ is quasi-invariant under this action. This leads to a unitary representation of Diff X in $L^2_\lambda(T\Gamma_X)$. A proof that this is irreducible for ergodic measures μ can be modelled on §3. The parameters of the representations just constructed are the measure μ in Γ_X and the measures

λ^γ in $T^\gamma \Gamma_X$.

It is easy to see how to construct representations of order 1 analogous to the elementary representations. We do not say much about representations of higher order, because difficulties in describing them arise even in the case of a finite number of particles (see [9]).

REMARK. The representations of Diff X listed in this appendix can be extended to representations of the cross-product $C^\infty(X)\cdot$ Diff X.

References

[1] A. M. Vershik, I. M. Gel'fand and M. I. Graev, Representations of the group $SL(2, R)$, where R is a ring of functions, Uspekhi Mat. Nauk **28**:5 (1973), 83–128. = Russian Math. Surveys **28**:5 (1973), 87–132.

[2] A. M. Vershik, I. M. Gel'fand and M. I. Graev, Irreducible representations of the group G^X and cohomology, Functsional. Anal. i Prilozhen. **8**:2 (1974), 67–69. MR **50** # 530. = Functional Anal. Appl. **8** (1974), 151–153.

[3] D. B. Anosov and A. B. Katok, New examples in smooth ergodic theory. Ergodic diffeomorphisms, Trudy Moskov. Mat. Obshch. **23** (1970), 3–36.

[4] A. Weil, L'intégration dans les groupes topologiques et ses applications, Actual. Sci. Ind. 869, Hermann & Cie, Paris 1940. MR **3** # 198.
Translation: *Integrirovanie v topologicheskikh gruppakh i ego prilozheniya,* Izdat. Inost. Lit., Moscow 1950.

[5] I. M. Gel'fand and N.Ya. Vilenkin, *Obobshchennye funktsii, vyp.4. Nekotorye primeneniya garmonicheskogo analiza. Oskashchennye gil'bertovy prostranstva* Gos. Izdat. Fiz.-Mat. Lit., Moscow 1961. MR **26** #4173.
Translation: Generalized functions, vol.4, Some applications of harmonic analysis, Equipped Hilbert spaces, Academic Press, New York-London 1964.

[6] A. M. Vershik, Description of invariant measures for the actions of some infinite-dimensional groups, Dokl. Akad. Nauk SSSR **218** (1974), 749–752. = Soviet Math. Dokl. **15** (1974), 1396–1400.

[7] R. L. Dobrushin, R. A. Minlos and Yu. M. Sukhov, *Prilozhenie k knige Ryuelya: Statisticheskaya mekhanika* (Supplement to Ruelle's book *Statistical mechanics*). Mir, Moscow 1971.

[8] R. S. Ismagilov, Unitary representations of the group of diffeomorphisms of the circle, Funktsional Anal. i Prilozhen. **5**:3 (1971), 45–53. = Functional. Anal. Appl. **5** (1971), 209–216.

[9] A. A. Kirillov, Unitary representations of the group of diffeomorphisms and some of its subgroups, Preprint IPM, No.82 (1974).

[10] A. A. Kirillov, Dynamical systems, factors, and group representations, Uspekhi Mat. Nauk **22**:5 (1967), 67–80. = Russian Math. Surveys **22**:5 (1967), 63–75.

[11] V. A. Rokhlin, On the fundamental ideal of measure theory, Mat. Sb. **25** (1949), 107–150. MR **11** # 18

[12] D. Ruelle, Statistical mechanics. Rigorous results, W. A. Benjamin Inc., Amsterdam 1969.
Translation: *Statisticheskaya mekhanika. Strogie rezul'taty,* Mir, Moscow 1971.

[13] S. V. Fomin, On measures invariant under a certain group of transformations, Izv. Akad. Nauk SSSR Ser. Mat. **14** (1950), 261–274. MR **12** # 33.

[14] G. Goldin, Non-relativistic current algebras as unitary representations of groups, J. Mathematical Phys. **12** (1971), 462–488. MR **44** # 1330.

[15] G. Goldin, K. J. Grodnik, R. Powers and D. Sharp, Non-relativistic current algebra in the N/V limit, J. Mathematical Phys. **15** (1974), 88–100.

[16] A. Guichardet, Symmetric Hilbert spaces and related topics, Lecture Notes in Math. **261**, Springer-Verlag, Berlin-Heidelberg-New York 1972.

[17] J. Kerstan, K. Mattes and J. Mecke, Unbegrenzt teilbare Punktprozesse, Berlin 1974.

[18] D. Knutson, λ-rings and the representation theory of the symmetric group, Lecture Notes in Math. **308**, Springer-Verlag, Berlin-Heidelberg-New York, 1973.

[19] R. Menikoff, The hamiltonian and generating functional for a non-relativistic local current algebra, J. Mathematical Phys. **15** (1974), 1138–1152. MR **49** # 10285.

[20] R. Menikoff, Generating functionals determining representations of a non-relativistic local current algebra in the N/V limit, J. Mathematical Phys. **15** (1974), 1394–1408.

[21] J. Moser, On the volume elements on a manifold, Trans. Amer. Math. Soc. **120** (1965), 286–294. MR **32** # 409

[22] R. S. Ismagilov, Unitary representations of the group of diffeomorphisms of the space $\mathbf{R}^n$, $n \geqslant 2$, Funktsional. Anal. i Prilozhen. **9**:2 (1975), 71–72.
= Functional Anal. Appl. **9** (1975), 144–145.

Received by the Editors, 15 May 1975

Translated by A. West

AN INTRODUCTION TO THE PAPER 'SCHUBERT CELLS AND COHOMOLOGY OF THE SPACES G/P'

Graeme Segal

It is well known that a generic invertible matrix can be factorized as the product of an upper triangular and a lower triangular matrix. A more precise statement is that any invertible $n \times n$ matrix g can be written in the form $b_1 w b_2$, where b_1 and b_2 are upper triangular and w is a permutation matrix. Here w is uniquely determined by g, though b_1 and b_2 are not. The matrices g for which w is the order-reversing permutation $i \mapsto n - i + 1$ form a dense open set in $\mathrm{GL}_n(\mathbf{C})$.

This double-coset decomposition $\mathrm{GL}_n(\mathbf{C}) = \bigcup_w BwB$, where B denotes the upper triangular matrices and w runs through the permutation matrices, has an analogue for any connected affine algebraic group G. The role of B is played by a Borel subgroup (i.e. a maximal soluble subgroup), and the role of the permutation matrices by the Weyl group $W = N(H)/H$, where $H \cong (\mathbf{C}^\times)^l$ is a maximal algebraic torus in G and $N(H)$ is its normalizer. The decomposition is nowadays called the *Bruhat decomposition*; but Gelfand had earlier recognized its importance in his work on the representations of the classical groups.

It is best to think of the decomposition as the decomposition of the homogeneous space $X = G/B$ into the orbits of the left action of B. The space X plays a central role in representation theory. It turns out that it is a complex projective algebraic variety, and that the orbits of B are algebraic affine spaces $\mathbf{C}^m$ of various dimensions, the "Bruhat cells". The closures of the cells are algebraic subvarieties which in general have singularities. It is important that the maximal compact subgroup K of G acts transitively on X, so that X has an alternative description as K/T, where $T = K \cap B$ is a maximal torus of K.

If $G = \mathrm{GL}_n(\mathbf{C})$ then X is the *flag manifold*: a flag in $\mathbf{C}^n$ is an increasing sequence of subspaces

$$F = (F_1 \subset F_2 \subset \ldots \subset F_n = \mathbf{C}^n)$$

with $\dim(F_k) = k$. For $\mathrm{GL}_n(\mathbf{C})$ acts transitively on the set of all flags, and B is the isotropy group of the standard flag $\mathbf{C} \subsetneq \mathbf{C}^2 \subset \ldots \subset \mathbf{C}^n$. In this case the cells are indexed by permutations w of $\{1, 2, \ldots, n\}$, and we can take as a representative point in the cell X_w the flag F^w such that F_k^w is spanned by $\{e_{w(1)}, e_{w(2)}, \ldots, e_{w(k)}\}$, where $\{e_1, \ldots, e_n\}$ is the standard basis of $\mathbf{C}^n$. X_w can be defined by "Schubert conditions": it consists precisely of the flags F such that $\dim(F_k \cap \mathbf{C}^m) = \nu_{km}$, where

$$\nu_{km} = \operatorname{card}\{i : i \leqslant k,\ w(i) \leqslant m\}.$$

The dimension of X_w is the *length* $l(w)$ of w, defined by

$$l(w) = \sum_{i=1}^{n} |w(i) - i|.$$

Alternatively $l(w)$ is the number of pairs (i, j) such that $i < j$ but $w(i) > w(j)$. In fact if N_w is the subgroup of B consisting of matrices (a_{ij}) with diagonal elements $a_{ii} = 1$ and such that $a_{ij} = 0$ unless $i \leqslant j$ and $w(i) \geqslant w(j)$ then the map $N_w \to X_w$ given by $g \mapsto g \cdot F^w$ is an isomorphism of algebraic spaces.

An analogous discussion can be carried out for the other classical groups. If $\mathbf{C}^n$ has a non-degenerate bilinear form $\langle\, ,\rangle$, and G is the group of automorphisms of $\mathbf{C}^n$ which preserve the form, then G/B can be identified with the set of flags F such that $F_k^\perp = F_{n-k}$ for $k = 1, \ldots, n$.

Returning to the general case, we have a topological cell decomposition of X into cells $\{X_w\}_{w \in W}$ which are all of even dimension. The homology groups of X are therefore free abelian with the classes $[X_w]$ of the cells as a natural basis.

On the other hand there is a completely different way of describing the cohomology ring of X. For every algebraic homomorphism $\lambda: B \to \mathbf{C}^\times$, or equivalently for every character λ of the compact torus T, there is a holomorphic line bundle E_λ on X. Associating to λ the first Chern class $c_\lambda = c_1(E_\lambda)$ of E_λ gives an isomorphism $\hat{T} \to H^2(X: \mathbf{Z})$. ($\hat{T}$ is the lattice of *weights* of G: in the paper it is called $\mathfrak{h}^*_{\mathbf{Z}}$.) The classes c_λ generate the cohomology ring of X multiplicatively over the rationals, and $H^*(X; \mathbf{Q}) \cong R/J$, where R is the polynomial algebra over $\mathbf{Q}$ generated by the c_λ, and J is the ideal generated by the homogeneous W-invariant polynomials of positive degree. (When $G = \mathrm{GL}_n(\mathbf{C})$ there are n obvious line bundles $E_1, \ldots, E_n$ on X: the fibre of E_i at a flag F is F_i/F_{i-1}. The classes $x_i = c(E_i)$ span $H^2(X; \mathbf{Z})$. The elementary symmetric functions in the x_i vanish because they are the Chern classes of $E_1 \oplus \ldots \oplus E_n$, which is a trivial bundle.)

It is natural to ask for the relation between these descriptions of the homology and the cohomology. In other words if p is a homogeneous polynomial of degree k in the c_λ, and w is an element of W length k, what is the value $\langle p, [X_w] \rangle$ of p on the cell X_w? One can also ask how to express the cohomology class Poincare dual to a cell X_w as a polynomial in the Chern classes.

To answer these questions it is enough in principle to determine the cap-product $c_\lambda \cap [X_w] \in H_{2k-2}(X; \mathbf{Z})$ for each $\lambda \in \hat{T}$ and each $w \in W$ of length k. The paper uses a simple and very attractive geometrical argument to do this. One begins by observing that by linearity it is enough to consider weights λ which are in the interior of the positive Weyl chamber. In that case X can be embedded as a projective algebraic variety in $P(V_\lambda)$, the projective space of the irreducible representation V_λ of G with highest weight λ, as the orbit under G of the highest weight vector f_λ. (V_λ is the dual of the space of holomorphic sections of E_λ.) The cohomology class $c_\lambda = c_1(E_\lambda)$ is then the class dual to the intersection $X \cap \Pi$, where Π is a hyperplane in $P(V_\lambda)$; and the cap-product with c_λ can be interpreted as the geometric operation of intersection with Π. This is amenable to calculation because of the following properties of the embedding $X \to P(V_\lambda)$:

(i) the centre of the cell X_w maps to the point of $P(V_\lambda)$ represented by the weight vector $f_w \in V_\lambda$ of weight $w\lambda$,

(ii) $\bar{X}_w$ is precisely the intersection of $\bar{X}$ with a linear subspace of $P(V_\lambda)$, and

(iii) the boundary $\bar{X}_w - X_w$ of X_w is $\bar{X}_w \cap \Pi_w$, where Π_w is the hyperplane perpendicular of f_w.

Now let us recall that the Weyl group W - which we are regarding as a group of automorphisms of the lattice $\hat{T}$ – is generated by the reflections σ_γ in the hyperplanes of $\hat{T}$ perpendicular to the roots γ of G. If $w \in W$ has length k it turns out that the $(k-1)$-dimensional cells in the boundary of X_w are precisely the $X_{w\sigma_\gamma}$ such that $l(w\sigma_\gamma) = k-1$. Thus the cap-product $c_\lambda \in [X_w]$ is necessarily of the form $\sum_\gamma n_\gamma [X_{w\sigma_\gamma}]$, where n_γ is a positive integer. To determine n_γ one must calculate the order to which the linear form $\langle f_w, \rangle$, when regarded as a function on $\bar{X}_w$, vanishes on the cell $X_{w\sigma_\gamma}$. That is easy to do because the formula

$$t \mapsto w\sigma_\gamma \exp(t\, E_{-\gamma}) f_e,$$

where $E_{-\gamma}$ is the standard element of $\mathfrak{g}$ in the $(-\gamma)$ root-space, defines a holomorphic curve in $\bar{X}_w$ which passes through the centre $f_{w\sigma_\gamma} = w\sigma_\gamma f_e$ of $X_{w\sigma_\gamma}$ when $t = 0$, and is transversal to $X_{w\sigma_\gamma}$. We calculate

$$\langle f_w, w\sigma_\gamma \exp(t\, E_{-\gamma}) f_e \rangle = \langle \sigma_\gamma f_e, \exp(t\, E_{-\gamma}) f_e \rangle = 0(t^{n_\gamma}),$$

where $n_\gamma = \langle \lambda, H_\gamma \rangle$, H_γ being the co-root associated to γ, i.e. the element of the dual lattice to $\hat{T}$ characterized by the property

$$\sigma_\gamma(\chi) = \chi - \langle \chi, H_\gamma \rangle \gamma$$

for all $\chi \in \hat{T}$.

The formula

$$c_\lambda \cap [X_w] = \sum_\lambda \langle \lambda, H_\gamma \rangle [X_{w\sigma_\gamma}]$$

gives us the pairing between homology and cohomology in the form

$$\langle c_{\lambda_1} c_{\lambda_2} \ldots c_{\lambda_n}, [X_w] \rangle = \Sigma \langle \lambda_1, H_{\gamma_1} \rangle \ldots \langle \lambda_k, H_{\gamma_1} \rangle,$$

where the sum is over all strings $\gamma_1, \ldots, \gamma_k$ of positive roots such that

$$w = \sigma_{\gamma_1} \sigma_{\gamma_2} \ldots \sigma_{\gamma_k}.$$

I shall not describe here the elegant algebraic formulations the authors derive from this.

It ought, however, to be mentioned that the methods apply equally well not only to the space G/B, but to G/P for every parabolic subgroup P of G. The most obvious case of this is the Grassmannian $\mathrm{Gr}_{k,n}$ of k-dimensional subspaces of $\mathbf{C}^n$, which is $\mathrm{GL}_n(\mathbf{C})/P$, where P is the appropriate group of echelon matrices.

(In terms of compact groups $\mathrm{Gr}_{k,n} = U_n/U_k \times U_{n-k}$.) The analogue of $\mathrm{Gr}_{k,n}$ for the orthogonal groups is the Grassmannian of *isotropic* k-dimensional subspaces of $\mathbf{C}^n$ for some non-degenerate quadratic form on $\mathbf{C}^n$: this space can be identified with $O_n/U_k \times O_{n-2k}$. When $k = 1$ it is a complex projective quadric hypersurface.

SCHUBERT CELLS AND COHOMOLOGY OF THE SPACES G/P

I. N. Bernstein, I. M. Gel'fand, S. I. Gel'fand

We study the homological properties of the factor space G/P, where G is a complex semisimple Lie group and P a parabolic subgroup of G. To this end we compare two descriptions of the cohomology of such spaces. One of these makes use of the partition of G/P into cells (Schubert cells), while the other consists in identifying the cohomology of G/P with certain polynomials on the Lie algebra of the Cartan subgroup H of G. The results obtained are used to describe the algebraic action of the Weyl group W of G on the cohomology of G/P.

Contents

Introduction

Let G be a linear semisimple algebraic group over the field $\mathbf{C}$ of complex numbers and assume that G is connected and simply-connected. Let B be a Borel subgroup of G and $X = G/B$ the fundamental projective space of G.

The study of the topology of X occurs, explicitly or otherwise, in a large number of different situations. Among these are the representation theory of semisimple complex and real groups, integral geometry and a number of problems in algebraic topology and algebraic geometry, in which analogous spaces figure as important and useful examples. The study of the homological properties of G/P can be carried out by two well-known methods. The first of these methods is due to A. Borel [1] and involves the identification of the cohomology ring of X with the quotient ring of the ring of polynomials on the Lie algebra $\mathfrak{h}$ of the Cartan subgroup

$H \subset G$ by the ideal generated by the W-invariant polynomials (where W is the Weyl group of G). An account of the second method, which goes back to the classical work of Schubert, is in Borel's note [2] (see also [3]); it is based on the calculation of the homology with the aid of the partition of X into cells (the so-called Schubert cells). Sometimes one of these approaches turns out to be more convenient and sometimes the other, so naturally we try to establish a connection between them. Namely, we must know how to compute the correspondence between the polynomials figuring in Borel's model of the cohomology and the Schubert cells. Furthermore, it is an interesting problem to find in the quotient ring of the polynomial ring a symmetrical basis dual to the Schubert cells. These problems are solved in this article. The techniques developed for this purpose are applied to two other problems. The first of these is the calculation of the action of the Weyl group on the homology of X in a basis of Schuberts cells, which turns out to be very useful in the study of the representations of the Chevalley groups.

We also study the action of W on X. This action is not algebraic (it depends on the choice of a compact subgroup of G). The corresponding action of W on the homology of X can, however, be specified in algebraic terms. For this purpose we use the trajectories of G in $X \times X$, and we construct explicitly the correspondences on X (that is, cycles in $X \times X$) that specify the action of W on $H_*(X, \mathbf{Z})$. The study of such correspondences forms the basis of many problems in integral geometry.

At the end of the article, we generalize our results to the case when B is replaced by an arbitrary parabolic subgroup $P \subset G$. When $G = GL(n)$ and G/P is the Grassmann variety, analogous results are to be found in [4].

B. Kostant has previously found other formulae for a basis of $H^*(X, \mathbf{Z})$, $X = G/B$, dual to the Schubert cells. We would like to express our deep appreciation to him for drawing our attention to this series of problems and for making his own results known to us.

The main results of this article have already been announced in [13].

We give a brief account of the structure of this article. At the beginning of § 1 we introduce our notation and state the known results on the homology of $X = G/B$ that are used repeatedly in the paper. The rest of § 1 is devoted to a statement of our main results.

In § 2 we introduce an ordering on the Weyl group W of G that arises naturally in connection with the geometry of X, and we investigate its properties.

§ 3 is concerned with the ring R of polynomials on the Lie algebra $\mathfrak{h}$ of the Cartan subgroup $H \subset G$. In this section we introduce the functionals D_w on R and the elements P_w in R and discuss their properties.

In § 4 we prove that the elements D_w introduced in § 3 correspond to the Schubert cells of X.

§ 5 contains generalizations and applications of the results obtained, in particular, to the case of manifolds $X(P) = G/P$, where P is an arbitrary parabolic subgroup of G. We also study in § 5 the correspondences on X and in particular, we describe explicitly those correspondences that specify the action of the Weyl group W on the cohomology of X. Finally, in this section some of our results are put in the form in which they were earlier obtained by B. Kostant, and we also interpret some of them in terms of differential forms on X.

§ 1. Notation, preliminaries, and statement of the main results

We introduce the notation that is used throughout the article.

G is a complex semisimple Lie group, which is assumed to be connected and simply-connected;

B is a fixed Borel subgroup of G;

$X = G/B$ is a fundamental projective space of G;

N is the unipotent radical of B;

H is a fixed maximal torus of G, $H \subset B$;

$\mathfrak{G}$ is the Lie algebra of G; $\mathfrak{h}$ and $\mathfrak{N}$ are the subalgebras of $\mathfrak{G}$ corresponding to H and N;

$\mathfrak{h}^*$ is the space dual to $\mathfrak{h}$;

$\Delta \subset \mathfrak{h}^*$ is the root system of $\mathfrak{h}$ in $\mathfrak{G}$;

Δ_+ is the set of positive roots, that is, the set of roots of $\mathfrak{h}$ in $\mathfrak{N}$, $\Delta_- = -\Delta_+$, $\Sigma \subset \Delta_+$ is the system of simple roots;

W is the Weyl group of G; if $\gamma \in \Delta$, then $\sigma_\gamma : \mathfrak{h}^* \to \mathfrak{h}^*$ is an element of W, a reflection in the hyperplane orthogonal to γ. For each element[1] $w \in W = \text{Norm}(H)/H$, the same letter is used to denote a representative of w in Norm $(H) \subset G$.

$l(w)$ is the length of an element $w \in W$ relative to the set of generators $\{\sigma_\alpha, \alpha \in \Sigma\}$ of W, that is, the least number of factors in the decomposition

$$w = \sigma_{\alpha_1}\sigma_{\alpha_2} \ldots \sigma_{\alpha_l}, \quad \alpha_i \in \Sigma. \tag{1}$$

A decomposition (1), with $l = l(w)$, is called reduced; $s \in W$ is the unique element of maximal length, $r = l(s)$;

$N_- = sNs^{-1}$ is the subgroup of G "opposite" to N.

For any $w \in W$ we put $N_w = w N_- w^{-1} \cap N$.

HOMOLOGY AND COHOMOLOGY OF THE SPACE X. We give at this point two descriptions of the homological structure of X. The first of these (Proposition 1.2) makes use of the decomposition of X into cells, while the second (Proposition 1.3) involves the realization of two-dimensional cohomology classes as the Chern classes of one-dimensional bundles.

We recall (see [5]) that $N_w = w N_- w^{-1} \cap N$ is a unipotent subgroup of

---

[1] Norm H is the normalizer of H in G.

G of (complex) dimension $l(w)$.

1.1. PROPOSITION (see [5]), *Let $o \in X$ be the image of B in X. The locally closed subvarieties $X_w = Nwo \subset X$, $w \in W$, yield a decomposition of X into N-orbits. The natural mapping $N_w \to X_w$ ($n \mapsto nwo$) is an isomorphism of algebraic varieties.*

Let $\overline{X}_w$ be the closure[1] of X_w in X, $[\overline{X}_w] \in H_{2l(w)}(\overline{X}_w, \mathbf{Z})$ the fundamental cycle of the complex algebraic variety $\overline{X}_w$ and $s_w \in H_{2l(w)}(X, \mathbf{Z})$ the image of $[\overline{X}_w]$ under the mapping induced by the embedding $\overline{X}_w \hookrightarrow X$.

1.2. PROPOSITION (see [2]). *The elements s_w form a free basis of $H_*(X, \mathbf{Z})$.*

We now turn to the other approach to the description of the cohomology of X. For this purpose we introduce in $\mathfrak{h}$ the root system $\{H_\gamma, \gamma \in \Delta\}$ dual to Δ. (This means that $\sigma_\gamma \chi = \chi - \chi(H_\gamma)\gamma$ for all $\chi \in \mathfrak{h}^*$, $\gamma \in \Delta$). We denote by $\mathfrak{h}_{\mathbf{Q}} \subset \mathfrak{h}$ the vector space over $\mathbf{Q}$ spanned by the H_γ. We also set $\mathfrak{h}^*_{\mathbf{Z}} = \{\chi \in \mathfrak{h}^* \mid \chi(H_\gamma) \in \mathbf{Z}$ for all $\gamma \in \Delta\}$ and $\mathfrak{h}^*_{\mathbf{Q}} = \mathfrak{h}^*_{\mathbf{Z}} \otimes_{\mathbf{Z}} \mathbf{Q}$.

Let $R = S(\mathfrak{h}^*_{\mathbf{Q}})$ be the algebra of polynomial functions on $\mathfrak{h}_{\mathbf{Q}}$ with rational coefficients. We extend the natural action of W on $\mathfrak{h}^*$ to R. We denote by I the subring of W-invariant elements in R and set $I_+ = \{f \in I \mid f(0) = 0\}$, $J = I_+R$.

We construct a homomorphism $\alpha: R \to H^*(X, \mathbf{Q})$ in the following way. First let $\chi \in \mathfrak{h}^*_{\mathbf{Z}}$. Since G is simply-connected, there is a character $\theta \in \mathrm{Mor}(H, \mathbf{C}^*)$ such that $\theta(\exp h) = \exp \chi(h)$, $h \in \mathfrak{h}$. We extend θ to a character of B by setting $\theta(n) = 1$ for $n \in N$. Since $G \to X$ is a principal fibre space with structure group B, this θ defines a one-dimensional vector bundle E_χ on X. We set $\alpha_1(\chi) = c_\chi$, where $c_\chi \in H^2(X, \mathbf{Z})$ is the first Chern class of E_χ. Then α_1 is a homomorphism of $\mathfrak{h}^*_{\mathbf{Z}}$ into $H^2(X, \mathbf{Z})$, which extends naturally to a homomorphism of rings $\alpha: R \to H^*(X, \mathbf{Q})$.

Note that W acts on the homology and cohomology of X. Namely, let $K \subset G$ be a maximal compact subgroup such that $T = K \cap H$ is a maximal torus in K. Then the natural mapping $K/T \to X$ is a homeomorphism (see [1]). Now W acts on the homology and cohomology of X in the same way as on K/T.

1.3. PROPOSITION ([1], [8]). (i) *The homomorphism α commutes with the action of W on R and $H^*(X, \mathbf{Q})$.*

(ii) Ker $\alpha = J$, *and the natural mapping $\bar{\alpha}: R/J \to H^*(X, \mathbf{Q})$ is an isomorphism.*

In the remainder of this section we state the main results of this article.

The integration formula. We have given two methods of describing the

[1] As X_w is a locally closed variety, its closure in the Zariski topology is the same as in the ordinary topology.

cohomological structure of X. One of the basic aims of this article is to establish a connection between these two approaches. By this we understand the following. Each Schubert cell $s_w \in H_*(X, \mathbf{Z})$ gives rise to a linear functional $\hat{D}_w$ on R according to the formula

$$\hat{D}_w(f) = \langle s_w, \alpha(f)\rangle$$

(where $\langle\ ,\ \rangle$ is the natural pairing of homology and cohomology). We indicate an explicit form for $\hat{D}_w$.

For each root $\gamma \in \Delta$, we define an operator $A_\gamma : R \to R$ by the formula

$$A_\gamma f = \frac{f - \sigma_\gamma f}{\gamma}$$

(that is, $A_\gamma f(h) = [f(h) - f(\sigma_\gamma h)]/\gamma(h)$ for all $h \in \mathfrak{h}$). Then we have the following proposition.

PROPOSITION. *Let* $w = \sigma_{\alpha_1} \ldots \sigma_{\alpha_l}$, $\alpha_i \in \Sigma$. If $l(w) < l$, *then* $A_{\alpha_1} \ldots A_{\alpha_l} = 0$. *If* $l(w) = l$, *then the operator* $A_{\alpha_1} \ldots A_{\alpha_l}$ *depends only on* w *and not on the representation of* w *in the form* $w = \sigma_{\alpha_1} \ldots \sigma_{\alpha_l}$; *we put* $A_w = A_{\alpha_1} \ldots A_{\alpha_l}$.

This proposition is proved in §3 (Theorem 3.4).

The functional $\hat{D}_w$ is easily described in terms of the A_w: we define for each $w \in W$ another functional D_w on R by the formula $D_w f = A_w f(0)$. The following theorem is proved in §4 (Theorem 4.1).

THEOREM. $D_w = \hat{D}_w$ *for all* $w \in W$.

We can give another more explicit description of D_w (and thus of $\hat{D}_w$). To do this, we write $w_1 \xrightarrow{\gamma} w_2$, $w_1, w_2 \in W$, $\gamma \in \Delta_+$, to express the fact that $w_1 = \sigma_\gamma w_2$ and $l(w_2) = l(w_1) + 1$.

THEOREM. *Let* $w \in W$, $l(w) = l$.

(i) *If* $f \in R$ *is a homogeneous polynomial of degree* $k \neq l$, *then* $\hat{D}_w(f) = 0$.

(ii) If $\chi_1, \ldots, \chi_l \in \mathfrak{h}^*_{\mathbf{Q}}$, then $\hat{D}_w(\chi_1 \ldots \chi_l) = \sum \chi_1(H_{\gamma_1}) \ldots \chi_l(H_{\gamma_l})$, where the sum is taken over all chains of the form

$$e = w_0 \xrightarrow{\gamma_1} w_1 \xrightarrow{\gamma_2} \ldots \xrightarrow{\gamma_l} w_l = w^{-1}$$

(see Theorem 3.12 (i), (v)).

The next theorem describes the basis of $H^*(X, \mathbf{Q})$ dual to the basis $\{s_w \mid w \in W\}$ of $H^*(X, \mathbf{Z})$. We identify the ring $\bar{R} = R/J$ with $H^*(X, \mathbf{Q})$ by means of the isomorphism $\bar{\alpha}$ of Proposition 1.3. Let $\{P_w \mid w \in W\}$ be the basis of $\bar{R}$ dual to the basis $\{s_w \mid w \in W\}$ of $H_*(X, \mathbf{Z})$. To specify P_w, we note that the operators $A_w : R \to R$ preserve the ideal $J \subset R$ (lemma 3.3 (v)), and so the operators $\bar{A}_w : \bar{R} \to \bar{R}$ are well-defined.

THEOREM. (i) *Let* $s \in W$ *be the element of maximal length*, $r = l(s)$ *Then* $P_s = \rho^r/r!$ (mod J) $= |W|^{-1} \prod_{\gamma \in \Delta_+} \gamma$ (mod J), (*where* $\rho \in \mathfrak{h}^*_{\mathbf{Q}}$ *is half* *the sum of the positive roots and $|W|$ is the order of W)*

(ii) *If $w \in W$, then $P_w = \overline{A}_{w^{-1}s}P_s$* (see Theorem 3.15, Corollary 3.16, Theorem 3.14(i)).

Another expression for the P_w has been obtained earlier by B. Kostant (see Theorem 5.9).

The following theorem gives a couple of important properties of the P_w.

THEOREM (i). *Let $\chi \in \mathfrak{h}^*_{\mathbf{Q}}$, $w \in W$. Then $\chi \cdot P_w = \sum_{w \xrightarrow{\gamma} w'} w\chi(H_\gamma)\, P_{w'}$* (see Theorem 3.14 (ii)).

(ii) *Let $\mathcal{P}: H_*(X, \mathbf{Q}) \to H^*(X, \mathbf{Q})$ be the Poincaré duality. Then* $\mathcal{P}(s_w) = \overline{\alpha}(P_{ws})$ (see Corollary 3.19).

THE ACTION OF THE WEYL GROUP. The action of W on $H^*(X, \mathbf{Q})$ can easily be described using the isomorphism $\overline{\alpha}$: $R/J \to H^*(X, \mathbf{Q})$, but we are interested in the problem of describing the action of W on the basis $\{s_w\}$ of $H_*(X, \mathbf{Q})$.

THEOREM. *Let $\alpha \in \Sigma$, $w \in W$. Then $\sigma_\alpha s_w = -s_w$ if $l(w\sigma_\alpha) = l(w) - 1$ and $\sigma_\alpha s_w = -s_w + \sum_{w' \xrightarrow{\gamma} w\sigma_\alpha} w'\alpha(H_\gamma)\, s_{w'}$, if $l(w\sigma_\alpha) = l(w) + 1$* (see Theorem 3.12 (iv)).

In §5 we consider some applications of the results obtained. To avoid overburdening the presentation, we do not make precise statements at this point. We merely mention that Theorem 5.5 appears important to us, in which a number of results is generalized to the case of the varieties $X(P) = G/P$ (P being an arbitrary parabolic subgroup of G), and also Theorem 5.7, in which we investigate certain correspondences on X.

§2. The ordering on the Weyl group and the mutual disposition of the Schubert cells

2.1 DEFINITION (i) *Let $w_1, w_2 \in W$, $\gamma \in \Delta_+$. Then $w_1 \xrightarrow{\gamma} w_2$ indicates the fact that $\sigma_\gamma w_1 = w_2$ and $l(w_2) = l(w_1) + 1$.*

(ii) *We put $w < w'$ if there is a chain*

$$w = w_1 \to w_2 \to \ldots \to w_k = w'.$$

It is helpful to picture W in the form of a directed graph with edges drawn in accordance with Definition 2.1 (i).

Here are some properties of this ordering.

2.2 LEMMA. *Let $w = \sigma_{\alpha_1} \ldots \sigma_{\alpha_l}$ be the reduced decomposition of an element $w \in W$. We put $\gamma_i = \sigma_{\alpha_1} \ldots \sigma_{\alpha_{i-1}}(\alpha_i)$. Then the roots $\gamma_1, \ldots, \gamma_l$ are distinct and the set $\{\gamma_1, \ldots, \gamma_l\}$ coincides with $\Delta_+ \cap w\Delta_-$.*

This lemma is proved in [6].

2.3 COROLLARY. (i) *Let $w = \sigma_{\alpha_1} \ldots \sigma_{\alpha_l}$ be the reduced decomposition and let $\gamma \in \Delta_+$ be a root such that $w^{-1}\gamma \in \Delta_-$. Then for some* i

$$\sigma_\gamma \sigma_{\alpha_1} \ldots \sigma_{\alpha_i} = \sigma_{\alpha_1} \ldots \sigma_{\alpha_{i-1}}. \tag{2}$$

(ii) *Let* $w \in W$, $\gamma \in \Delta_+$. *Then* $l(w) < l(\sigma_\gamma w)$, *if and only if* $w^{-1}\gamma \in \Delta_+$.

PROOF (i) From Lemma 2.2 we deduce that $\gamma = \sigma_{\alpha_1} \ldots \sigma_{\alpha_{i-1}} (\alpha_i)$ *for some* i, and (2) follows.

(ii) If $w^{-1}\gamma \in \Delta_-$, then by (2) $\sigma_\gamma w = \sigma_{\alpha_1} \ldots a_{\alpha_{i-1}} \sigma_{\alpha_{i+1}} \ldots \sigma_{\alpha_l}$, that is $l(\sigma_\gamma w) < l(w)$. Interchanging w and $\sigma_\gamma w$, we see that if $w^{-1}\gamma \in \Delta_+$, then $l(w) < l(\sigma_\gamma w)$.

2.4 LEMMA. *Let* $w_1, w_2 \in W$, $\alpha \in \Sigma$, $\gamma \in \Delta_+$, *and* $\gamma \neq \alpha$. *Let* $\gamma' = \sigma_\alpha \gamma$. If

(3)

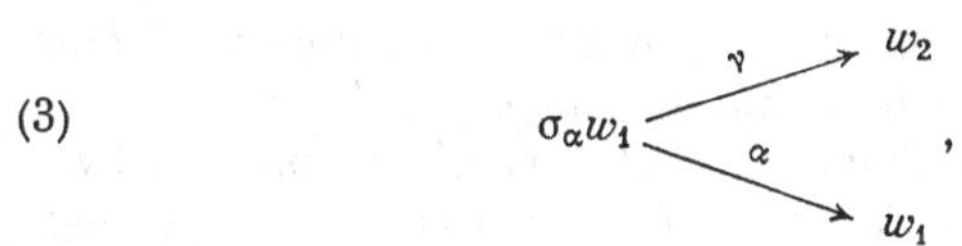

then

(4)

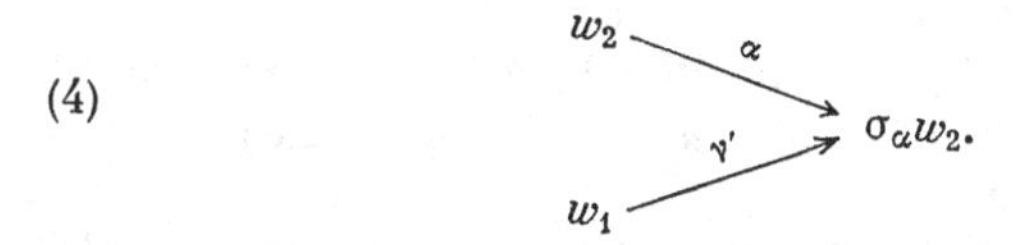

Conversely, (3) *follows from* (4).

PROOF. Since $\alpha \in \Sigma$ and $\gamma \neq \alpha$, we have $\gamma' = \sigma_\alpha \gamma \in \Delta_+$. It is therefore sufficient to show that $l(\sigma_\alpha w_2) > l(w_2) = l(w_1)$. This follows from Corollary 2.3, because $\sigma_\alpha w_2 = \sigma_{\gamma'} w_1$ and $(\sigma_\alpha w_2)^{-1}\gamma' = w_2^{-1}\sigma_\alpha\gamma' = w_2^{-1}\gamma \in \Delta_-$ by (3). The second assertion of the lemma is proved similarly.

2.5 LEMMA. *Let* $w, w' \in W$, $\alpha \in \Sigma$ *and assume that* $w < w'$. *Then*

a) *either* $\sigma_\alpha w \leqslant w'$ *or* $\sigma_\alpha w < \sigma_\alpha w'$,

b) *either* $w \leqslant \sigma_\alpha w'$ *or* $\sigma_\alpha w < \sigma_\alpha w'$.

PROOF a) Let

$$w = w_1 \to w_2 \to \ldots \to w_k = w'.$$

We proceed by induction on k. If $\sigma_\alpha w < w$ or $\sigma_\alpha w = w_2$, the assertion is obvious. Let $w < \sigma_\alpha w$, $\sigma_\alpha w \neq w_2$. Then $\sigma_\alpha w < \sigma_\alpha w_2$ by Lemma 2.4. We obtain a) by applying the inductive hypothesis to the pair (w_2, w').

b) is proved in a similar fashion.

2.6. COROLLARY. *Let* $\alpha \in \Sigma$, $w_1 \xrightarrow{\alpha} w_1'$, $w_2 \xrightarrow{\alpha} w_2'$. *If one of the elements* w_1, w_1' *is smaller (in the sense of the above ordering) than one of* w_2, w_2', *then* $w_1 \leqslant w_2 < w_2'$ *and* $w_1 < w_1' \leqslant w_2'$.

The property in Lemma 2.5 characterizes the ordering $<$. More precisely, we have the following proposition:

2.7 PROPOSITION. *Suppose that we are given a partial ordering* $w \dashv w'$ *on* W *with the following properties:*

a) *If* $\alpha \in \Sigma$, $w \in W$ *with* $l(\sigma_\alpha w) = l(w) + 1$, *then* $w \dashv \sigma_\alpha w$.

b) *If* $w \dashv w'$, $\alpha \in \Sigma$, *then either* $\sigma_\alpha w \dashv w'$ *or* $\sigma_\alpha w \dashv \sigma_\alpha w'$.

Then $w \dashv w'$ *if and only if* $w \leqslant w'$.

PROOF. Let s be the element of maximal length in W. It follows from a) that $e \dashv w \dashv s$ for all $w \in W$.

I. We prove that $w \leqslant w'$ implies that $w \dashv w'$. We proceed by induction on $l(w')$. If $l(w') = 0$, then $w' = e$, $w = e$ and so $w \dashv w'$. Let $l(w') > 0$ and let $\alpha \in \Sigma$ be a root such that $l(\sigma_\alpha w') = l(w') - 1$. Then by Lemma 2.5 a), either $\sigma_\alpha w \leqslant \sigma_\alpha w'$ or $w \leqslant \sigma_\alpha w'$.

(i) $w \leqslant \sigma_\alpha w' \Rightarrow w \dashv \sigma_\alpha w'$ (by the inductive hypothesis), $\Rightarrow w \dashv w'$. (using a)).

(ii) $\sigma_\alpha w \leqslant \sigma_\alpha w' \Rightarrow \sigma_\alpha w \dashv \sigma_\alpha w'$ (by the inductive hypothesis), $\Rightarrow$ either $w \dashv \sigma_\alpha w'$ or $w \dashv w'$ (applying b) to the pair $(\sigma_\alpha w, \sigma_\alpha w')$), $\Rightarrow w \dashv w'$.

II. We now show that $w \dashv w'$ implies that $w \leqslant w'$. We proceed by backward induction on $l(w)$. If $l(w) = r = l(s)$, then $w = s$, $w' = s$, and so $w \leqslant w'$. Let $l(w) < r$ and let α be an element of Σ such that $l(\sigma_\alpha w) = l(w) + 1$. By b) either $\sigma_\alpha w \dashv w'$ or $\sigma_\alpha w \dashv \sigma_\alpha w'$.

(i) $\sigma_\alpha w \dashv w' \Rightarrow \sigma_\alpha w \leqslant w'$ (by the inductive hypothesis) $\Rightarrow w \leqslant w'$.

(ii) $\sigma_\alpha w \dashv \sigma_\alpha w' \Rightarrow \sigma_\alpha w \leqslant \sigma_\alpha w' \Rightarrow w \leqslant w'$ (by Corollary 2.6). Proposition 2.7 is now proved.

2.8 PROPOSITION. *Let $w \in W$ and let $w = \sigma_{\alpha_1} \ldots \sigma_{\alpha_l}$ be the reduced decomposition of w.*

a) *If $1 \leqslant i_1 < i_2 < \ldots < i_k \leqslant l$ and*

$$w' = \sigma_{\alpha_{i_1}} \ldots \sigma_{\alpha_{i_k}}, \tag{5}$$

then $w' \leqslant w$.

b) *If $w' < w$, then w' can be represented in the form* (5) *for some indexing set $\{i_j\}$.*

c) *If $w' \to w$, then there is a unique index i, $1 \leqslant i \leqslant l$, such that*

$$w' = \sigma_{\alpha_1} \ldots \sigma_{\alpha_{i-1}} \sigma_{\alpha_{i+1}} \ldots \sigma_{\alpha_l}. \tag{6}$$

PROOF. Let us prove c). Let $w' \to w$. Then by Lemma 2.2 there is at least one index i for which (6) holds. Now suppose that (6) holds for two indices i, j, $i < j$. Then $\sigma_{\alpha_{i+1}} \ldots \sigma_{\alpha_j} = \sigma_{\alpha_i} \ldots \sigma_{\alpha_{j-1}}$. Thus, $\sigma_{\alpha_i} \ldots \sigma_{\alpha_j} = \sigma_{\alpha_{i+1}} \ldots \sigma_{\alpha_{j-1}}$, which contradicts the assumption that the decomposition $w = \sigma_{\alpha_1} \ldots \sigma_{\alpha_l}$ is reduced.

b) follows at once from c) if we take into account the fact that the decomposition (6) is reduced. We now prove a) by induction on l. We treat two cases separately.

(i) $i_1 > 1$. Then by the inductive hypothesis $w' \leqslant \sigma_{\alpha_2} \ldots \sigma_{\alpha_l}$, that is, $w' \leqslant \sigma_{\alpha_1} w < w$.

(ii) $i_1 = 1$. Then, by the inductive hypothesis, $\sigma_{\alpha_1} w' = \sigma_{\alpha_{i_2}} \ldots \sigma_{\alpha_{i_k}} \leqslant \sigma_{\alpha_1} w = \sigma_{\alpha_2} \ldots \sigma_{\alpha_l}$. By Corollary 2.6, $w' \leqslant w$.

Proposition 2.8 yields an alternative definition of the ordering on W (see [7]). The geometrical interpretation of this ordering is very interesting and useful in what follows.

2.9 THEOREM. *Let V be a finite-dimensional representation of a Lie algebra $\mathfrak{G}$ with dominant weight λ. Assume that all the weights $w\lambda$ $w \in W$, are distinct and select for each w a non-zero vector $f_w \in V$ of weight $w\lambda$. Then*

$$w' \leqslant w \Leftrightarrow f_{w'} \in U(\mathfrak{N}) f_w$$

(where $U(\mathfrak{N})$ is the enveloping algebra of the Lie algebra $\mathfrak{N}$).

PROOF. For each root $\gamma \in \Delta$ we fix a root vector $E_\gamma \in \mathfrak{G}$ in such a way that $[E_\gamma, E_{-\gamma}] = H_\gamma$. Denote by $\mathfrak{A}_\gamma$ the subalgebra of $\mathfrak{G}$, generated by E_γ, $E_{-\gamma}$, and H_γ. $\mathfrak{A}_\gamma$ is isomorphic to the Lie algebra $sl_2(\mathbf{C})$. Let $w' \xrightarrow{\gamma} w$ and let $\widetilde{V}$ be the smallest $\mathfrak{A}_\gamma$-invariant subspace of V containing $f_{w'}$.

2.10 LEMMA. *Let $n = w'\lambda(H_\gamma) \in \mathbf{Z}$, $n > 0$. The elements $\{E^i_{-\gamma} f_{w'} \mid i = 0, 1, \ldots, n\}$ form a basis of $\widetilde{V}$. Put $\tilde{f} = E^n_{-\gamma} f_{w'}$. Then $E_{-\gamma}\tilde{f} = 0$, $E^n_\gamma \tilde{f} = c' f_{w'}$ $(c' \neq 0)$ and $f_w = c\tilde{f}$ $(c \neq 0)$.*

PROOF. By Lemma 2.2, $w'^{-1}\gamma \in \Delta_+$, hence $E_\gamma f_{w'} = = cE_\gamma w' f_e = cw' E_{w'^{-1}\gamma} f_e = 0$, that is, $f_{w'}$ is a vector of dominant weight relative to $\mathfrak{A}_\gamma$. All the assertions of the lemma, except the last, follow from standard facts about the representations of the algebra $\mathfrak{A}_\gamma \cong sl_2(\mathbf{C})$. Furthermore, $\tilde{f}$ and f_w are two non-zero vectors of weight $w\lambda$ in V, and since the multiplicity of $w\lambda$ in V is equal to 1, these vectors are proportional. The lemma is now proved.

To prove Theorem 2.9 we introduce a partial ordering on W by putting $w' \dashv w$ if $f_{w'} \in U(\mathfrak{N}) f_w$. Since all the weights $w\lambda$ are distinct, the relation $\dashv$ is indeed an ordering; we show that it satisfies conditions a) and b) of Proposition 2.7.

a) Let $\alpha \in \Sigma$ and $l(\sigma_\alpha w) = l(w) + 1$. Then $w \xrightarrow{\alpha} \sigma_\alpha w$, and by Lemma 2.10, $f_w \in U(\mathfrak{N}) f_{\sigma_\alpha w}$, that is, $w \dashv \sigma_\alpha w$.

b) Let $w \dashv w'$. We choose an $\alpha \in \Sigma$ such that $w \xrightarrow{\alpha} \sigma_\alpha w$. Replacing w' by $\sigma_\alpha w'$, if necessary, we may assume that $\sigma_\alpha w' \to w'$. We prove that $\sigma_\alpha w \dashv w'$, that is, $f_{\sigma_\alpha w} \in U(\mathfrak{N}) f_{w'}$. It follows from Lemma 2.10 that $E_{-\alpha} f_{w'} = 0$ and $f_{\sigma_\alpha w} = cE^n_{-\alpha} f_w$. Let $\mathfrak{P}_\alpha$ be the subalgebra of $\mathfrak{G}$ generated by $\mathfrak{N}$, $\mathfrak{h}$ and $\mathfrak{A}_\alpha$. Since $w \dashv w'$, $f_w \in U(\mathfrak{N}) f_{w'}$ and so $f_{\sigma_\alpha w} = cE^n_{-\alpha} f_w = X f_{w'}$, where $X \in U(\mathfrak{P}_\alpha)$. Any element X of $U(\mathfrak{P}_\alpha)$ can be represented in the form $X = \sum_{i=1}^{k} Y_i Y_i' + \widetilde{Y} E_{-\alpha}$, where $Y_i \in U(\mathfrak{N})$, $Y_i' \in U(\mathfrak{h})$, $\widetilde{Y} \in U(\mathfrak{P}_\alpha)$. Therefore, $f_{\sigma_\alpha w} = \sum Y_i Y_i' f_{w'} = \sum c_i Y_i f_{w'} \in U(\mathfrak{N}) f_{w'}$ and Theorem 2.9 is proved.

We use Theorem 2.9 to describe the mutual disposition of the Schubert cells.

2.11. THEOREM (Steinberg [7]). *Let $w \in W$, $X_w \subset X$ a Schubert cell, and $\overline{X}_w$ its closure. Then $X_{w'} \subset \overline{X}_w$ if and only if $w' \leqslant w$.*

To prove this theorem, we give a geometric description of the variety X_w.

Let V be a finite-dimensional representation of G with regular dominant weight λ (that is, all the weights $w\lambda$ distinct). As above, we choose for each $w \in W$ a non-zero vector $f_w \in V$ of weight $w\lambda$. We consider the space $P(V)$ of lines in V; if $f \in V$, $f \neq 0$, then we denote by $[f] \in P(V)$ a line passing through f. Since λ is regular, the stabilizer of the point $[f_e] \in P(V)$ under the natural action of G on $P(V)$ is B. The G-orbit of $[f_e]$ in $P(V)$ is therefore naturally isomorphic to $X = G/B$. In what follows, we regard X as a subvariety of $P(V)$.

For each $w \in W$ we denote by ϕ_w the linear function on V given by $\phi_w(f_w) = 1$, $\phi_w(f) = 0$ if $f \in V$ is a vector of weight distinct from $w\lambda$.

2.12 LEMMA. *Let $f \in V$ and $[f] \in X$. Then*

$$[f] \in X_w \Leftrightarrow f \in U(\mathfrak{N})\, f_w,\ \varphi(f) \neq 0.$$

PROOF. We may assume that $f = gf_e$ for some $g \in G$.

Let $[f] \in X_w$, that is, $g \in NwB$. Then $f = c_1 \exp(Y)wf_e$ for some $Y \in \mathfrak{N}$, hence $f \in U(\mathfrak{N})f_w$, and $\phi_w(f) \neq 0$.

On the other hand, it is clear that for each $f \in V$ there is at most one $w \in W$ such that $f \in U(\mathfrak{N})\, f_w$ and $\phi_w(f) \neq 0$. The Lemma now follows from the fact that $X = \bigcup_{w \in W} X_w$.

We now prove Theorem 2.11

a) Let $X_{w'} \subset \overline{X}_w$. Then $[f_{w'}] \in \overline{X}_w$, and by Lemma 2.12, $f_{w'} \in U(\mathfrak{N})f_w$. So $w' \leqslant w$, by Theorem 2.9.

b) To prove the converse it is sufficient to consider the case $w' \xrightarrow{\gamma} w$. Let $n = w\lambda(H_\gamma) \in \mathbf{Z}$. Just as in the proof of Theorem 2.9, a) we can show that $n > 0$, $E_\gamma^n f_w = cf_{w'}$ and $E_\gamma^{n+1} f_w = 0$.

Therefore $\lim_{t \to \infty} t^{-n} \exp(tE_\gamma)\, f_w = \frac{c}{n!} f_{w'}$, that is, $[f_{w'}] \in X_w$. Hence, $X_{w'} \subset \overline{X}_w$.

§ 3. Discussion of the ring of polynomials on $\mathfrak{h}$

In this section we study the rings R and $\overline{R}$. For each $w \in W$ we define an element $P_w \in \overline{R}$ and a functional D_w on R and investigate their properties. In the next section we shall show that the D_w correspond to Schubert cells, and that the P_w yield a basis, dual to the Schubert cell basis, for the cohomology of X.

3.1 DEFINITION. (i) *$R = \oplus R_i$ is the graded ring of polynomial functions on $\mathfrak{h}_\mathbf{Q}$ with rational coefficients. W acts on R according to the rule $wf(h) = f(w^{-1}h)$.*

(ii) *I is the subring of W-invariant elements in R,*

$$I_+ = \{f \in I \mid f(0) = 0\}.$$

(iii) *J is the ideal of R generated by I_+.*

(iv) $R = R/J$.

3.2 DEFINITION. *Let $\gamma \in \Delta$. We specify an operator A_γ on R by the rule*

$$A_\gamma f = \frac{f - \sigma_\gamma f}{\gamma}.$$

$A_\gamma f$ lies in R, since $f - \sigma_\gamma f = 0$ on the hyperplane $\gamma = 0$ in $\mathfrak{h}_Q$.

The simplest properties of the A_γ are described in the following lemma.

3.3 LEMMA.(i) $A_{-\gamma} = -A_\gamma$, $A_\gamma^2 = 0$.

(ii) $wA_\gamma w^{-1} = A_{w\gamma}$.

(iii) $\sigma_\gamma A_\gamma = - A_\gamma \sigma_\gamma = A_\gamma$, $\sigma_\gamma = -\gamma A_\gamma + 1 = A_\gamma \gamma - 1$.

(iv) $A_\gamma f = 0 \Leftrightarrow \sigma_\gamma f = f$.

(v) $A_\gamma J \subset J$.

(vi) *Let $\chi \in \mathfrak{h}_Q^*$. Then the commutator of A_γ with the operator of multiplication by χ has the form $[A_\gamma, \chi] = \chi(H_\gamma)\sigma_\gamma$.*

PROOF. (i) – (iv) are clear. To prove (v), let $f = f_1 f_2$, where $f_1 \in I_+$, $f_2 \in R$. It is then clear that $A_\gamma f = f_1 . A_\gamma f_2 \in J$. As to (vi), since $\sigma_\gamma \chi = \chi - \chi(H_\gamma)\gamma$, we have

$$[A_\gamma, \chi] f = A_\gamma(\chi f) - \chi A_\gamma(f) = \frac{1}{\gamma}(\chi f - \sigma_\gamma\chi \cdot \sigma_\gamma f - \chi f + \chi\sigma_\gamma f) =$$

$$= \frac{\chi - \sigma_\gamma \chi}{\gamma} \cdot \sigma_\gamma f = \chi(H_\gamma) \cdot \sigma_\gamma f.$$

The following property of the A_γ is fundamental in what follows.

3.4 THEOREM. *Let $\alpha_1, \ldots, \alpha_l \in \Sigma$, and put $w = \sigma_{\alpha_1} \ldots \sigma_{\alpha_l}$; $A_{(\alpha_1, \ldots, \alpha_l)} = A_{\alpha_1} \ldots A_{\alpha_l}$.*

a) *If $l(w) < l$, then $A_{(\alpha_1, \ldots, \alpha_l)} = 0$.*

b) *If $l(w) = l$, then $A_{(\alpha_1, \ldots, \alpha_l)}$ depends only on w and not on the set $\alpha_1, \ldots, \alpha_l$. In this case we put $A_w = A_{(\alpha_1, \ldots, \alpha_l)}$.*

The proof is by induction on l, the result being obvious when $l = 1$.

For the proof of a), we may assume by the inductive hypothesis that $l(\sigma_{\alpha_1} \ldots \sigma_{\alpha_{l-1}}) = l - 1$, consequently $l(\sigma_{\alpha_1} \ldots \sigma_{\alpha_{l-1}} \sigma_{\alpha_l}) = l - 2$. Then $\sigma_{\alpha_i} \sigma_{\alpha_{i+1}} \ldots \sigma_{\alpha_{l-1}} = \sigma_{\alpha_{i+1}} \ldots \sigma_{\alpha_{l-1}} \sigma_{\alpha_l}$ for some i (we have applied Corollary 2.3 to the case $w = \sigma_{\alpha_{l-1}} \ldots \sigma_{\alpha_1}$, $\gamma = \alpha_l$). We show that $A_{(\alpha_i, \ldots, \alpha_l)} = 0$.

Since $l - i < l$, the inductive hypothesis shows that $A_{\alpha_i} A_{\alpha_{i+1}} \ldots A_{\alpha_{l-1}} = A_{\alpha_{i+1}} \ldots A_{\alpha_{l-1}} A_{\alpha_l}$, and so by lemma 3.3 (i) $A_{\alpha_i} \ldots A_{\alpha_l} = A_{\alpha_{i+1}} \ldots A_{\alpha_l} A_{\alpha_l} = 0$.

To prove b), we introduce auxiliary operators $B_{(\alpha_1, \ldots, \alpha_l)}$, by setting

$$B_{(\alpha_1, \ldots, \alpha_l)} = \sigma_{\alpha_l} \ldots \sigma_{\alpha_1} A_{(\alpha_1, \ldots, \alpha_l)}.$$

We put $w_i = \sigma_{\alpha_l} \ldots \sigma_{\alpha_i}$. Then in view of Lemma 3.3 (ii, iii) we have

(7) $$B_{(\alpha_1,\ldots,\alpha_l)} = A_{\alpha_1}^{w_2} A_{\alpha_2}^{w_3} \ldots A_{\alpha_{l-1}}^{w_l} A_{\alpha_l}$$

(where A_γ^w stands for $wA_\gamma w^{-1}$).

3.5 LEMMA. *Let $\chi \in \mathfrak{h}_{\mathbb{Q}}^*$. The commutator of $B_{(\alpha_1,\ldots,\alpha_l)}$ with the operator of multiplication by χ is given by the following formula:*[1]

(8) $$[B_{(\alpha_1,\ldots,\alpha_l)}, \chi] = \sum_{i=1}^{l} \chi(w_{i+1}H_{\alpha_i})\, w_{i+1}w_i^{-1} B_{(\alpha_1,\ldots,\hat{\alpha}_i,\ldots,\alpha_l)}.$$

PROOF. We have

$$[B_{(\alpha_1,\ldots,\alpha_l)}, \chi] = [A_{\alpha_1}^{w_2} A_{\alpha_2}^{w_3} \ldots A_{\alpha_l}, \chi] =$$
$$= \sum_{i=1}^{l} A_{\alpha_1}^{w_2} A_{\alpha_2}^{w_3} \ldots [A_{\alpha_i}^{w_{i+1}}, \chi] \ldots A_{\alpha_l} = \sum_{i=1}^{l} T_i.$$

By Lemma 3.3 (ii, vi), $[A_{\alpha_i}^{w_{i+1}}, \chi] = \chi(w_{i+1}H_{\alpha_i})\, \sigma_{w_{i+1}\alpha_i}$.

Since $\sigma_{w_{i+1}\alpha_i} = w_{i+1}w_i^{-1}$, we have

$$T_i = \chi(w_{i+1}H_{\alpha_i}) A_{\alpha_1}^{w_2} \ldots A_{\alpha_{i-1}}^{w_i} w_{i+1}w_i^{-1} A_{\alpha_{i+1}}^{w_{i+2}} \ldots A_{\alpha_l}.$$

We want to move the term $w_{i+1}w_i^{-1}$ to the left. To do this we note that for $j < i$

$$A_{\alpha_j}^{w_{j+1}} w_{i+1}w_i^{-1} = w_{i+1}w_i^{-1} (A_{\alpha_j}^{w_{j+1}})^{w_i w_{i+1}^{-1}} = w_{i+1}w_i^{-1} A_{\alpha_j}^{w_i w_{i+1}^{-1} w_{j+1}} =$$
$$= w_{i+1}w_i^{-1} A_{\alpha_j}^{\sigma_{\alpha_l}\ldots\hat{\sigma}_{\alpha_i}\ldots\sigma_{\alpha_{j+1}}}.$$

Therefore,

$$T_i = \chi(w_{i+1}H_{\alpha_i})\, w_{i+1}w_i^{-1} A_{\alpha_1}^{\sigma_{\alpha_l}\ldots\hat{\sigma}_{\alpha_i}\ldots\sigma_{\alpha_2}} \ldots A_{\alpha_{i-1}}^{\sigma_{\alpha_l}\cdots\sigma_{\alpha_{i+1}}} A_{\alpha_{i+1}}^{\sigma_{\alpha_l}\ldots\sigma_{\alpha_{i+2}}} \ldots A_{\alpha_l}.$$

By (7), applied to the sequence or roots $(\alpha_1, \ldots, \hat{\alpha}_i, \ldots, \alpha_l)$, we have

$$T_i = \chi(w_{i+1}H_{\alpha_i})\, w_{i+1}w_i^{-1} B_{(\alpha_1,\ldots,\hat{\alpha}_i,\ldots,\alpha_l)},$$

and Lemma 3.5 is proved.

If $l(\sigma_{\alpha_1} \ldots \hat{\sigma}_{\alpha_i} \ldots \sigma_{\alpha_l}) < l - 1$, then $T_i = 0$ by the inductive hypothesis. If $l(\sigma_{\alpha_1} \ldots \hat{\sigma}_{\alpha_i} \ldots \sigma_{\alpha_l}) = l - 1$, then, putting $w' = \sigma_{\alpha_1} \ldots \hat{\sigma}_{\alpha_i} \ldots \sigma_{\alpha_l}$ and $\gamma = \sigma_{\alpha_1} \ldots \sigma_{\alpha_{i-1}}(\alpha_i)$, we see from Lemma 2.2 that $w' \xrightarrow{\gamma} w$, and also

$$\chi(w_{i+1}H_{\alpha_i}) = w'\chi(w'w_{i+1}H_{\alpha_i}) = w'\chi(\sigma_{\alpha_1} \cdots \sigma_{\alpha_{i-1}} H_{\alpha_i}) = w'\chi(H_\gamma)$$

and

$$w_{i+1}w_i^{-1} B_{(\alpha_1,\ldots,\hat{\alpha}_i,\ldots,\alpha_l)} = w_{i+1}w_i^{-1} w'^{-1} A_{(\alpha_1,\ldots,\hat{\alpha}_i,\ldots,\alpha_l)} = w^{-1} A_{(\alpha_1,\ldots,\hat{\alpha}_i,\ldots,\alpha_l)}.$$

[1] ^ indicates that the corresponding term must be omitted.

Using Proposition 2.8 c) and the inductive hypothesis, (8) can be rewritten in the following form:

$$[B_{(\alpha_1, \ldots, \alpha_l)}, \chi] = \sum_{w' \xrightarrow{\gamma} w} w'\chi(H_\gamma) w^{-1} A_{w'}.$$

The right-hand side of this formula does not depend on the representation of w in the form of a product $\sigma_{\alpha_1} \ldots \sigma_{\alpha_l}$. The proof of theorem 3.4 is thus completed by the following obvious lemma.

3.6. LEMMA. *Let B be an operator in R such that $B(1) = 0$ and $[B, \chi] = 0$ for all $\chi \in \mathfrak{h}_{\mathbf{Q}}^*$. Then $B = 0$.*

3.7. COROLLARY. *The operators A_w satisfy the following commutator relation:*

$$[w^{-1}A_w, \chi] = \sum_{w' \xrightarrow{\gamma} w} w'\chi(H_\gamma) w^{-1} A_{w'}.$$

We put $S_i = R_i^*$ (where $R_i \subset R$ is the space of homogeneous polynomials of degree i) and $S = \oplus S_i$. We denote by (,) the natural pairing $S \times R \to \mathbf{Q}$. Then W acts naturally on S.

3.8 DEFINITION. (i) *For any $\chi \in \mathfrak{h}_{\mathbf{Q}}^*$ we let χ^* denote the transformation of S adjoint to the operator of multiplication by χ in R.*

(ii) *We denote by F_γ: $S \to S$ the linear transformation adjoint to A_γ: $R \to R$.*

The next lemma gives an explicit description of the F_γ.

3.9 LEMMA. *Let $\gamma \in \Delta$. For any $D \in S$ there is a $\tilde{D} \in S$ such that $\gamma^* (\tilde{D}) = D$. If $\tilde{D}$ is any such operator, then $\tilde{D} - \sigma_\gamma \tilde{D} = F_\gamma(D)$, (in particular, the left-hand side of this equation does not depend on the choice of $\tilde{D}$).*

PROOF. The existence of $\tilde{D}$ follows from the fact that multiplication by γ is a monomorphism of R. Furthermore, for any $f \in R$ we have

$$(\tilde{D} - \sigma_\gamma \tilde{D}, f) = (\tilde{D}, f - \sigma_\gamma f) = (\tilde{D}, A_\gamma f \cdot \gamma) = (\gamma^* (\tilde{D}), A_\gamma f) = (D, A_\gamma f),$$

hence $\tilde{D} - \sigma_\gamma \tilde{D} = F_\gamma$.

REMARK. It is often convenient to interpret S as a ring of differential operators on $\mathfrak{h}$ with constant rational coefficients. Then the pairing (,) is given by the formula $(D, f) = (Df)(0)$, $D \in S$, $f \in R$. Also, it is easy to check that $\chi^*(D) = [D, \chi]$, where $\chi \in \mathfrak{h}_{\mathbf{Q}}^*$ and $D \in S$ are regarded as operators on R.

Theorem 3.4 and Corollary 3.7 can be restated in terms of the operators F_γ.

3.10 THEOREM. *Let $\alpha_1, \ldots, \alpha_l \in \Sigma$, $w = \sigma_{\alpha_1} \ldots \sigma_{\alpha_l}$.*

(i) *If $l(w) < l$, then $F_{\alpha_l} \ldots F_{\alpha_1} = 0$.*

(ii) *If $l(w) = l$, then $F_{\alpha_l} \ldots F_{\alpha_1}$ depends only on w and not on $\alpha_1, \ldots, \alpha_l$. In this case the transformation $F_{\alpha_l} \ldots F_{\alpha_1}$ is denoted by F_w. (Note that $F_w = A_w^*$).*

(iii) $$[\chi^*, F_w w] = \sum_{w' \xrightarrow{\gamma} w} w'\chi(H_\gamma) F_{w'} w.$$

3.11. DEFINITION. *We set* $D_w = F_w$ (1).

As we shall show in §4, the functionals D_w correspond to the Schubert cells in $H_*(X, \mathbf{Q})$ in the sense that $(D_w, f) = \langle s_w, \alpha(f)\rangle$ *for all* $f \in R$.

The properties of the D_w are listed in the following theorem.

3.12. THEOREM. (i) $D_w \in S_{l(w)}$.

(ii) *Let* $w \in W$, $\alpha \in \Sigma$. *Then*

$$F_\alpha D_w = \begin{cases} 0 & \text{if } l(w\sigma_\alpha) = l(w) - 1, \\ D_{w\sigma_\alpha} & \text{if } l(w\sigma_\alpha) = l(w) + 1. \end{cases}$$

(iii) *Let* $\chi \in \mathfrak{h}^*_{\mathbf{Q}}$. *Then*

$$\chi^*(D_w) = \sum_{w' \xrightarrow{\gamma} w} w'\chi(H_\gamma) D_{w'}.$$

(iv) *Let* $\alpha \in \Sigma$. *Then*

$$\sigma_\alpha D_w = \begin{cases} -D_w, & \text{if } l(w\sigma_\alpha) = l(w) - 1, \\ -D_w + \sum\limits_{w' \xrightarrow{\gamma} w\sigma_\alpha} w'\alpha(H_\gamma) D_{w'} & \text{if } l(w\sigma_\alpha) = l(w) + 1. \end{cases}$$

(v) *Let* $w \in W$, $l(w) = l$, $\chi_1, \ldots, \chi_l \in \mathfrak{h}^*_{\mathbf{Q}}$. *Then* $(D_w, \chi_1, \ldots, \chi_l) = \sum \chi_1(H_{\gamma_1}) \ldots \chi_l(H_l)$, *where the summation extends over all chains*

$$e \xrightarrow{\gamma_1} w_1 \xrightarrow{\gamma_2} w_2 \longrightarrow \ldots \xrightarrow{\gamma_l} w_l = w^{-1}.$$

PROOF. (i) and (ii) follow from the definition of D_w and Theorem 3.10 (i).

(iii) $\chi^*(D_w) = \chi^* F_w w(1) = [\chi^*, F_w w]$ (1) (since $\chi^*(1) = 0$), and (iii) follows from Theorem 3.10 (iii).

It follows from Lemma 3.3 (iii) that $\sigma_\alpha = \alpha^* F_\alpha - 1$. Thus, (iv) follows from (ii) and (iii).

(v) We put $\widetilde{D}_w = D_{w^{-1}}$. Then the $\widetilde{D}_w$ satisfy the relation

$$\chi^*(\widetilde{D}_w) = \sum_{w' \xrightarrow{\gamma} w} \chi(H_\gamma) \widetilde{D}_{w'}. \tag{9}$$

Since $(D, \chi f) = (\chi^*(D), f)$, (v) is a consequence of (9) by induction on l.

Let $\mathcal{H}$ be the subspace of S orthogonal to the ideal $J \subset R$. It follows from Lemma 3.3 (vi) that $\mathcal{H}$ is invariant with respect to all the F_γ. It is also clear that $1 \in \mathcal{H}$. Thus, $D_w \in \mathcal{H}$ for all $w \in H$.

3.13. THEOREM. *The functionals* D_w, $w \in W$, *form a basis for* $\mathcal{H}$.

PROOF. a) We first prove that the D_w are linearly independent. Let $s \in W$ be the element of maximal length and $r = l(s)$. Then, by Theorem 3.12 (v), $D_s(\rho^r) > 0$ and so $D_s \neq 0$. Now let $\sum c_w D_w = 0$ and let $\widetilde{w}$ be one of the elements of maximal length for which $c_w \neq 0$. Put $l = l(\widetilde{w})$.

There is a sequence $\alpha_1, \ldots, \alpha_{r-l}$ for which $\widetilde{w}\sigma_{\alpha_1} \ldots \sigma_{\alpha_{r-l}} = s$. Let $F = F_{\alpha_{r-l}} \ldots F_{\alpha_1}$. It follows from Theorem 3.10 that $FD_{\widetilde{w}} = D_s$ and $FD_w = 0$ if $l(w) \geqslant l$, $w \neq \widetilde{w}$. Therefore $F(\sum c_w D_w) = c_{\widetilde{w}} D_s \neq 0$.

b) We now show that the D_w span $\mathscr{H}$. It is sufficient to prove that if $f \in R$ and $(D_w, f) = 0$ for all $w \in W$, then $f \in J$. We may assume that f is a homogeneous element of degree k. For $k = 0$ the assertion is clear.

Now let $k > 0$ and assume that the result is true for all polynomials f of degree less than k. Then for all $\alpha \in \Sigma$ and $w \in W$, $(D_w, A_\alpha f) = (F_\alpha D_w, f) = 0$, by Theorem 3.10 (i) and (ii). By the inductive hypothesis, $A_\alpha f \in J$, that is, $f - \sigma_\alpha f = \alpha A_\alpha f \in J$. Hence for all $w \in W$, $f \equiv wf \pmod J$. Thus, $|W|^{-1} \sum_{w \in W} wf \equiv f \pmod J$. Since the left-hand side belongs to I_+, we see that $f \in J$. Theorem 3.13 is now proved.

The form (,) gives rise to a non-degenerate pairing between $\overline{R} = R/J$ and $\mathscr{H}$. Let $\{P_w\}$ be the basis of $\overline{R}$ dual to $\{D_w\}$. The following properties of the P_w are immediate consequences of Theorem 3.12.

3.14 THEOREM. (i) *Let* $w \in W$, $\alpha \in \Sigma$. *Then*

$$A_\alpha P_w = \begin{cases} 0 & \text{if } l(w\sigma_\alpha) = l(w) + 1, \\ P_{w\sigma_\alpha} & \text{if } l(w\sigma_\alpha) = l(w) - 1. \end{cases}$$

(ii) $\chi P_w = \sum_{w \xrightarrow{\gamma} w'} w\chi(H_\gamma) P_{w'}$ *for* $\chi \in \mathfrak{h}^*_{\mathbf{Q}}$.

(iii) *Let* $\alpha \in \Sigma$. *Then*

$$\sigma_\alpha P_w = \begin{cases} P_w & \text{if } l(w\sigma_\alpha) = l(w) + 1, \\ P_w - \sum_{w\sigma_\alpha \xrightarrow{\gamma} w'} w\alpha(H_\gamma) P_{w'} & \text{if } l(w\sigma_\alpha) = l(w) - 1. \end{cases}$$

From (i) it is clear that all the P_w can be expressed in terms of the P_s. More precisely, let $w = \sigma_{\alpha_1} \ldots \sigma_{\alpha_l}$, $l(w) = r - l$. Then

$$P_w = A_{\alpha_l} \ldots A_{\alpha_1} P_s.$$

To find an explicit form for the P_w it therefore suffices to compute the $P_s \in \overline{R}$.

3.15 THEOREM. $P_s = |W|^{-1} \prod_{\gamma \in \Delta_+} \gamma \pmod J$.

PROOF. We divide the proof into a number of steps. We fix an element $h \in \mathfrak{h}$ such that all the wh, $w \in W$, are distinct.

1. We first prove that there is a polynomial $Q \in R$ of degree r such that

$$(10) \qquad Q(sh) = 1, \; Q(wh) = 0 \; \text{for } w \neq s.$$

For each $w \in W$ we choose in R a homogeneous polynomial $\widetilde{P}_w$ of degree

$l(w)$ whose image in $\bar{R} = R/J$ is P_w. Since $\{P_w\}$ is a basis of R, any polynomial $f \in R$ can be written in the form $f = \sum \tilde{P}_w f_w$, where $f_w \in I$ (this is easily proved by induction on the degree of f). Now let $Q' \in R$ be an arbitrary polynomial satisfying (10) and let $Q' = \sum \tilde{P}_w g_w$, $g_w \in I$. It is clear that $Q = \sum_w g_w(h) \tilde{P}_w$ meets our requirements.

2. Let $\bar{Q}$ be the image of Q in $\bar{R}$, and let $\bar{Q} = \sum c_w P_w$ be the representation of $\bar{Q}$ in terms of the basis $\{P_w\}$ of R. We now prove that $c_s = (-1)^r \prod_{\gamma \in \Delta_+} (\gamma(h))^{-1}$.

To prove this we consider $A_s \bar{Q}$. On the one hand $A_s \bar{Q} = c_s$, by Theorem 3.13 (i); on the other hand, $A_s Q$ is a constant, since Q is a polynomial of degree r. Hence, $A_s Q = c_s$.

We now calculate $A_s Q$, Let $s = \sigma_{\alpha_1} \ldots \sigma_{\alpha_r}$ be the reduced decomposition. We put $w_i = \sigma_{\alpha_i} \ldots \sigma_{\alpha_1}$ (in particular, $w_0 = e$), $\gamma_i = w_{i-1}^{-1} \alpha_1$, $Q_i = A_{\alpha_{i+1}} \ldots A_{\alpha_r} Q$.

LEMMA. *Q_i is a polynomial of degree i,*

$$Q_i(w_i h) = (-1)^{r-i} \cdot \prod_{r \geqslant j > i} (\gamma_j(h))^{-1}$$

and $Q_i(wh) = 0$ if $w \not\geqslant w_i$.

PROOF. We prove the lemma by backward induction on i. For $i = r$ we have $w_r = s$, $Q_r = Q$, and the assertion of the lemma follows from the definition of Q.

We now assume the lemma proved for Q_i, $i > 0$. In the first place, it is clear that $Q_{i-1} = A_{\alpha_i} Q_i$ is a polynomial of degree $i - 1$.

Furthermore,

$$Q_{i-1}(wh) = A_{\alpha_i} Q_i(wh) = \frac{Q_i(wh) - Q_i(\sigma_{\alpha_i} wh)}{\alpha_i(wh)}.$$

If $w = w_{i-1}$, then $w < w_i$, $\sigma_{\alpha_i} w = w_i$ and $\alpha_i(w_{i-1} h) = (w_{i-1}^{-1} \alpha_i)(h) = -(w_i^{-1} \alpha_i)(h) = -\gamma_i(h)$. Therefore, using the inductive hypothesis, we have

$$Q_{i-1}(w_{i-1} h) = -\frac{Q_i(w_i h)}{\alpha_i(w_{i-1} h)} = (-1)^{r-i+1} \prod_{r \geqslant j > i-1} (\gamma_j(h))^{-1}.$$

But if $w \not\geqslant w_{i-1}$, Corollary 2.6 implies that $w \not\geqslant w_i$ and $\sigma_{\alpha_i} w \not\geqslant w_i$. So $Q_{i-1}(wh) = 0$, and the lemma is proved.

Note that by Lemma 2.2, as i goes from 1 to r, γ_i ranges over all the positive roots exactly once. Therefore

$$c_s = A_s Q = Q_0 = (-1)^r \prod_{\gamma \in \Delta_+} (\gamma(h))^{-1}.$$

3. Consider the polynomial $\mathrm{Alt}(Q) = \sum (-1)^{l(w)} wQ$; $\mathrm{Alt}(Q)$ is skew-symmetric, that is, $\sigma_\alpha \mathrm{Alt}(Q) = -\mathrm{Alt}(Q)$ for all $\gamma \in \Delta$. Therefore $\mathrm{Alt}(Q)$ is divisible

(in R) by $\prod_{\gamma\in\Delta_+}\gamma$. Since the degrees of Alt(Q) and $\prod_{\gamma\in\Delta_+}\gamma$ are equal (to r), $\mathrm{Alt}(Q) = \lambda \prod_{\gamma\in\Delta_+}\gamma$. Furthermore, $\mathrm{Alt}(Q)(h) = (-1)^r$, so that

(11) $$\mathrm{Alt}(Q) = (-1)^r \prod_{\gamma\in\Delta_+} (\gamma(h))^{-1} \prod_{\gamma\in\Delta_+}\gamma.$$

4. We put $\mathrm{Alt}(\bar{Q}) = \sum (-1)^{l(w)} w\bar{Q}$. By Theorem 3.14 (iii), $\mathrm{Alt}(P_s) = \sum(-1)^{l(w)} wP_s = |W|\, P_s$. Therefore $\mathrm{Alt}(\bar{Q}) = c_s\, |W|\, P_s$ + terms of smaller degree. Since Alt(Q) is a homogeneous polynomial of degree r, we have

(12) $$\mathrm{Alt}(\bar{Q}) = c_s |W| P_s.$$

By comparing (11) and (12) we find that

$$P_s = |W|^{-1} \prod_{\gamma\in\Delta_+} \gamma \,(\mathrm{mod}\, J).$$

The theorem is now proved.

3.16 COROLLARY. *Let ρ be half the sum of the positive roots. Then* $P_s = \rho^r/r!$ (mod J).

PROOF. For each $\chi \in \mathfrak{h}^*$ we consider the formal power series $\exp\chi$ on $\mathfrak{h}$ given by

$$\exp\chi = \sum_{n=0}^{\infty} \chi^n/n!.$$

Then we have (see [9])

$$\sum_{w\in W} (-1)^{l(w)} \exp(w\rho) = \prod_{\gamma\in\Delta_+} \left[\exp\frac{\gamma}{2} - \exp\left(-\frac{\gamma}{2}\right)\right].$$

Comparing the terms of degree r we see that

$$\frac{1}{r!}\sum (-1)^{l(w)} (w\rho)^r = \prod_{\gamma\in\Delta_+}\gamma.$$

If $\rho^r(\mathrm{mod}\, J) = \lambda P_s$, $\lambda \in \mathbf{C}$, then $(w\rho^r)$ (mod J) $= \lambda w P_s = \lambda(-1)^{l(w)} P_s$. Thus, $\frac{1}{|W|}\sum (-1)^{l(w)} (w\rho)^r = \lambda P_s\, (\mathrm{mod}\, J)$. The result now follows from Theorem 3.15.

To conclude this section we prove some results on products of the P_w in $\bar{R}$.

3.17. THEOREM. (i) *Let $\alpha \in \Sigma$, $w \in W$. Then*

$$P_{\sigma_\alpha} P_w = \sum_{w \xrightarrow{\gamma} w'} \chi_\alpha (H_{w^{-1}\gamma})\, P_{w'},$$

where $\chi_\alpha \in \mathfrak{h}^*_{\mathbf{Z}}$ is the fundamental dominant weight corresponding to the root α (that is, $\chi_\alpha(H_\beta) = 0$ for $\alpha \neq \beta \in \Sigma$, $\chi_\alpha(H_\alpha) = 1$).

(ii) *Let $w_1, w_2 \in W$, $l(w_1) + l(w_2) = r$. Then $P_{w_1} P_{w_2} = 0$ for*

$w_2 \neq w_1 s$, $P_{w_1} P_{w_1 s} = P_s$.

(iii) *Let* $w \in W$, $f \in \overline{R}$. *Then* $f P_w = \sum_{w' > w} c_{w'} P_{w'}$.

(iv) *If* $w_1 \nleq w_2 s$, *then* $P_{w_1} P_{w_2} = 0$.

PROOF. (i) By Theorem 3.12 (v), $P_{\sigma_\alpha} = \chi_\alpha \pmod{J}$. Therefore (i) follows from Theorem 3.14 (ii).

(ii) The proof goes by backward induction on $l(w_2)$. If $l(w_2) = r$, then $w_2 = s$, $w_1 = e$ and $P_{w_1} = 1$.

To deal with the general case we find the following simple lemma useful, which is an easy consequence of the definition of the A_γ.

3.18 LEMMA. *Let* $\gamma \in \Delta$, f, $g \in R$. *Then* $A_\gamma(A_\gamma f \cdot g) = A_\gamma f \cdot A_\gamma g$.

Thus, let $w_2 \in W$, $l(w_2) = l < r$, and choose $\alpha \in \Sigma$ so that $w_2 \overset{\alpha}{\to} \sigma_\alpha w_2$. We consider two cases separately.

A) $w_1 \overset{\alpha}{\nrightarrow} \sigma_\alpha w_1$. We observe that the following equation holds for any $w \in W$

$$l(ws) = r - l(w). \tag{13}$$

Since in our case $l(\sigma_\alpha w_2) = l + 1$ and $l(\sigma_\alpha w_1) = r - l + 1$, we see that $\sigma_\alpha w_1 s \neq \sigma_\alpha w_2$, and so $w_1 s \neq w_2$. On the other hand, $P_{w_2} = A_\alpha P_{\sigma_\alpha w_2}$ and $P_{w_1} = A_\alpha P_{\sigma_\alpha w_1}$ by Theorem 3.14 (i). Therefore, an application of Lemma 3.18 shows that

$$P_{w_1} P_{w_2} = A_\alpha P_{\sigma_\alpha w_1} \cdot A_\alpha P_{\sigma_\alpha w_2} = A_\alpha (P_{\sigma_\alpha w_1} \cdot A_\alpha P_{\sigma_\alpha w_2}) = A_\alpha (P_{\sigma_\alpha w_1} \cdot P_{w_2}).$$

Since $l(\sigma_\alpha w_1) + l(w_2) = r - l + 1 + l > r$, we have $P_{\sigma_\alpha w_1} P_{w_2} = 0$. Hence $P_{w_1} P_{w_2} = 0$ as well.

B) $\sigma_\alpha w_1 \overset{\alpha}{\to} w_1$. In this case, $P_{\sigma_\alpha w_1} = A_\alpha P_{w_1}$ and $P_{w_2} = A_\alpha P_{\sigma_\alpha w_2}$, by Theorem 3.14 (i). Again applying Lemma 3.18, we have

$$A_\alpha (P_{w_1} P_{w_2}) \doteq A_\alpha (P_{w_1} \cdot A_\alpha P_{\sigma_\alpha w_2}) = A_\alpha P_{w_1} \cdot A_\alpha P_{\sigma_\alpha w_2} = \tag{14}$$

$$= A_\alpha (A_\alpha P_{w_1} \cdot P_{\sigma_\alpha w_2}) = A_\alpha (P_{\sigma_\alpha w_1} \cdot P_{\sigma_\alpha w_2}).$$

Since the P_w form a basis of $\overline{R}$, any element f of degree r in $\overline{R}$ has the form $f = \lambda P_s$, $\lambda \in \mathbf{C}$. Furthermore, $A_\alpha P_s = P_{\sigma_\alpha s} \neq 0$. But $\deg P_{w_1} P_{w_2} = \deg P_{\sigma_\alpha w_1} \cdot P_{\sigma_\alpha w_2} = r$. Therefore (14) is equivalent to

$$P_{w_1} P_{w_2} = P_{\sigma_\alpha w_1} P_{\sigma_\alpha w_2}.$$

Applying the inductive hypothesis to the pair $(\sigma_\alpha w_1, \sigma_\alpha w_2)$, we obtain part (ii) of the theorem.

(iii) is an immediate consequence of Theorem 3.14 (ii).

(iv) follows from (ii) and (iii).

We define the operator $\mathcal{P}: \overline{R} \to \mathcal{H}$ of Poincaré duality by the formula $(\mathcal{P}f)(g) = D_s(fg)$, f, $g \in \overline{R}$, $\mathcal{P}f \in \mathcal{H}$.

3.19. COROLLARY. $\mathcal{P} P_w = D_{ws}$.

§4. Schubert cells

We prove in this section that the functionals D_w, $w \in W$ introduced in §3 correspond to Schubert cells s_w, $w \in W$.

Let $s_w \in H_*(X, \mathbf{Q})$ be a Schubert cell. It gives rise to a linear functional on $H^*(X, \mathbf{Q})$, which, by means of the homomorphism α: $R \to H^*(X, \mathbf{Q})$ (see Theorem 1.3), can be regarded as a linear functional on R. This functional takes the value 0 on all homogeneous components P_k with $k \neq l(w)$, and thus determines an element $\hat{D}_w \in S_{l(w)}$.

4.1. THEOREM. $\hat{D}_w = D_w$ (cf. Definition 3.11).

This theorem is a natural consequence of the next two propositions.

PROPOSITION 1. $\hat{D}_e = 1$, *and for any* $\chi \in \mathfrak{h}^*_{\mathbf{Z}}$

$$\chi^* (\hat{D}_w) = \sum_{w' \xrightarrow{\gamma} w} w'\chi(H\gamma)\, \hat{D}_{w'}. \tag{15}$$

PROPOSITION 2. *Suppose that for each* $w \in W$ *we are given an element* $\hat{D}_w \in S_{l(w)}$, *with* $\hat{D}_e = 1$, *for which* (15) *holds for any* $\chi \in \mathfrak{h}^*_{\mathbf{Z}}$. *Then* $\hat{D}_w = D_w$.

Proposition 2 follows at once from Theorem 3.12 (iii) by induction on $l(w)$.

We turn now to the proof of Proposition 1.

We recall (see [10]) that for any topological space Y there is a bilinear mapping

$$H^i(Y, \mathbf{Q}) \times H_j(Y, \mathbf{Q}) \xrightarrow{\cap} H_{j-i}(Y, \mathbf{Q})$$

(the cap-product). It satisfies the condition:

$$1.\ \langle c \cap y, z\rangle = \langle y, c \cdot z\rangle \tag{16}$$

for all $y \in H_j(Y, \mathbf{Q})$, $z \in H^{j-i}(Y, \mathbf{Q})$, $c \in H^i(Y, \mathbf{Q})$.

2. Let f: $Y_1 \to Y_2$ be a continuous mapping. Then

$$f_*(f^*c \cap y) = c \cap f_* y \tag{17}$$

for all $y \in H_j(Y_1, \mathbf{Q})$, $c \in H^i(Y_2, \mathbf{Q})$.

By virtue of (17) we have for any $\chi \in \mathfrak{h}^*_{\mathbf{Z}}$, $f \in R$

$$(\chi^*(\hat{D}_w), f) = (\hat{D}_w, \chi f) = \langle s_w, \alpha_1(\chi)\alpha(f)\rangle = \langle s_w \cap \alpha_1(\chi), \alpha(f)\rangle.$$

Therefore (15) is equivalent to the following geometrical fact.

PROPOSITION 3. For all $\chi \in \mathfrak{h}^*_{\mathbf{Z}}$

$$s_w \cap \alpha_1(\chi) = \sum_{w' \xrightarrow{\gamma} w} w'\chi(H_\gamma)\, s_{w'}. \tag{18}$$

We restrict the fibering E_χ to $\bar{X}_w \subset X$ and let $c_\chi \in H^2(\bar{X}_w, \mathbf{Q})$ be the first Chern class of E_χ. By (17) and the definition of the homomorphism $\alpha_1: \mathfrak{h}^*_{\mathbf{Z}} \to H^2(X, \mathbf{Q})$, it is sufficient to prove that

(19) $$s_w \cap c_\chi = \sum_{w' \xrightarrow{\gamma} w} w'\chi(H_\gamma)\, s_{w'}.$$

in $H_{2l(w)-2}(\bar{X}_w, \mathbf{Q})$.

To prove (19), we use the following simple lemma, which can be verified by standard arguments involving relative Poincaré duality.

4.2 LEMMA. *Let Y be a compact complex analytic space of dimension n, such that the codimension of the space of singularities of Y is greater than 1. Let E be an analytic linear fibering on Y, and $c \in H^2(Y, \mathbf{Q})$ the first Chern class of E. Let μ be a non-zero analytic section of E and $\sum m_i Y_i = \operatorname{div} \mu$ the divisor of μ. Then $[Y] \cap c = \sum m_i [Y_i] \in H_{2n-2}(Y, \mathbf{Q})$, where $[Y]$ and $[Y_i]$ are the fundamental classes of Y and Y_i.*

Let $w \in W$, and let $X_w \subset X$ be the corresponding Schubert cell. From Lemma 4.2 and Theorem 2.11 it is clear that to prove Proposition 3 it is sufficient to verify the following facts.

4.3. PROPOSITION. *Let $w' \xrightarrow{\gamma} w$. Then $\bar{X}_w$ is non-singular at points $x \in X_{w'}$.*

4.4. PROPOSITION. *There is a section μ of the fibering E_χ over $\bar{X}_w$ such that*

$$\operatorname{div} \mu = \sum_{w' \xrightarrow{\gamma} w} w'\chi(H_\gamma)\, \bar{X}_{w'}.$$

To verify these facts we use the geometrical description of Schubert cells given in 2.9. We consider a finite-dimensional representation of G on a space V with regular dominant weight λ, and we realize X as a subvariety of $P(V)$. For each $w \in W$ we fix a vector $f_w \in V$ of weight $w\lambda$.

PROOF OF PROPOSITION 4.3. For a root $\gamma \in \Delta_+$ we construct a three-dimensional subalgebra $\mathfrak{A}_\gamma \subset \mathfrak{G}$ (as in the proof of Theorem 2.9). Let i: $SL_2(\mathbf{C}) \to G$ be the homomorphism corresponding to the embedding $\mathfrak{A}_\gamma \to \mathfrak{G}$. Consider in $SL_2(\mathbf{C})$ the subgroups $B' = \left\{\begin{pmatrix} a & b \\ 0 & a^{-1}\end{pmatrix}\right\}$, $H' = \left\{\begin{pmatrix} a & 0 \\ 0 & a^{-1}\end{pmatrix}\right\}$ and $N'_- = \left\{\begin{pmatrix} 1 & 0 \\ x & 1\end{pmatrix}\right\}$ and the element $\sigma = \begin{pmatrix} 0 & 1 \\ -1 & 0\end{pmatrix}$. We may assume that $i(H') \subset H$, $i(B') \subset B$.

Let $\tilde{V}$ be the smallest $\mathfrak{A}_\gamma$-invariant subspace of V containing $f_{w'}$. It is clear that $\tilde{V}$ is invariant under $i(SL_2(\mathbf{C}))$, and that the stabilizer of the line $[f_{w'}]$ is B'. This determines a mapping δ: $SL_2(\mathbf{C})/B' \to X$. The space $SL_2(\mathbf{C})/B'$ is naturally identified with the projective line $\mathbf{P}^1$. Let $o, \infty \in \mathbf{P}^1$ be the images of $e, \sigma \in SL_2(\mathbf{C})$.

We define a mapping ξ: $N_{w'} \times \mathbf{P}^1 \to X$ by the rule

$$(x, z) \mapsto x \cdot \delta(z).$$

4.5. LEMMA. *The mapping ξ has the following properties:*

(i) $\xi(N_{w'} \times \{o\}) = X_{w'}$, $\xi(N_{w'} \times (\mathbf{P}^1 \setminus o)) \subset X_w$.

(ii) *The restriction of* ξ *to* $(N_{w'} \times \mathbf{P}^1 \setminus \infty))$ *is an isomorphism onto a certain open subset of* $\overline{X}_w$.

Proposition 4.3 clearly follows from this lemma.

PROOF OF LEMMA 4.5. The first assertion of (i) follows at once from the definition of $X_{w'}$. Since the cell X_w is invariant under N, the proof of the second assertion of (i) is reduced to showing that $\delta(z) \cdot \in X_w$ for $z \in \mathbf{P}^1 \setminus o$. Let $h \in SL_2(\mathbf{C})$ be an inverse image of z. Then h can be written in the form $h = b_1 \sigma b_2$, where $b_1, b_2 \in B'$. It is clear that $i(b_2)f_{w'} = c_1 f_{w'}$ and $i(\sigma)f_{w'} = c_2 f_w$, where c_1, c_2 are constants. Therefore $i(h)f_{w'} = c_1 c_2 i(b_1)f_w$, that is, $\delta(z) \in X_w$.

To prove (ii), we consider the mapping

$$w'^{-1} \circ \xi \colon N_{w'} \times (\mathbf{P}^1 \setminus \infty) \to X.$$

The space $\mathbf{P}^1 \setminus \infty$ is naturally isomorphic to the one-parameter subgroup $N'_- \subset SL_2(\mathbf{C})$.

The mapping $\xi\colon N_{w'} \times N'_- \to X$ is given by the rule

$$\xi(n, n_1) = ni(n_1)\,[f_{w'}],\ n \in N_{w'},\ n_1 \in N'.$$

Thus,

$$w'^{-1} \circ \xi(n, n_1) = (w'^{-1} n w')\,(w'^{-1} i\,(n_1)\,w')\,[f_e].$$

We now observe that $w'^{-1} N_{w'} w' \subset N_-$ (by definition of $N_{w'}$), and $w'^{-1} i(N')w' \in N_-$ (since $w'^{-1}\ \gamma \in \Delta_+$). Furthermore, the intersection of the tangent spaces to these subgroups consists only of 0, because $N_{w'} \subset N$, $i(N'_-) \subset N_-$. The mapping $N_- \to X$ ($n \mapsto n[f_e]$) is an isomorphism onto an open subset of X. Therefore (ii) follows from the next simple lemma, which is proved in [5], for example.

4.6. LEMMA. *Let* N_1 *and* N_2 *be two closed algebraic subgroups of a unipotent group* N *whose tangent spaces at the unit element intersect only in* 0. *Then the product mapping* $N_1 \times N_2 \to N$ *gives an isomorphism of* $N_1 \times N_2$ *with a closed subvariety of* N.

This completes the proof of Proposition 4.3.

PROOF OF PROPOSITION 4.4. Any element of $\mathfrak{h}^*_{\mathbf{Z}}$ has the form $\chi = \lambda - \lambda'$, where λ, λ' are regular dominant weights. In this case, $E_\chi = E_\lambda \otimes E_{\lambda'}^{-1}$, and it is therefore sufficient to find a section μ with the required properties in the case $\chi = \lambda$.

We consider the space $P(V)$, where V is a representation of G with dominant weight λ. Let η_V be the linear fibering on $P(V)$ consisting of pairs (P, ϕ), where ϕ is a linear functional on the line $P \subset V$. Then $E_\lambda = i^*(\eta_V)$, where $i\colon X \to P(V)$ is the embedding described in § 2.

The linear functional ϕ_w on V (see the proof of Theorem 2.11) yields a section of the bundle η. We shall prove that the restriction of μ to this section on $\overline{X}_w$ is a section of the fibering E_λ having the requisite properties.

By Lemma 2.12, $\mu(x) \neq 0$ for all $x \in X_w$. The support of the divisor div μ is therefore contained in $\overline{X}_w \setminus X_w = \bigcup_{w' \xrightarrow{-\gamma} w} \overline{X}_{w'}$.

Since $\bar{X}_{w'}$ is an irreducible variety, we see that $\operatorname{div}\mu = \sum_{w' \xrightarrow{\gamma} w} a_\gamma \bar{X}_{w'}$, where $a_\gamma \in \mathbf{Z}$, $a_\gamma \geqslant 0$. It remains to show that $a_\gamma = w'\chi(H_\gamma)$.

In view of Lemma 4.5 (i) and (ii), the coefficient a_γ is equal to the multiplicity of zero of the section $\delta^*(\mu)$ of the fibering $\delta^*(E_\lambda)$ on $\mathbf{P}^1$ at the point o, that is, the multiplicity of zero of the function $\psi(t) = \phi_{w'}((\exp tE_{-\gamma})f_{w'})$ for $t = 0$. It follows from Lemma 2.10 that $\psi(t) = ct^n$, hence $a_\gamma = n = w'\chi(H_\gamma)$. This completes the proof of Proposition 4.4 and with it of Theorem 4.1.

§ 5. Generalizations and supplements

1. Degenerate flag varieties. We extend the results of the previous sections to spaces $X(P) = G/P$, where P is an arbitrary parabolic subgroup of G. For this purpose we recall some facts about the structure of parabolic subgroups $P \subset G$ (see [7]).

Let Θ be some subset of Σ, and Δ_Θ the subset of Δ_+ consisting of linear combinations of elements of Θ. Let G_Θ be the subgroup of G generated by H together with the subgroups $N_\gamma = \{\exp tE_\gamma \mid t \in \mathbf{C}\}$ for $\gamma \in \Delta_\Theta \cup -\Delta_\Theta$, and let N_Θ be the subgroup of N generated by the N_γ for $\gamma \in \Delta_+ \backslash \Delta_\Theta$. Then G_Θ is a reductive group normalizing N_Θ, and $P_\Theta = G_\Theta N_\Theta$ is a parabolic subgroup of G containing B.

It is well known (see [7], for example) that every parabolic subgroup $P \subset G$ is conjugate in G to one of the subgroups P_Θ. We assume in what follows that $P = P_\Theta$, where Θ is a fixed subset of Σ. Let W_Θ be the Weyl group of G_Θ. It is the subgroup of W generated by the reflections σ_α, $\alpha \in \Theta$.

We describe the decomposition of $X(P)$ into orbits under the action of B.

5.1. PROPOSITION. (i) *$X(P) = \bigcup_{w \in W} Bwo$, where $o \in X(P)$ is the image of P in G/P.*

(ii) *The orbits Bw_1o and Bw_2o are identical if $w_1 w_2^{-1} \in W_\Theta$ and otherwise are disjoint.*

(iii) *Let W_Θ^1 be the set of $w \in W$ such that $w\Theta \subset \Delta_+$. Then each coset of W/W_Θ contains exactly one element of W_Θ^1. Furthermore, the element $w \in W_\Theta^1$ is characterized by the fact that its length is less than that of any other element in the coset wW_Θ.*

(iv) *If $w \in W_\Theta^1$, then the mapping $N_w \to X(P)$ $(n \to nwo)$ is an isomorphism of N_w with the subvariety $Bwo \subset X(P)$.*

PROOF. (i)–(ii) follow easily from the Bruhat decomposition for G and G_Θ. The proof of (iii) can be found in [7], for example, and (iv) follows at once from (iii) and Proposition 1.1.

Let $w \in W_\Theta^1$, $X_w(P) = Bwo$, let $\bar{X}_w(P)$ be the closure of $X_w(P)$ and $[\bar{X}_w(P)] \in H_{2l(w)}(\bar{X}_w(P), \mathbf{Z})$ its fundamental class. Let $s_w(P) \in H_{2l(w)}(X(P), \mathbf{Z})$ be the image of $[\bar{X}_w(P)]$ under the mapping induced by the embedding $\bar{X}_w(P) \hookrightarrow X(P)$. The next proposition is an analogue of Proposition 1.2.

5.2. PROPOSITION ([2]). *The elements $s_w(P)$, $w \in W_\Theta^1$, form a free basis in $H_*(X(P), \mathbf{Z})$.*

5.3. COROLLARY. *Let α_P: $X \to X(P)$ be the natural mapping. Then $(\alpha_P)_* s_w = 0$ if $w \notin W_\Theta^1$, $(\alpha_p)_* s_w = s_w(P)$ if $w \in W_\Theta^1$.*

5.4. COROLLARY. $(\alpha_P)_*$: $H_*(X, \mathbf{Z}) \to H_*(X(P), \mathbf{Z})$ *is an epimorphism, and* $(\alpha_P)^*$: $H^*(X(P), \mathbf{Z}) \to H^*(X, \mathbf{Z})$ *is a monomorphism.*

5.5. THEOREM. (i) $\mathrm{Im}(\alpha_P)^* \subset H^*(X, \mathbf{Z}) = \bar{R}$ *coincides with the set of W_Θ-invariant elements of $\bar{R}$.*

(ii) $P_w \in \mathrm{Im}(\alpha_P)^*$ *for* $w \in W_\Theta^1$ *and* $\{(\alpha_P)^{*-1} P_w\}_{w \in W_\Theta^1}$ *is the basis in $H^*(X(P), \mathbf{Z})$ dual to the basis $\{s_w(P)\}_{w \in W_\Theta^1}$ in $H_*(X(P), \mathbf{Z})$.*

PROOF. Let $w \in W_\Theta^1$. Since $\langle P_w, s_{w_1} \rangle = 0$ for $w_1 \notin W_\Theta^1$, P_w is orthogonal to $\mathrm{Ker}(\alpha_P)_*$, that is, $P_w \in \mathrm{Im}(\alpha_P)^*$. Now (ii) follows from the fact that $\langle (\alpha_P)^* P_w, s_{w'}(P) \rangle = \langle P_w, s_{w'} \rangle$ for $w, w' \in W_\Theta^1$. To prove (i), it is sufficient to verify that the P_w, $w \in W_\Theta^1$, form a basis for the space of W_Θ-invariant elements of $\bar{R}$. We observe that an element $f \in \bar{R}$ is W_Θ-invariant if and only if $A_\alpha f = 0$ for all $\alpha \in \Theta$. Since $w \in W_\Theta^1$ if and only if $l(w\sigma_\alpha) = l(w) + 1$ for all $\alpha \in \Theta$, (i) follows from Theorem 3.14(i).

2. CORRESPONDENCES. Let Y be a non-singular oriented manifold. An arbitrary element $z \in H_*(Y \times Y, \mathbf{Z})$ is called a correspondence on Y. Any such element z gives rise to an operator z_*: $H_*(Y, \mathbf{Z}) \to H_*(Y, \mathbf{Z})$, according to

$$z_*(c) = (\pi_2)_*((\pi_1)^*(\mathcal{P}c) \cap z), \quad c \in H_*(Y, \mathbf{Z}),$$

where π_1, π_2: $Y \times Y \to Y$ are the projections onto the first and second components, and $\mathcal{P}$ is the Poincaré duality operator. We also define an operator z^*: $H^*(Y, \mathbf{Z}) \to H^*(Y, \mathbf{Z})$ by setting $z^*(\xi) = \mathcal{P}[(\pi_1)_*((\pi_2)^*(\xi) \cap z)]$, $\xi \in H^*(Y, \mathbf{Z})$. It is clear that z_* and z^* are adjoint operators.

Let z be assigned to a (possibly singular) submanifold $Z \subset Y \times Y$, in such a way that z is the image of the fundamental cycle $[Z] \in H_*(Z, \mathbf{Z})$ under the mapping induced by the embedding $Z \hookrightarrow Y \times Y$. Then

$$z_*(c) = (\rho_2)_* ([Z] \cap (\rho_1)^* \mathcal{P}c),$$

where ρ_1, ρ_2: $Z \to Y$ are the restrictions of π_1, π_2 to Z.

If, in this situation, ρ_1: $Z \to Y$ is a fibering and c is given by a submanifold $C \subset Y$, then the cycle

$$[Z] \cap (\rho_1)^* \mathcal{P}c$$

is given by the submanifold $\rho_1^{-1}(C) \subset Z$.

We want to study correspondences in the case $Y = X = G/B$.

5.6. DEFINITION. *Let $w \in W$. We put $Z_w = \{(gwo, go)\} \subset X \times X$ and denote by z_w the correspondence $z_w = [\bar{Z}_w] \subset H_*(X \times X, \mathbf{Z})$.*

5.7. THEOREM. $(z_w)_* = F_w$.

PROOF. We calculate $(z_w)_* (s_{w'})$.

Since the variety $\overline{Z}_w$ is G-invariant and G acts transitively on X, the mapping $\rho_1 : \overline{Z}_w \to X$ is a fibering. Thus,

$$(z_w)_* (s_{w'}) = (\rho_2)_* [\rho_1^{-1} (\overline{X}_{w'})].$$

It is easily verified that $\rho_1^{-1} (\overline{X}_{w'}) = \overline{\pi_1^{-1} (X_{w'}) \cap Z_w}$. We put $Y = \pi_1^{-1} (X_{w'}) \cap Z_w \subset X \times X$. Then

$$Y = \{(nw'o,\ nw'bwo) \mid n \in N,\ b \in B\}. \tag{20}$$

Since the dimension of the fibre of $\rho_1 : Z_w \to X$ is equal to $2l(w)$, we see that $\dim Y = 2l(w) + 2l(w')$. It is clear from (20) that

$$\rho_2(Y) = \{nw'bwo \mid n \in N,\ b \in B\} = Bw'Bwo.$$

It is well known (see [6]. Ch. IV, § 2.1 Lemma 1) that

$$Bw'Bwo = Bw'wo \cup \Big(\bigcup_{l(w_1)<l(w)+l(w')} Bw_1o\Big).$$

Thus, two cases can arise.

a) $l(w'w) < l(w') + l(w)$. In this case, $\dim \rho_2 (Y) < 2l(w') + 2l(w)$, and so $(z_w)_*(s_{w'}) = (\rho_2)_*[Y] = 0$.

b) $l(w'w) = l(w') + l(w)$. In this case, $\rho_2 (Y) = X_{w'w} + X'$, where $\dim X' < \dim X_{w'w} = 2l(w') + 2l(w)$. Thus, $(\rho_2)_*[Y] = [\overline{X}_{w'w}]$, that is, $(z_w)_*(s_{w'}) = s_{w'w}$. Comparing the formulae obtained with 3.12 (ii), we see that $(z_w)_* = F_w$.

5.8. COROLLARY. $z_w = \sum s_{w's} \otimes s_{w'w}$, *where the summation extends over those* $w' \in W$ *for which* $l(w'w) = l(w) + l(w')$.

In § 1 we have defined an action of W on $H_*(X, \mathbf{Z})$. This definition depended on the choice of a compact subgroup K. Using Theorem 5.7 we can find explicitly the correspondences giving this action.

In fact, it follows from Lemma 3.3 (iii) that $\sigma_\alpha = \alpha^* F_\alpha - 1$ for any $\alpha \in \Sigma$. The transformation F_α is given by the correspondence Z_{σ_α}. The operator α^* can also be given by a correspondence: if $U_\alpha = \Sigma c_i U_i$ is a divisor in X giving the cycle $\mathscr{P}(\alpha) \in H_{2r-2}(X, \mathbf{Z})$ (for example, $U_\alpha = \sum_{\beta \in \Sigma} \alpha(H_\beta) X_{\sigma_\beta}$), then the cycle $\widetilde{U}_\alpha = \Sigma c_i \widetilde{U}_i$, where $\widetilde{U}_i = \{(x, x) \mid x \in U_i\} \subset X \times X$, determines the correspondence that gives the operator α^*. The operator σ_α in $H_*(X, \mathbf{Z})$ is therefore given by the correspondence $\widetilde{U}_\alpha * Z_{\sigma_\alpha} - 1$ (where $*$ denotes the product of correspondences, as in [11]). Using the geometrical realization of the product of correspondences (see [11]), we can explicitly determine the correspondence S_α that gives the transformation $1 + \sigma_\alpha$ in $H_*(X, \mathbf{Z})$, namely, $S_\alpha = \Sigma c_i \hat{U}_i$ where $\hat{U}_i = \{(x, y) \in X \times X \mid x \in U_i,\ \tilde{x}^{-1}\tilde{y} \in P_{\{\alpha\}}\}$. In this expression, $\tilde{x}, \tilde{y} \in G$ are arbitrary representatives of x, y, and $P_{\{\alpha\}}$ is the parabolic subgroup corresponding to the root α.

3. B. Kostant has described the P_w in another way. We state his result.

Let $h \in \mathfrak{h}^*_{\mathbf{Q}}$ be an element such that $\alpha(h) > 0$ for all $\alpha \in \Sigma$. Let $J_h = \{f \in R \mid f(wh) = 0 \text{ for all } w \in W\}$ be an ideal of R.

5.9. THEOREM. (i) *Let $w \in W$, $l(w) = l$. There is a polynomial $Q_w \in R$ of degree l such that*

(21) $\qquad Q_w(wh) = 1,\ Q_w(w'h) = 0 \quad \text{if} \quad l(w') \leqslant l(w),\ \ w' \neq w.$

The Q_w are uniquely determined by (21) *to within elements of J_h.* (ii) *Let Q_w^0 be the form of highest degree in the polynomial Q_w. The image of Q_w^0 in $\bar{R}$ is equal to* $\prod_{\gamma \in \Delta_- \cap w^{-1}\Delta_+} (\gamma(h))^{-1} \cdot P_w$.

The proof is analogous to that of Theorem 3.15.

4. We choose a maximal compact subgroup $K \subset G$ such that $K \cap B \subset H$ (see § 1). The cohomology of X can be described by means of the K-invariant closed differential forms on X. For let $\chi \in \mathfrak{h}^*_{\mathbf{Z}}$, and let E_χ be the corresponding one-dimensional complex G-fibering on X. Let $\tilde{\omega}_\chi$ be the 2-form on X which is the curvature form of the connection associated with the K-invariant metric on E_χ (see [12]). Then the class of the form $\omega_\chi \frac{1}{2\pi i} \tilde{\omega}_\chi$ is $c_\chi \in H^2(X, \mathbf{Z})$. The mapping $\chi \to \omega_\chi$ extends to a mapping $\theta\colon R \to \Omega^*_{ev}(X)$, where Ω^*_{ev} is the space of differential forms of even degree on X. One can prove the following theorem, which is a refinement of Proposition 1.3 (ii) and Theorem 3.17.

5.10. THEOREM (i) Ker $\theta = J$, *that is, θ induces a homomorphism of rings $\bar{\theta}\colon \bar{R} \to \Omega^*_{ev}(X)$.* (ii) *Let $w_1, w_2 \in W$, $w_1 \not\leqslant w_2 s$. Then the restriction of the form $\bar{\theta}(P_{w_1})$ to X_{w_2} is equal to* 0. (iii) *Let $w_1, w_2 \in W$, $w_1 \not\leqslant w_2 s$. Then $\bar{\theta}(P_{w_1})\,\bar{\theta}(P_{w_2}) = 0$.*

References

[1] A. Borel, Sur la cohomologie des espaces fibrés principaux et des espaces homogènes des groupes de Lie compacts, Ann. of Math. (2) **57** (1953), 115–207; MR 14 #490. Translation: in '*Rassloennye prostranstva*', Inost. lit., Moscow 1958

[2] A. Borel, Kählerian coset spaces of semisimple Lie groups, Proc. Nat. Acad, Sci. U.S.A. **40** (1954), 1147–1151; MR 17 #1108.

[3] B. Kostant, Lie algebra cohomology and generalized Schubert cells, Ann. of Math. (2) 77 (1963), 72–144; MR 26 #266.

[4] G. Horrocks, On the relations of S-functions to Schubert varieties, Proc. London Math. Soc. (3) 7 (1957), 265–280; MR 19 #459.

[5] A. Borel, Linear algebraic groups, Benjamin, New York, 1969; MR #4273. Translation: *Lineinye algebraicheskie gruppy*, 'Mir', Moscow 1972.

[6] N. Bourbaki, Groupes et algebras de Lie, Ch. 1–6, Éléments de mathématique, 26, 34, 36, Hermann & Cie, Paris, 1960–72. Translation: *Gruppyi algebry Li*, 'Mir', Moscow 1972.

[7] R. Steinberg, Lectures on Chevalley groups, Yale University Press, New Haven, Conn. 1967.

[8] M. F. Atiyah and F. Hirzebruch, Vector bundles and homogeneous spaces, Proc.

Sympos. Pure Math.; Vol. III, 7–38, Amer. Math. Soc., Providence, R. I., 1961; MR 25 # 2617.
= Matematika **6**: 2 (1962), 3–39.

[9] P. Cartier, On H. Weyl's character formula, Bull. Amer. Math. Soc. **67** (1961), 228–230; MR 26 # 3828.
= Matematika **6**: 5 (1962), 139–141.

[10] E. H. Spanier, Algebraic topology, McGraw-Hill, New York 1966; MR 35 # 1007
Translation: *Algebraicheskaya topologiya,* 'Mir', Moscow 1971.

[11] Yu.I. Manin, Correspondences, motives and monoidal transformations, Matem. Sb. 77 (1968), 475–507; MR 41 # 3482
= Math. USSR–Sb. **6** (1968), 439–470.

[12] S. S. Chern, Complex manifolds, Instituto de Fisca e Matemática, Recife 1959; MR 22 # 1920.
Translation: *Kompleksnye mnogoobraziya*, Inost. Lit., Moscow 1961.

[13] I. N. Bernstein, I. M. Gel'fand and S. I. Gel'fand, Schubert cells and the cohomology of flag spaces, Funkts. analiz **7**: 1 (1973), 64–65.

Received by the Editors 13 March 1973

Translated by D. Johnson

FOUR PAPERS ON PROBLEMS IN LINEAR ALGEBRA

Claus Michael Ringel

This volume contains four papers on problems in linear algebra. They form part of a general investigation which was started with the famous paper [Q] on the four subspace problem. The r subspace problem asks for the determination of the possible positions of r subspaces in a vector space, or, equivalently, of the indecomposable representations of the following oriented graph

with $r + 1$ vertices. For $r \geqslant 5$, this problem seems to be rather hard to attack, however one may try to obtain at least partial results dealing with special kinds of representations. Also, the r subspace problem can be used as a test problem for more elaborate problems in linear algebra. This seems to be the case for some of the investigations published in this volume, they have been generalized recently to the case of arbitrary oriented graphs [M, S].

Three of the four papers deal with the r subspace problem. (We should remark that there is a rather large overlap of [F] and [I, II]. However, the main argument of [F], the proof given in section 7, is not repeated in [I, II], whereas [I, II] give the details for the complete irreducibility of the representations $\rho_{t,l}$ which only was announced in [F]. We also recommend the survey given by Dlab [8].) Given r subspaces $E_1, \ldots, E_r$ of a finite-dimensional vector space V, we obtain a lattice homomorphism ρ from the free modular lattice D^r with r generators $e_1, \ldots, e_r$ into the lattice $L(V)$ of all subspaces of V given by $\rho(e_i) = E_i$. Such a lattice homomorphism is called a representation of D^r. In [F], Gelfand and Ponomarev introduce a set of indecomposable representations $\rho_{t,l}$ with $0 \leqslant t \leqslant r$ and $l \in \mathbf{N}$, which we will call the preprojective representations (in [F], the representations $\rho_{t,l}$ with $1 \leqslant t \leqslant r$ are called representations of the first kind, those of the form $\rho_{o,l}$ representations of the second kind; in [I, II] there may arise some confusion: $\rho_{t,l}$ is denoted by $\rho^+_{t,l}$, whereas the symbol $\rho_{t,l}$ used in [I, II] stands for the same type of representation but with a shift of the indices, see Proposition 8.2 in [II]). For the construction of the preprojective representations, we refer to section 1.4 of

[F] : one first defines a finite set $A_t(r, l)$ (which later we will identify with a set of paths in some oriented graph), considers the vector space with basis the set $A_t(r, l)$, and also a subspace $Z_t(r, l)$ generated by certain sums of the canonical base elements of $A_t(r, l)$. The residue classes of the canonical base elements of $A_t(r, l)$ in $V_{t,l} = A_t(r, l)/Z_t(r, l)$ will be denoted by ξ_α (with $\alpha \in A_t(r, l)$). Now, the representation $\rho_{t,l}$ is given by the vector space $V_{t,l}$ together with a certain r tuple of subspaces of $V_{t,l}$, all being generated by some of the generators ξ_α. Note that this implies that $\rho_{t,l}$ is defined over the prime field k_0 of k. (Gelfand and Ponomarev usually assume that the characteristic of k is zero, thus $k_0 = \mathbb{Q}$. However, all results and proofs remain valid in general.)

The main result concerning these representations $\rho_{t,l}$ asserts that in case $\dim V_{t,l} > 2$, the representation $\rho_{t,l}$ is completely irreducible. This means that the image of D^r under the lattice homomorphism $\rho_{t,l}$: $D^r \to L(V_{t,l})$ is the set of all subspaces of $V_{t,l}$ defined over the prime field k_0, thus $\rho_{t,l}(D^r)$ is a projective geometry over k_0. The first essential step in the proof of this result is to show that the subspaces $k\xi_\alpha$ are of the form $\rho(e_\alpha)$ for some $e_\alpha \in D^r$. (In [F], this is only announced, but it is an immediate consequence of theorem 8.1 in [II].)

The second step is to show that any subspace of $V_{t,l}$ which is defined over the prime field, lies in the lattice of subspaces generated by the $k\xi_\alpha$ provided $\dim V_{t,l} > 2$. Combining both assertions, we conclude that $\rho_{t,l}$ is completely irreducible unless $\dim V_{t,l} \leqslant 2$. The proof of the second step occupies section 9 of [II]. Here, one considers the following situation: there is given a set $R = \{\xi_\alpha \mid \alpha\}$ of non-zero vectors of a vector space V (= $V_{t,l}$), with the following properties:

(1) R generates V

(2) R is indecomposable (there is no proper direct decomposition $V = V' \oplus V''$ with $R = (R \cap V') \cup (R \cap V'')$), and

(3) R is defined over the prime field (there exists a basis of V such that any $\xi_\alpha \in R$ is a linear combination of the base vectors with coefficients in the prime field k_0).

Then it is shown that the lattice of subspaces of V generated by the one-dimensional subspaces $k\xi_\alpha$, is isomorphic to the lattice of subspaces of k_0^n, with $n = \dim V$.

Perhaps we should add that the representations ρ: $D^r \to L(V)$ with V being generated by the one-dimensional subspaces of the form $\rho(a)$, $a \in D^r$, seem to be of special interest. In this case, the one-dimensional subspaces of the form $\rho(a)$, $a \in D^r$ determine completely $\rho(D^r)$. (Namely, let $b \in D^r$, and U the subspace generated by all one-dimensional subspaces of the form $\rho(x)$, $x \in D^r$, satisfying $\rho(x) \subseteq \rho(b)$, and choose $x_1, \ldots, x_s$ such that

$$\rho(b) \subseteq U \oplus \rho(x_1) \oplus \ldots \oplus \rho(x_s) = U \oplus \rho(\sum_{i=1}^{s} x_i). \text{ Thus, } \rho(b) = U \oplus (\rho(\sum_{i=1}^{s} x_i) \cap \rho(b)).$$

Assume, U is a proper subspace of $\rho(b)$. Then there exists $t \leqslant s$ with

$\rho(\sum_{i=1}^{t-1} x_i) \cap \rho(b) = 0$, whereas $\rho(\sum_{i=1}^{t} x_i) \cap \rho(b)$ is non-zero, and therefore one-dimensional. This however implies that $\rho(\sum_{i=1}^{t} x_i) \cap \rho(b) = \rho(b \sum_{i=1}^{t} x_i)$ is contained in U, a contradiction. Thus $\rho(b) = U$.). For $r \geqslant 4$, there always are indecomposable representations which do not have this property.

In the case $r = 4$, we may give the complete list of all lattices of the form $\rho(D^4)$, where ρ is an indecomposable representation. Besides the projective geometries over any prime field, and of arbitrary finite dimension $\neq 1$, and the lattice , we obtain all the lattices $S(n, 4)$ introduced by Day, Herrmann and Wille in [6]. Let us just copy $S(14, 4)$ and note that any interval $[c_n, c_m]$ is again of the form $S(n - m, 4)$.

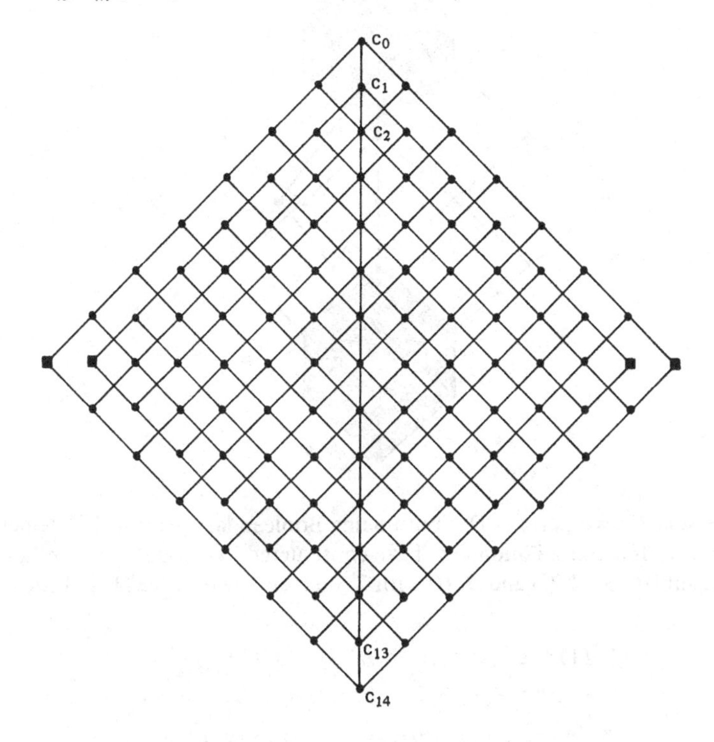

(In fact, in case either $\rho: D^4 \to L(V)$ or its dual is preprojective and dim $V > 2$, we have seen above that $\rho(D^4)$ is the full projective geometry over the prime field. If neither ρ nor its dual is preprojective, ρ is said to be regular. If ρ is regular and non-homogeneous, say of regular length n (see [9]), then $\rho(D^4) \approx S(n, 4)$, whereas for ρ homogeneous, we have

$\rho(D^4) \approx$.)

Gelfand and Ponomarev use the representations $\rho_{t,l}$ of D^r in order to get some insight into the structure of D^r. The existence of a free modular lattice with a given set of generators is easily established, however the mere existence result does not say anything about the internal structure of D^r. In fact, it has been shown by Freese [14] that for $r \geqslant 5$, the word problem in D^r is unsolvable. The free modular lattice D^3 in 3 generators e_1, e_2, e_3 was first described by Dedekind [7], it looks as follows:

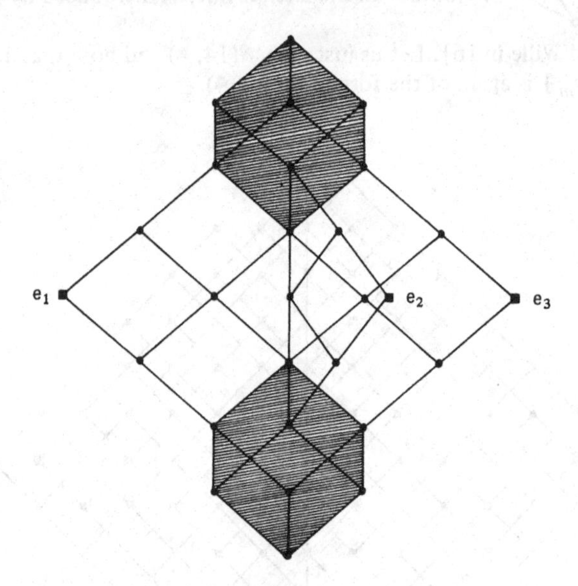

We have shaded two parts of D^3, both being Boolean lattices with 2^3 elements. For $r \geqslant 4$, Gelfand and Ponomarev have constructed two countable families of Boolean sublattices $B^+(l)$ and $B^-(l)$ with 2^r elements, where $l \in \mathbf{N}$, and such that

$$B^-(1) < B^-(2) < \ldots < B^-(l) < B^-(l+1) < \ldots$$

and

$$\ldots B^+(l+1) < B^+(l) < \ldots < B^+(2) < B^+(1),$$

called the lower and the upper cubicles, respectively. Let $B^- = \bigcup_{l \in \mathbf{N}} B^-(l)$, and

$$B^+ = \bigcup_{l \in \mathbf{N}} B^+(l).$$

The elements of these cubicles have an important property: they are perfect. This notion has been introduced by Gelfand and Ponomarev in [F] for the following property: a is said to be perfect if $\rho(a)$ is either O or V for any indecomposable representation $\rho\colon D^r \to L(V)$. This means that for any representation, the image of a is a direct summand. For any perfect element a, let $N_k(a)$ be the set of all indecomposable representations $\rho\colon D^r \to L(V)$, with V a finite dimension vector space over the field k and which satisfy $\rho(a) = 0$. It is shown in [F] that for $a \in B^+$, the set $N_k(a)$ is finite and contains only preprojective representations. Dually, for $a \in B^-$, the set $N_k(a)$ contains all but a finite number of indecomposable representations, and all indecomposable representations not in $N_k(a)$ are preinjective (the representations dual to preprojective ones are called preinjective).

In dealing with perfect elements it seems to be convenient to work modulo linear equivalence. Two elements $a, a' \in D^r$ are said to be linear equivalent provided $\rho(a) = \rho(a')$ for any representation $\rho\colon D^r \to L(V)$. Of course, any element linear equivalent to a perfect element is also perfect. Up to linear equivalence, one has $B^- < B^+$ and Gelfand and Ponomarev have conjectured that, up to linear equivalence, all perfect elements belong to $B^- \cup B^+$. However, this has to be modified. Herrmann [19] has pointed out that there are additional perfect elements arising from the different characteristics of fields. For example, for any prime number p, and $m \geqslant 2$, there is some perfect element $d_{pm} \in D^r$ such that $N_k(d_{pm})$ contains all representations $\rho_{t,l}$ with $l < m$, and, in case the characteristic of k is p, then, in addition, the representation $\rho_{o,m}$, and nothing else. Thus, it is even more convenient to work in the free p-linear lattice D^r_p, the quotient of D^r modulo p-linear equivalence where p is either zero or a prime. Here, two elements, $a, a' \in D^r$ are said to be p-linear equivalent provided $\rho(a) = \rho(a')$ for any representation ρ in a vector space over a field of characteristic p.

The modified conjecture now asserts that any perfect element is p-linearly equivalent to an element in $B^- \cup B^+$. This indeed is true, as we want to show. Thus, assume there exists a perfect element $a \in D^r$ which is not p-linear equivalent to an element of $B^- \cup B^+$. Gelfand and Ponomarev have shown that then $N(a) = N_k(a)$ contains all preprojective representations and no preinjective representation. In a joint paper [10] with Dlab, we have shown that for $r \geqslant 5$, the set $N(a)$ either contains only the preprojective representations or else all but the preinjective representations. The elements $x \in D^r$ are given by lattice polynomials in the variables $e_1, \ldots, e_r$. Of course, there will be many different lattice polynomials which define the same element x. A lattice polynomial with minimal number of occurrences of variables defining x will be called a reduced expression of x and this number of variables in a reduced expression will be called the complexity $c(x)$ of x. Now, let $\rho\colon D^r \to L(V)$ be a representation, U a one-dimensional subspace of V, and

$\rho' \colon D^r \to L(V/U)$ the induced representation, with $\rho'(e_i) = (\rho(e_i) + U)/U$ for the generators e_i, $1 \leqslant i \leqslant r$. We claim that for $x \in D^r$, we have

$$\dim \rho'(x) \leqslant c(x) - 1 + \dim \rho(x).$$

[For the proof, we consider instead of ρ' the representation $\rho'' \colon D^r \to L(V)$ with $\rho''(x)$ the full inverse image of $\rho'(x)$ under the projection $V \to V/U$, thus $\dim \rho''(x) = 1 + \dim \rho'(x)$, for $x \in D^r$. Also note that $\rho(x) \subseteq \rho''(x)$ for all x. By induction on $c(x)$, we show the formula

$$\dim \rho''(x) - \dim \rho(x) \leqslant c(x).$$

Since $\dim U = 1$, this clearly is true for $x = e_i$, with $\rho''(e_i) = \rho(e_i) + U$. Now assume the formula being valid both for x_1 and x_2. For $x = x_1 + x_2$ with $c(x) = c(x_1) + c(x_2)$, we have

$$\begin{aligned} \dim \rho''(x) &= \dim \rho''(x_1 + x_2) \leqslant \dim \rho(x_1 + x_2) + c(x_1) + c(x_2) \\ &= \dim \rho(x) + c(x). \end{aligned}$$

Similarly, for $x = x_1 x_2$ with $c(x) = c(x_1) + c(x_2)$, we have

$$\begin{aligned} \dim \rho''(x) &= \dim \rho''(x_1 x_2) = \dim \rho''(x_1) + \dim \rho''(x_2) - \dim \rho''(x_1 + x_2) \\ &\leqslant \dim \rho(x_1) + c(x_1) + \dim \rho(x_2) + c(x_2) - \dim \rho(x_1 + x_2) \\ &= \dim \rho(x_1 x_2) + c(x_1) + c(x_2) = \dim \rho(x) + c(x). \end{aligned}$$

This finishes the proof.]

It is now sufficient to find a preprojective representation $\rho \colon D^r \to L(V)$ with $\dim V > c(a)$ and a one-dimensional subspace U of V such that the induced representation ρ' in V/U has no preprojective direct summand. Namely, our considerations above imply that $\dim \rho'(a) \leqslant c(a) - 1 < \dim V/U$, due to the fact that $\rho(a) = 0$, and therefore there exists at least one indecomposable representation σ in $N(a)$ which is not preprojective. As a consequence, in case $r \geqslant 5$, we know that $N(a)$ contains all but the preinjective representations. By duality, we similarly show that $N(a)$ contains only the preprojective representations, thus we obtain a contradiction. So, let us construct a suitable preprojective representation with the properties mentioned above. In fact, instead of considering representations of D^r, we will work inside the abelian category of representations of the oriented graph (*). We denote by $P_{t,l} = (V_{t,l}; \rho_{t,l}(e_1), \ldots, \rho_{t,l}(e_r))$ the graph representation corresponding to $\rho_{t,l}$. Take any homomorphism $\varphi \colon P_{0,1} \to P_{0,2}$ such that $R = \operatorname{Cok} \varphi$ is regular (that is, has no non-zero preprojective or preinjective direct summand. For example, there always exists such a φ with R being the direct sum of two indecomposable representations of dimension types $(1; 1, 1, 0, \ldots, 0)$ and $(r-3; 0, 0, 1, 1, \ldots, 1)$.) Now apply Φ^{-i} for $i \in \mathbf{N}$. We obtain exact sequences

$$0 \to P_{0,i+1} \xrightarrow{\Phi^{-i}(\varphi)} P_{0,i+2} \to \Phi^{-i} R \to 0,$$

thus, the inclusion

$$\varphi_i = \Phi^{-i}(\varphi) \circ \ldots \circ \Phi^{-1}(\varphi) \circ \varphi\colon P_{0,1} \to P_{0,i+2}$$

has regular cokernel (extensions of regular representations being regular, again). We now only have to choose i such that $\dim V_{0,i+2} > c(a)$. This finishes the proof in case $r \geqslant 5$. (For $r = 4$, we again take $\varphi\colon P_{0,1} \to P_{0,2}$ with $\operatorname{Cok} \varphi$ being the direct sum of two representations of dimension types $(1; 1, 1, 0, 0)$ and $(1, 0, 0, 1, 1)$, and form φ_i. The indecomposable summands of $\operatorname{Cok} \varphi_i$ all belong to one component C of the Auslander–Reiten quiver, thus we conclude as above that $C \subseteq N(a)$. By duality, one similarly shows that there are representations in C which do not belong to $N(a)$, so again we obtain a contradiction. Note that in case $r = 4$, the conjecture has been solved before by Herrmann [19].)

We consider now the general problem of representations of an oriented graph (Γ, Λ). We do not recall the definition of the category $L(\Gamma, \Lambda)$ of representations of (Γ, Λ) over some fixed field k, nor the typical examples, but just refer to the first two pages of [BGP]. We only note that $L(\Gamma, \Lambda)$ can also be considered as the category of modules over the path algebra $k(\Gamma, \Lambda)$, see [17], and $k(\Gamma, \Lambda)$ is a finite-dimensional k-algebra if and only if (Γ, Λ) does not have oriented cycles. In [15], Gabriel had shown that (Γ, Λ) has only finitely many indecomposable representations if and only if Γ is the disjoint union of graphs of the form A_n, D_n, E_6, E_7 and E_8 (they are depicted on the third page of [BGP]). It turned out that in case Γ is of the form A_n, D_n, E_6, E_7 or E_8, the indecomposable representations of (Γ, Λ), with Λ an arbitrary orientation, are in one-to-one correspondence to the positive roots of Γ. It is the aim of the paper [BGP] to give a direct proof of this fact. It introduces appropriate functors which produce all indecomposable representations from the simple ones in the same way as the canonical generators of the Weyl group produce all positive roots from the simple ones. We later will come back to these functors and their various generalizations.

Given a finite graph Γ, let E_Γ be the $\mathbb{Q}$-vector space of functions $\Gamma_0 \to \mathbb{Q}$, an element of E_Γ being written as a tuple $x = (x_\alpha)$ indexed by the elements $\alpha \in \Gamma_0$. For $\beta \in \Gamma_0$, we denote its characteristic function by $\bar{\beta}$ (thus $\bar{\beta}_\alpha = 0$ for $\alpha \neq \beta$, and $\bar{\beta}_\beta = 1$). Any representation V of (Γ, Λ) gives rise to an element $\dim V$ in E_Γ, its dimension type. For any orientation Λ of Γ, and any $\beta \in \Gamma_0$, there is a unique simple representation L_β of dimension type $\dim L_\beta = \bar{\beta}$. In case there are no oriented cycles in (Γ, Λ), we obtain in this way all simple representations of (Γ, Λ), thus, in this case, E_Γ may be identified with the rational Grothendieck group $G_0(\Gamma, \Lambda) \otimes_{\mathbb{Z}} \mathbb{Q}$ (here, $G_0(\Gamma, \Lambda)$ is the factor group of the free abelian group with basis the set of all representations of (Γ, Λ)

modulo all exact sequences) with dim being the canonical map (sending a representation to the corresponding residue class). On E_Γ, there is defined a quadratic form B. In fact, for any orientation Λ of Γ, we may consider the (non-symmetric) bilinear form B_Λ on E_Γ given by

$$B_\Lambda(x, y) = \sum_{l \in \Gamma_0} x_\alpha y_\alpha - \sum_{l \in \Gamma_1} x_{\alpha(l)} y_{\beta(l)}$$

and B is the corresponding quadratic form $B(x) = B_\Lambda(x, x)$. Note that B is positive definite if and only if Γ is the disjoint union of graphs of the form A_n, D_n, E_6, E_7 and E_8, and in these cases, the root system for Γ is by definition just the set of solutions of the equation $B(x) = 1$.

For k algebraically closed and B being positive definite we will outline a direct proof that dim: $L(\Gamma, \Lambda) \to E_\Gamma$ induces a bijection between the indecomposable representations of (Γ, Λ) and the positive roots. There is the following algebraic-geometric interpretation of B due to Tits [15]: The representations of (Γ, Λ) of dimension type x may be considered as the algebraic variety

$$m^x(\Gamma, \Lambda) = \prod_{l \in \Gamma_1} \operatorname{Hom}(k^{\alpha(l)}, k^{\beta(l)}),$$

and there is an obvious action on it by the algebraic group

$$G^x = \prod_{\alpha \in \Gamma_0} \operatorname{GL}(\alpha, k)/\Delta$$

with Δ being the multiplicative group of k diagonally embedded as group of scalars. Clearly

$$B(x) = \dim G^x + 1 - \dim m^x(\Gamma, \Lambda).$$

Using this interpretation, Gabriel has shown in [16] that it only remains to prove that the endomorphism ring of any indecomposable representation is k. So assume V is indecomposable, and that there are non-zero nilpotent endomorphisms. Then V contains a subrepresentation U with $\operatorname{End}(U) = k$ and $\operatorname{Ext}^1(U, U) \neq 0$. [Namely, let $0 \neq \varphi$ be an endomorphism with image S of smallest possible length, thus $\varphi^2 = 0$, and let $W = \bigoplus_{i=1}^{r} W_i$ be the kernel of φ, with all W_i indecomposable. Now $S \subseteq W$, thus the projection of S into some W_i must be non-zero. Since S was an image of a non-zero endomorphism of smallest length, we see that S embeds into this W_i. We may assume $i = 1$. Thus there is an inclusion $\iota: S \to W_1$. If W_1 has non-zero nilpotent endomorphisms, we use induction. Otherwise $\operatorname{End}(W_1) = k$. Also, $\operatorname{Ext}^1(W_1, W_1) \neq 0$, since on the one hand $\operatorname{Ext}^1(S, W_1) \neq 0$ due to the exact sequence

$$0 \to \bigoplus_{i=1}^{r} W_i \to V \to S \to 0$$

and, on the other hand, the inclusion ι gives rise to a surjection $\mathrm{Ext}^1(\iota, W)$. Here we use that $L(\Gamma, \Lambda)$ is a hereditary category]. The bilinear form B_Λ has the following homological interpretation [25]:

$$B_\Lambda(\dim V, \dim V') = \dim_k \mathrm{Hom}(V, V') - \dim_k \mathrm{Ext}^1(V, V'),$$

for all representations V, V'. Consequently, the existence of a representation U satisfying $\mathrm{End}(U) = k$, $\mathrm{Ext}^1(U, U) \neq 0$ would imply that

$$B(\dim U) = B_\Lambda(\dim U, \dim U) \leqslant 0,$$

contrary to the assumption that B is positive definite. This finishes the proof.

For any finite connected graph Γ without loops, Kac [21, 22] gave a purely combinatorial definition of its root system Λ. Note that Λ is a subset of E_Γ containing the canonical base vectors $\bar{\beta}$, for $\beta \in \Gamma_0$, and being stable under the Weyl group W, the group generated by the reflections σ_β along $\bar{\beta}$ with respect to B. The set Δ can also be interpreted in terms of root spaces of certain (usually infinite dimensional) Lie algebras [21]. Denote by Δ_+ the set of roots with only non-negative coordinates with respect to the canonical basis. Then Δ is the union of Δ_+ and $\Delta_- = -\Delta_+$. In case Γ is of type A_n, D_n, E_6, E_7 or E_8, the root system is finite and coincides with the set of solutions of $B(x) = 1$. Otherwise the root system is infinite and will contain besides certain solutions of $B(x) = 1$ also some solutions of $B(x) \leqslant 0$. The elements x of the root system which satisfy $B(x) = 1$ are called real roots, they are precisely the elements of the W-orbits of the canonical base elements. The remaining elements of the root system are called imaginary roots, and Kac has determined a fundamental domain for this set, the fundamental chamber.

Now, one has the following results (at least if k is either finite or algebraically closed): For any finite graph Γ without loops, and any orientation Λ, the set of dimension types of indecomposable modules is precisely the set Δ_+ of positive roots. For any positive real root x, there exists precisely one indecomposable representation V of (Γ, Λ) with $\dim V = x$. For any positive imaginary root x, the maximal dimension μ_x of an irreducible component in the set of isomorphism classes of indecomposable representations of dimension x is precisely $1 - B(x, x)$. (Note that the subset of indecomposable representations in $m^x(\Gamma, \Lambda)$ is constructible, and G^x-invariant, thus we can decompose it as a finite disjoint union of G^x-invariant subsets each of which admits a geometric quotient. By definition, μ_x is the maximum of the dimensions of these quotients.) In particular, we see that the number of indecomposable representations (or of the maximal dimension of families of indecomposable representations) of (Γ, Λ) does not depend on the orientation Λ. For Γ of the form A_n, D_n, E_6, E_7 or E_8, this is Gabriel's theorem (of course, there are no imaginary roots). For Γ of the form $\tilde{A}_n, \tilde{D}_n, \tilde{E}_6, \tilde{E}_7$, or $\tilde{E}_8$, the so called tame cases, these results have been shown by Donovan–Freislich [13] and Nazarova [23], see also [9]; in fact, in these cases one obtains a full classification of all indecomposable representations; also, it is possible in these cases to describe completely the rational invariants of the action of G^x on $m^x(\Gamma, \Lambda)$, for any dimension type x, see [27]. Of course, the oriented graphs of finite or tame representation type are rather special ones. It has been known since some time that the remaining (Γ, Λ) are wild: there always is a full exact subcategory of $L(\Gamma, \Lambda)$ which is equivalent to the category $\mathcal{M}_{k\langle X, Y\rangle}$ of $k\langle X, Y\rangle$-modules ($k\langle X, Y\rangle$ being the polynomial ring in two non-commuting indeterminates). In this situation, the results above are due to Kac [21, 22]. Note that this solves all the conjectures of Bernstein–Gelfand–Ponomarev formulated in [BGP]. However, there remain many open questions concerning wild graphs (Γ, Λ). One does not expect to obtain a complete classification of the indecomposable representations of such a graph, but one would like to have some more knowledge about certain classes of representations. For example, there does not yet exist a combinatorial description of the set of those roots which are dimension types of representations V with $\mathrm{End}(V) = k$.

We have mentioned above that the root system Δ of Γ is stable under the Weyl group W and that any W-orbit of Δ contains either one of the base vectors $\bar{\beta}$ (with $\beta \in \Gamma_0$) or an element of the fundamental chamber. One therefore tries to find operations which associate to an indecomposable representation V of (Γ, Λ) with Λ an orientation, and a Weyl group element $w \in W$ a new indecomposable representation of (Γ, Λ'), where Λ' is a possibly different orientation of Γ. By now, several such operations are known (see [BGP, 21, 28]), the first one being the reflection functors F_β^-, F_β^+ introduced by Bernstein, Gelfand and Ponomarev in [BGP]. Here, for the definition of F_β^+, the vertex β is supposed to be a sink, thus the simple representation L_β with dimension vector $\dim L_\beta = \bar{\beta}$ is projective. This concept has been generalized by Auslander, Platzeck and Reiten [1] dealing with any finite dimensional algebra A (or even an artin algebra) with a simple projective module L. For this, we need the Auslander–Reiten translates τ, τ^{-1}. Recall that τX_A is defined for any A-module X_A: let $P_1 \xrightarrow{p} P_0 \to X_A \to 0$ be a minimal projective resolution of X_A, then $\mathrm{Tr}\, X_A$ is by definition the cokernel of the map $\mathrm{Hom}(p, A_A)$ and $\tau X = D\, \mathrm{Tr}\, X$, $\tau^{-1} X = \mathrm{Tr}\, D\, X$, with D the usual duality with respect to the base field k. So assume L is a simple projective A-module, let P be the direct sum of one copy of each of the indecomposable projective modules different from L, and $B = \mathrm{End}(P \oplus \tau^{-1} L)$. The functor considered by Auslander, Platzeck and Reiten is $F = \mathrm{Hom}_A(P \oplus \tau^{-1} L, -)$ from the category $\mathcal{M}_A$ of A-modules to $\mathcal{M}_B$. The functor induces an equivalence of the full subcategory T of $\mathcal{M}_A$ of all modules which do not have L as a direct summand and a certain full subcategory of $\mathcal{M}_B$. Note that $P \oplus \tau^{-1} L$ is a tilting module in the sense of [18], except in the trivial case of L being, in addition, injective. (A tilting module T_A is defined by the following three properties:

(1) proj. dim. $T_A \leqslant 1$, (2) there exists an exact sequence $0 \to A_A \to T' \to T'' \to 0$, with T', T'' being direct sums of direct summands of T_A, and (3) $\mathrm{Ext}^1(T_A, T_A) = 0$. Now, if L_A is simple projective and not injective, the middle term Y of the Auslander–Reiten sequence

$$0 \to L \to Y \to \tau^{-1}L \to 0$$

starting with L is projective. This sequence shows, on the one hand, that proj. dim. $\tau^{-1}L = 1$. On the other hand, it also gives an exact sequence of the form needed in (2). Finally, $\mathrm{Ext}^1_A(P \oplus \tau^{-1}L, P \oplus \tau^{-1}L) \approx$ $\approx D\,\mathrm{Hom}(P \oplus \tau^{-1}L, L) = 0$, since any non-zero homomorphism from a module to L is a split epimorphism.)

A certain composition of the reflection functors F^+_β (or F^-_β, respectively) is of particular interest, the Coxeter functor Φ^+ (or Φ^-). An explicit calculation for the r-subspace situation is given in [F], in the special case of the 4-subspace problem it had been defined before in [Q]. The Coxeter functors are endofunctors of $L(\Gamma, \Lambda)$, they are only defined in case (Γ, Λ) does not have oriented cycles (non-oriented cycles are allowed, see [9]). Note that the assignment of an orientation Λ without oriented cycles is equivalent to the choice of a partial ordering of Γ_0 (let $\alpha \leqslant \beta$ iff there exists an oriented path $\alpha = \alpha_0 \leftarrow \alpha_1 \leftarrow \dots$ $\dots \leftarrow \alpha_m = \beta$), and also to the choice of a Coxeter transformation: this is a Weyl group element of the form $c = \sigma_{\alpha_n} \dots \sigma_{\alpha_1}$ with $\alpha_1, \dots, \alpha_n$ being the elements of Γ_0 in some fixed ordering (take an ordering of Γ which refines the given partial ordering). So assume from now on that (Γ, Λ) is a connected oriented graph without oriented cycles, and let c be the corresponding Coxeter element. The Coxeter functors Φ^+ and Φ^- defined in [BGP] have the following properties: if V is an indecomposable representation of (Γ, Λ), then either V is projective and then $\Phi^+(V) = 0$, or else V is not projective, and then $\Phi^+(V)$ again is indecomposable, $\Phi^-\Phi^+(V) \approx V$ and $\dim \Phi^+(V) = c \dim V$. Thus the Coxeter functor Φ^+ realizes the action of the Coxeter transformation on the set of all representations without non-zero projective direct summands. The usefulness of the Coxeter functors seems to have its origin in their relation to the Auslander–Reiten translation τ. Namely, Gabriel ([17], Prop. 5.3, see also [1,5]) has shown that τ can be identified with $C^+ \circ T$, where T is the functor which maps the representation (V, f) to $(V, -f)$. In particular, for Γ being a tree, we can identify τ with C^+ itself.

In order to explain the value of the Auslander–Reiten translation τ (and therefore of the Coxeter functors), we have to recall the definition of the Auslander–Reiten quiver of a finite dimensional algebra A. Its vertices are the isomorphism classes $[X]$ of the indecomposable A-modules X, and, if X, Y are indecomposable modules, then there is an arrow $[X] \to [Y]$ iff there exists an irreducible map $X \to Y$ (a map f is said to be irreducible provided it is neither a split monomorphism nor a split epimorphism, and for any factorization $f = f'' \circ f'$, we have that f' is a split monomorphism or f'' is a split epimorphism [2]). Now, the Auslander–Reiten quiver is a translation quiver with respect to τ: if X is indecomposable and not projective, then there exists an irreducible map $Y \to X$ iff there exists an irreducible map $\tau X \to Y$.

For the finite dimensional hereditary algebras A, the structure of the Auslander–Reiten quiver is known. We will recall this result in the special case of $A = k(\Gamma, \Lambda)$. First, we need some notation. Define $\mathbf{Z}(\Gamma, \Lambda)$ as follows: its vertices are the elements of $\Gamma_0 \times \mathbf{Z}$, and for any arrow ${}^i\circ \xleftarrow{\alpha} \circ^j$, there are arrows $(i, z) \xrightarrow{(\alpha, z)} (j, z)$ and $(j, z) \xrightarrow{(\alpha^*, z)} (i, z+1)$, for all $z \in \mathbf{Z}$, see [24] and also [17, 29]. Note that in case Γ is a tree, $\mathbf{Z}(\Gamma, \Lambda)$ does not depend on the orientation Λ and just may be denoted by $\mathbf{Z}\Gamma$. If $I \subseteq \mathbf{Z}$, let $I(\Gamma, \Lambda)$ be the full subgraph of all vertices (i, z) with $i \in I$. In particular, we will have to consider $\mathbf{N}(\Gamma, \Lambda)$ and $\mathbf{N}^-(\Gamma, \Lambda)$, where $\mathbf{N} = \{1, 2, 3, \ldots\}$ and $\mathbf{N}^- = \{-1, -2, -3, \ldots\}$. Also, denote by A_∞ the following infinite graph

$$\circ - \circ - \circ \ldots \circ - \circ \ldots$$

The result is as follows: in case Γ is of the form A_n, D_n, E_6, E_7 or E_8, the Auslander–Reiten quiver of $k(\Gamma, \Lambda)$ is a finite full connected subquiver of $\mathbf{Z}\Gamma$. (In case D_n with $n \equiv 0(2)$, the Auslander–Reiten quiver of $k(\Gamma, \Lambda)$ is $[1, n-1]$ (Γ, Λ), in case of E_7 or E_8, it is $[1,9]$ (Γ, Λ) or $[1,15]$ (Γ, Λ), respectively; in the remaining cases, it is slightly more difficult to describe, see [17, 29]). In all other cases, the Auslander–Reiten quiver of $k(\Gamma, \Lambda)$ has infinitely many components, all but two being quotients of $\mathbf{Z}A_\infty$ (see [26]), the remaining two being of the form $\mathbf{N}(\Gamma, \Lambda)$ and $\mathbf{N}^-(\Gamma, \Lambda)$. The component of the form $\mathbf{N}(\Gamma, \Lambda)$ contains the indecomposable projective modules: in fact, the indecomposable projective module P_i corresponding to the vertex $i \in \Gamma_0$ appears as indexed by $(i, 1)$, and the module indexed by (i, z), $z \in \mathbf{N}$, is just $\Phi^{-z+1}(P_i)$, this component is called the preprojective component. Similarly, the component of the form $\mathbf{N}^-(\Gamma, \Lambda)$ is called the preinjective component, it contains the indecomposable injective module J_i corresponding to $i \in \Gamma_0$ as indexed by $(i, -1)$, and the module indexed by $(i, -z)$, $z \in \mathbf{N}$, is just $\Phi^{+z-1}(J_i)$.

Let us consider in more detail a preprojective component $\mathscr{P}$, and the modules belonging to $\mathscr{P}$; they will be called preprojective modules. In case Γ is of type A_n, D_n, E_6, E_7, or E_8, we let $\mathscr{P}$ denote the full Auslander–Reiten quiver; in any case, we note that an indecomposable representation of (Γ, Λ) is said to be preprojective iff it is of the form $\Phi^{-z}P$, with P indecomposable projective and $z \geqslant 0$. (A general theory of preprojective modules has been developed by Auslander and Smalø, see [3]). For an indecomposable preprojective representation X, there are only finitely many indecomposable modules Y such that $\mathrm{Hom}(Y, X) \neq 0$, all of them are preprojective again, and any non-invertible homomorphism $Y \to X$ is a sum of compositions of irreducible maps. In particular, if X, Y are indecomposable and preprojective and

Hom(X, Y) $\neq 0$, then there is an oriented path $[X] \to \ldots \to [Y]$ in $\mathscr{P}$. In fact, the complete categorical structure of the full subcategory of preprojective modules can be read off from the combinatorial description of $\mathscr{P}$ as a translation quiver: the category of all preprojective modules is equivalent to the quotient category $\langle\rangle \mathscr{P}$ of the path category of $\mathscr{P}$ modulo the so called mesh relations (see [4, 24, 17]). Note that the category $\langle\rangle \mathscr{P}$ allows to reconstruct all the modules in $\mathscr{P}$. Namely, any module X_A is isomorphic to $\mathrm{Hom}({}_AA_A, X_A)$, thus, if $A_A = \bigoplus_{i \in \Gamma_0} P_i^{n_i}$, then X_A can be identified with $\bigoplus_{i \in \Gamma_0} \mathrm{Hom}(P_i, X)^{n_i}$, and $\mathrm{Hom}(P_i, X)$ can be calculated inside $\langle\rangle \mathscr{P}$, since both P_i, X are preprojective.

Starting from the preprojective component $\mathscr{P}$ of $k(\Gamma, \Lambda)$, one may define a (usually infinite-dimensional) algebra Π as follows: Take the direct sum of all homomorphism spaces $\mathrm{Hom}((j, 1), (t, l))$ in $\langle\rangle \mathscr{P}$ and define the product of two residue classes $\bar{w}$, $\bar{w}'$ of paths w: $(j, 1) \to \ldots \to (t, l)$ and w': $(j', 1) \to \ldots \to (t', l')$ as follows: in case $t = j'$, let $\bar{w}\,\bar{w}'$ be the residue class of the composed path $\tau^{-l+1}(w') \circ w$: $(j, 1) \to \ldots \to (t', l + l' - 1)$, and 0 otherwise. There is a purely combinatorial description of Π in terms of (Γ, Λ) due to Gelfand and Ponomarev, see [R]. Let $\hat{\Gamma}$ be obtained from (Γ, Λ) by adding to each arrow α: $i \to j$ an additional arrow α^*: $j \to i$. We clearly can identify Π with the factor algebra of the path algebra $k\hat{\Gamma}$ modulo the ideal generated by the element $\sum_{\alpha \in \Gamma_1} \alpha \cdot \alpha^* + \sum_{\alpha \in \Gamma_1} \alpha^* \cdot \alpha$. Note that this description is independent of the choice of the orientation Λ. Also, we see from both descriptions that Π contains as a subalgebra $k(\Gamma, \Lambda)$, thus we may consider Π as a right $k(\Gamma, \Lambda)$-module, and the first description now shows that the $k(\Gamma, \Lambda)$-module $\Pi_{k(\Gamma,\Lambda)}$ decomposes as the direct sum of all preprojective representations of (Γ, Λ) each occurring with multiplicity one, and therefore is called the preprojective algebra of Γ. (For the proper generalisation to the case of a species, we refer to [11]. We also should note the slight deviation of the preprojective algebra from the model algebra defined in [M], which reduces to the algebra A^r given in [I, II] in the case of the r-subspace situation. Namely, here the constant paths have square zero, whereas they are idempotents in Π. Now, in Π the sum of the constant paths is the identity element. In order also to have an identity element, Gelfand and Ponomarev add to the direct sum of all preprojective modules an additional one-dimensional space $k\epsilon$. There is a change of definition proposed in [S], using the constant paths as idempotents as in Π, but adding again an additional identity element.) Since Π is the direct sum of the preprojective representations of (Γ, Λ), it follows that Π is finite dimensional if and only if Γ is of the form A_n, D_n, E_6, E_7, or E_8. In [12], the tame cases $\tilde{A}_n$, $\tilde{D}_n$, $\tilde{E}_6$, $\tilde{E}_7$ and $\tilde{E}_8$ have been characterized by the fact that the Gelfand–Kirillov dimension of Π is 1, whereas it is ∞ for the wild cases.

Let us return to the special case of the r subspace graph (*), with $r \geq 4$. The description above gives that the preprojective component $\mathscr{P}$ is of the form

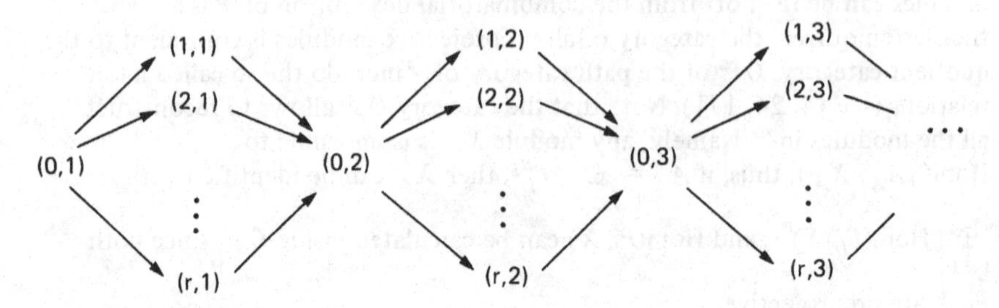

If we denote the arrows in the following way:

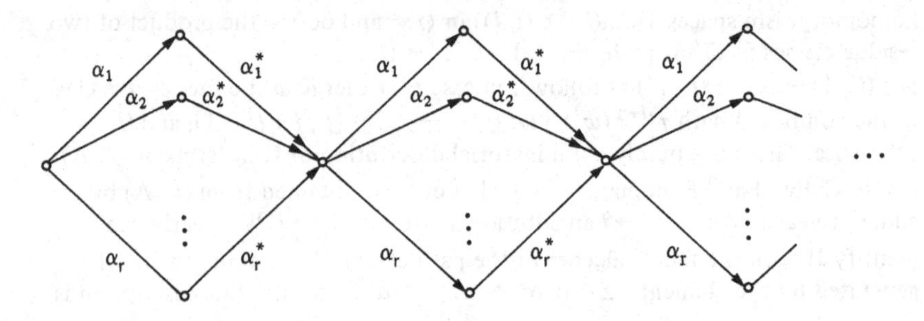

then the mesh relations are as follows: $\alpha_i\ \alpha_i^* = 0$ for all i, and $\sum_i \alpha_i^*\ \alpha_i = 0$.

Thus, if we want to determine the total space of the representation labelled (t, l), we have to calculate Hom$((0, 1), (t, l))$ inside the category $\langle\rangle\mathscr{P}$, and this amounts to the calculation of all possible paths from $(0, 1)$ to (t, l), taking this as the basis of a vector space and factoring out the mesh relations. However, taking from the beginning into account the relations $\alpha_i\ \alpha_i^* = 0$, we just as well may work with the vector space generated by the set $A_t(r, l)$ and factoring out the remaining mesh relations. This shows that we obtain as total space the vector space $V_{t,l}$. Similarly, the r different subspaces of the representation labelled (t, l) are given by the various Hom$((j, 1), (t, l))$, $1 \leq j \leq r$, again calculated in $\langle\rangle\mathscr{P}$, and therefore coincide with the subspaces $\rho_{t,l}(e_j)$. In this way, we obtain directly the description of the preprojective representations of D^r given by Gelfand and Ponomarev (and a direct proof of Proposition 8.2 in [F]).

Finally, let us note in which way the preprojective component of D^r determines the lattice B^+ of perfect elements belonging to the upper cubicles. For any perfect element a, we have denoted by $N(a)$ the set of indecomposable representation ρ satisfying $\rho(a) = 0$. We claim that for $a \in B^+$, the set $N(a)$ is a finite, predecessor closed subset of $\mathscr{P}$ (an element x is said to be a predecessor of y in case there is an oriented path $x \to \ldots \to y$. For the proof, we first note that clearly $N(a) \cap \mathscr{P}$ is predecessor closed, since for indecomposable representations ρ, ρ' with $\mathrm{Hom}(\rho, \rho') \neq 0$, and a perfect, $\rho' \in N(a)$ implies $\rho \in N(a)$. Since not all of $\mathscr{P}$ is contained in $N(a)$, it obviously follows that $N(a) \cap \mathscr{P}$ is finite. However, any complete slice of $\mathscr{P}$ generates all representations outside of $\mathscr{P}$, thus taking a complete slice of $\mathscr{P}$ outside of $N(a) \cap \mathscr{P}$, we easily see that no indecomposable representation outside of $\mathscr{P}$ can belong to $N(a)$, thus $N(a) \subseteq \mathscr{P}$.) Thus N determines a map from B^+ to the set of all finite, predecessor closed subsets of $\mathscr{P}$. This map is bijective and order-reversing, thus B^+ is anti-isomorphic to the lattice of finite, predecessor closed subsets of $\mathscr{P}$.

References

[Q] Gelfand, Ponomarev: Problems in linear algebra and classification of quadruples in a finite dimensional vector space. Coll. Math. Soc. Bolyai 5, Tihany (1970), 163–237.

[BGP] Bernstein, Gelfand, Ponomarev: Coxeter functors and Gabriel's theorem. Uspekhi Mat. Nauk **28** (1973), Russian Math. Surveys **28** (1973), 17–32, also in this volume.

[F] Gelfand, Ponomarev: Free modular lattices and their representations. Uspekhi Math. Nauk **29** (1974), 3–58. Russian Math. Surveys **29** (1974), 1–56, also in this volume.

[I] Gelfand, Ponomarev: Lattices, representations and algebras connected with them. I. Uspekhi Math. Nauk **31** (1976), 71–88. Russian Math. Surveys **31** (1976), 67–85, also in this volume.

[II] Gelfand, Ponomarev: Lattices, representations and algebras connected with them. II. Uspechi Math. Nauk **32** (1977), 85–106. Russian Math. Surveys **32** (1977), 91–114, also in this volume.

[M] Gelfand, Ponomarev: Model algebras and representations of graphs. Funkc. Anal. i Pril. **13**.3 (1979), 1–12. Funct. Anal. Appl. **13** (1979), 157–166.

[R] Rojter: Gelfand–Ponomarev algebra of a quiver. Abstract, 2nd ICRA (Ottawa 1979).

[S] Gelfand, Ponomarev: Representations of graphs. Perfect subrepresentations. Funkc. Anal. i Pril. **14**.3 (1980), 14–31. Funct. Anal. Appl. **14** (1980), 177–190.

[1] Auslander, Platzek, Reiten: Coxeter functors without diagrams. Trans. Amer. Math. Soc. **250** (1979), 1–46.

[2] Auslander, Reiten: Representation theory of artin algebras III, IV, V. Comm. Algebra **3** (1975), 239–294; **5** (1977), 443–518; **5** (1977), 519–554.

[3] Auslander, Smalø: Preprojective modules over artin algebras. J. Algebra (to appear).

[4] Bautista: Irreducible maps and the radical of a category. Preprint.

[5] Brenner, Butler: The equivalence of certain functors occurring in the representation theory of artin algebras and species. J. London Math. Soc. **14** (1976), 183–187.

[6] Day, Herrmann, Wille: On modular lattices with four generators. Algebra Universalis **3** (1972), 317–323.

[7] Dedeking: Über die von drei Moduln erzeugte Dualgruppe. Math. Ann. **53** (1900), 371–403.

[8] Dlab: Structure des treillis linéaires libres. Seminaire Dubreil. Springer LNM **795** (1980), 10–34.

[9] Dlab, Ringel: Indecomposable representations of graphs and algebras. Mem. Amer. Math. Soc. **173** (1976).

[10] Dlab, Ringel: Perfect elements in the free modular lattices. Math. Ann. **247** (1980), 95–100.

[11] Dlab, Ringel: The preprojective algebra of a modulated graph. Springer LNM **832** (1980), 216–131.

[12] Dlab, Ringel: Eigenvalues of Coxeter transformations and the Gelfand–Kirillov dimension of the preprojective algebra. Proc. Amer. Math. Soc. (to appear).

[13] Donovan, Freislich: The representation theory of finite graphs and associated algebras. Carleton Lecture Notes **5** (1973).

[14] Freese: Free modular lattices. Trans. AMS **261** (1980), 81–91.

[15] Gabriel: Unzerlegbare Darstellungen I. Manuscripta Math. **6** (1972), 71–103.

[16] Gabriel: Indecomposable representations II. Symposia Math. Inst. Naz. Alta Mat. **11** (1973), 81–104.

[17] Gabriel: Auslander–Reiten sequences and representation finite algebras. Springer LNM **831** (1980), 1–71.

[18] Happel, Ringel: Tilted algebras. Trans. Amer. Math. Soc. (to appear).

[19] Herrmann: Rahmen und erzeugende Quadrupeln in modularen Verbänden. To appear in Algebra Universalis.

[20] Hutchinson: Embedding and unsolvability theorems for modular lattices. Algebra Universalis **7** (1977), 47–84.

[21] Kac: Infinite root systems, representations of graphs and invariant theory. Inv. Math. **56** (1980), 57–92. part II: preprint.

[22] Kac: Some remarks on representations of quivers and infinite root systems. Springer LNM **832** (1980), 311–327.

[23] Nazarova: Representations of quivers of infinite type. Izv. Akad. Nauk. SSSR. Ser. Mat. **37** (1973), 752–791.

[24] Riedtmann: Algebren, Darstellungsköcher, Überlagerungen und Zurück. Comment. Math. Helv. **55** (1980), 199–224.

[25] Ringel: Representations of *K*-species and bimodules. J. Algebra **41** (1976), 269–302.

[26] Ringel: Finite dimensional algebras of wild representation type. Math. Z. **161** (1978), 235–255.

[27] Ringel: The rational invariants of tame quivers. Inv. Math. **58** (1980), 217–239.

[28] Ringel: Reflection functors for hereditary algebras. J. London Math. Soc. (2) **21** (1980), 465–479.

[29] Ringel: Tame algebras. Springer LNM **831** (1980), 137–287.

COXETER FUNCTORS AND GABRIEL'S THEOREM

I. N. Bernstein, I. M. Gel'fand, and V. A. Ponomarev

It has recently become clear that a whole range of problems of linear algebra can be formulated in a uniform way, and in this common formulation there arise general effective methods of investigating such problems. It is interesting that these methods turn out to be connected with such ideas as the Coxeter–Weyl group and the Dynkin diagrams.

We explain these connections by means of a very simple problem. We assume no preliminary knowledge. We do not touch on the connections between these questions and the theory of group representations or the theory of infinite-dimensional Lie algebras. For this see [3]–[5].

Let Γ be a finite connected graph; we denote the set of its vertices by Γ_0 and the set of its edges by Γ_1 (we do not exclude the cases where two vertices are joined by several edges or there are loops joining a vertex to itself). We fix a certain orientation Λ of the graph Γ; this means that for each edge $l \in \Gamma_1$ we distinguish a starting-point $\alpha(l) \in \Gamma_0$ and an end-point $\beta(l) \in \Gamma_0$.

With each vertex $\alpha \in \Gamma_0$ we associate a finite-dimensional linear space V_α over a fixed field K. Furthermore, with each edge $l \in \Gamma_1$ we associate a linear mapping f_l: $V_{\alpha(l)} \to V_{\beta(l)}$ ($\alpha(l)$ and $\beta(l)$ are the starting-point and end-point of the edge l). We impose no relations on the linear mappings f_l. We denote the collection of spaces V_α and mappings f_l by (V, f).

DEFINITION 1. Let (Γ, Λ) be an oriented graph. We define a *category* $\mathscr{L}(\Gamma, \Lambda)$ in the following way. An *object* of $\mathscr{L}(\Gamma, \Lambda)$ is any collection (V, f) of spaces V_α ($\alpha \in \Gamma_0$) and mappings f_l ($l \in \Gamma_1$). A morphism φ: $(V, f) \to (W, g)$ is a collection of linear mappings φ_α: $V_\alpha \to W_\alpha$ ($\alpha \in \Gamma_0$) such that for any edge $l \in \Gamma_1$ the following diagram

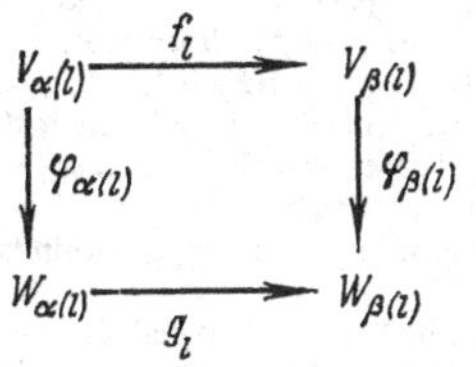

$$\begin{array}{ccc} V_{\alpha(l)} & \xrightarrow{f_l} & V_{\beta(l)} \\ \Big\downarrow \varphi_{\alpha(l)} & & \Big\downarrow \varphi_{\beta(l)} \\ W_{\alpha(l)} & \xrightarrow[g_l]{} & W_{\beta(l)} \end{array}$$

is commutative, that is, $\varphi_{\beta(l)} f_l = g_l \varphi_{\alpha(l)}$.

Many problems of linear algebra can be formulated in these terms. For example, the question of the canonical form of a linear transformation $f\colon V \to V$ is connected with the diagram

The classification of a pair of linear mappings $f_1\colon V_1 \to V_2$ and $f_2\colon V_1 \to V_2$ leads to the graph

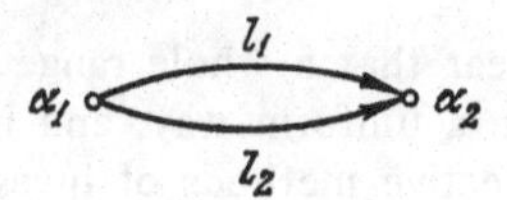

A very interesting problem is that of the classification of quadruples of subspaces in a linear space, which corresponds to the graph

This last problem contains several problems of linear algebra.[1]

Let (Γ, Λ) be an oriented graph. The direct sum of the objects (V, f) and (U, g) in $\mathscr{L}(\Gamma, \Lambda)$ is the object (W, h), where $W_\alpha = V_\alpha \oplus U_\alpha$, $h_l = f_l \oplus g_l$ $(\alpha \in \Gamma_0, l \in \Gamma_1)$.

We call a non-zero object $(V, f) \in \mathscr{L}(\Gamma, \Lambda)$ indecomposable if it cannot be represented as the direct sum of two non-zero objects. The simplest indecomposable objects are the irreducible objects L_α $(\alpha \in \Gamma_0)$, whose structure is as follows: $(L_\alpha)_\gamma = 0$ for $\gamma \neq \alpha$, $(L_\alpha)_\alpha = K$, $f_l = 0$ for all $l \in \Gamma_1$.

It is clear that each object (V, f) of $\mathscr{L}(\Gamma, \Lambda)$ is isomorphic to the direct sum of finitely many indecomposable objects.[2]

In many cases indecomposable objects can be classified.[3]

In his article [1] Gabriel raised and solved the following problem: to find all graphs (Γ, Λ) for which there exist only finitely many non-isomorphic indecomposable objects $(V, f) \in \mathscr{L}(\Gamma, \Lambda)$. He made the following

---

[1] Let us explain how the problem of the canonical form of a linear operator $f\colon V \to V$ reduces to that of a quadruple of subspaces. Consider the space $W = V \oplus V$ and in it the graph of f, that is, the subspace E_4 of pairs $(\xi, f\xi)$, where $\xi \in V$. The mapping f is described by a quadruple of subspaces in W, namely $E_1 = V \oplus 0$, $E_2 = 0 \oplus V$, $E_3 = \{(\xi, \xi) \mid \xi \in V\}$($E_3$ is the diagonal) and $E_4 = \{(\xi, f\xi) \mid \xi \in V\}$ – the graph of f. Two mappings f and f' are equivalent if and only if the quadruples corresponding to them are isomorphic. In fact, E_1 and E_2 define "coordinate planes" in W, E_3 establishes an identification between them, and then E_4 gives the mapping.

[2] It can be shown that such a decomposition is unique to within isomorphism (see [6], Chap. II, 14, the Krull–Schmidt theorem).

[3] We believe that a study of cases in which an explicit classification is impossible is by no means without interest. However, we should find it difficult to formulate precisely what is meant in this case by a "study" of objects to within isomorphism. Suggestions that are natural at first sight (to consider the subdivision of the space of objects into trajectories, to investigate versal families, to distinguish "stable" objects, and so on) are not, in our view, at all definitive.

surprising observation. For the existence of finitely many indecomposable objects in $\mathscr{L}(\Gamma, \Lambda)$ it is necessary and sufficient that Γ should be one of the following graphs:

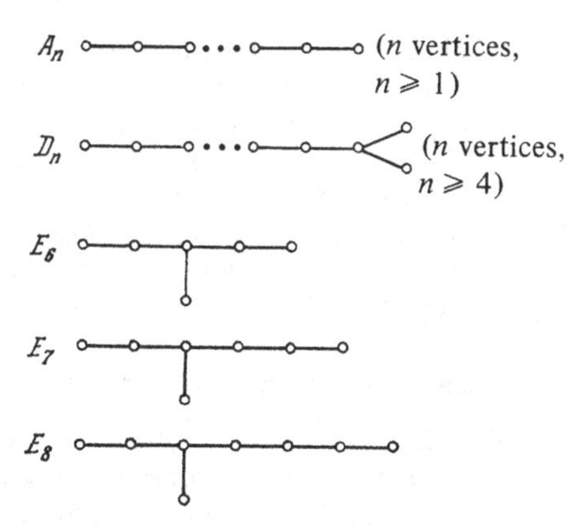

(this fact does not depend on the orientation Λ).The surprising fact here is that these graphs coincide exactly with the Dynkin diagrams for the simple Lie groups.[1]

However, this is not all. As Gabriel established, the indecomposable objects of $\mathscr{L}(\Gamma, \Lambda)$ correspond naturally to the positive roots, constructed according to the Dynkin diagram Γ.

In this paper we try to remove to some extent the "mystique" of this correspondence. Whereas in Gabriel's article the connection with the Dynkin diagrams and the roots is established a posteriori, we give a proof of Gabriel's theorem based on exploiting the technique of roots and the Weyl group. We do not assume the reader to be familiar with these ideas, and we give a complete account of the necessary facts.

An essential role is played in our proof by the functors defined below, which we call Coxeter functors (the name arises from the connection of these functors with the Coxeter transformations in the Weyl group). For the particular case of a quadruple of subspaces these functors were introduced in [2] (where they were denoted by Φ^+ and Φ^-). Essentially, our paper is a synthesis of Gabriel's idea on the connection between the categories $\mathscr{L}(\Gamma, \Lambda)$ with the Dynkin diagrams and the ideas of the first part of [2], where with the help of the functors Φ^+ and Φ^- the "simple" indecomposable objects are separated from the more "complicated" ones.

---

[1] More precisely, Dynkin diagrams with single arrows.

We hope that this technique is useful not only for the solution of Gabriel's problem and the classification of quadruples of subspaces, but also for the solution of many other problems (possibly, not only problems of linear algebra).

Some arguments on Gabriel's problem, similar to those used in this article, have recently been expressed by Roiter. We should also like to draw the reader's attention to the articles of Roiter, Nazarova, Kleiner, Drozd and others (see [3] and the literature cited there), in which very effective algorithms are developed for the solution of problems in linear algebra. In [3], Roiter and Nazarova consider the problem of classifying representations of ordered sets; their results are similar to those of Gabriel on the representations of graphs.

§ 1. Image functors and Coxeter functors

To study indecomposable objects in the category $\mathcal{L}(\Gamma, \Lambda)$ we consider "image functors", which construct for each object $V \in \mathcal{L}(\Gamma, \Lambda)$ some new object (in another category); here an indecomposable object goes either into an indecomposable object or into the zero object. We construct such a functor for each vertex α at which all the edges have the same direction (that is, they all go in or all go out). Furthermore, we construct the "Coxeter functors" Φ^+ and Φ^-, which take the category $\mathcal{L}(\Gamma, \Lambda)$ into itself.

For each vertex $\alpha \in \Gamma_0$ we denote by Γ^α the set of edges containing α. If Λ is some orientation of the graph Γ, we denote by $\sigma_\alpha \Lambda$ the orientation obtained from Λ by changing the directions of all edges $l \in \Gamma^\alpha$.

We say that a vertex α is (–)-accessible (with respect to the orientation Λ) if $\beta(l) \neq \alpha$ for all $l \in \Gamma_1$ (this means that all the edges containing α start there and that there are no loops in Γ with vertex at α). Similarly we say that the vertex β is (+)-accessible if $\alpha(l) \neq \beta$, for all $l \in \Gamma_1$.

DEFINITION 1.1 1) Suppose that the vertex β of the graph Γ is (+)-accessible with respect to the orientation Λ. From an object (V, f) in $\mathcal{L}(\Gamma, \Lambda)$ we construct a new object (W, g) in $\mathcal{L}(\Gamma, \sigma_\beta \Lambda)$.

Namely, we put $W_\gamma = V_\gamma$ for $\gamma \neq \beta$.

Next we consider all the edges $l_1, l_2, \ldots, l_k$ that end at β (that is, all edges of Γ^β). We denote by W_β the subspace in the direct sum $\bigoplus_{i=1}^{k} V_{\alpha(l_i)}$ consisting of the vectors $v = (v_1, \ldots, v_k)$ (here $v_i \in V_{\alpha(l_i)}$) for which $f_{l_1}(v_1) + \ldots + f_{l_k}(v_k) = 0$. In other words, if we denote by h the mapping $h: \bigoplus_{i=1}^{k} V_{\alpha(l_i)} \to V_\beta$ defined by the formula

$h(v_1, v_2, \ldots, v_k) = f_{l_1}(v_1) + \ldots + f_{l_k}(v_k)$, then $W_\beta = \operatorname{Ker} h$.

We now define the mappings g_l. For $l \notin \Gamma^\beta$ we put $g_l = f_l$. If $l = l_j \in \Gamma^\beta$, then g_l is defined as the composition of the natural embedding of W_β in $\oplus V_{\alpha(l_i)}$ and the projection of this sum onto the term $V_{\alpha(l_j)} = W_{\alpha(l_j)}$. We note that on all edges $l \in \Gamma^\beta$ the orientation has been changed, that is, the resulting object (W, g) belongs to $\mathcal{L}(\Gamma, \sigma_\beta\Lambda)$. We denote the object (W, g) so constructed by $F_\beta^+(V, f)$.

2) Suppose that the vertex $\alpha \in \Gamma_0$ is $(-)$-accessible with respect to the orientation Λ. From the object $(V, f) \in \mathcal{L}(\Gamma, \Lambda)$ we construct a new object $F_\alpha^-(V, f) = (W, g) \in \mathcal{L}(\Gamma, \sigma_\alpha\Lambda)$. Namely, we put

$$W_\gamma = V_\gamma \text{ for } \gamma \neq \alpha$$
$$g_l = f_l \text{ for } l \notin \Gamma^\alpha$$

$W_\alpha = \bigoplus_{i=1}^{k} V_{\beta(l_i)}/\operatorname{Im} \widetilde{h}$, where $\{l_1, \ldots, l_k\} = \Gamma^\alpha$, and the mapping

$\widetilde{h}: V_\alpha \to \bigoplus_{i=1}^{k} V_{\beta(l_i)}$ is defined by the formula $\widetilde{h}(v) = (f_{l_1}(v), \ldots, f_{l_k}(v))$.

If $l \in \Gamma^\alpha$, then the mapping $g_l: W_{\beta(l)} \to W_\alpha$ is defined as the composition of the natural embedding of $W_{\beta(l)} = V_{\beta(l)}$ in $\bigoplus_{i=1}^{k} V_{\beta(l_i)}$ and the projection of this direct sum onto W_α.

It is easy to verify that F_β^+ (and similarly F_α^-) is a functor from $\mathcal{L}(\Gamma, \Lambda)$ into $\mathcal{L}(\Gamma, \sigma_\beta\Lambda)$ (or $\mathcal{L}(\Gamma, \sigma_\alpha\Lambda)$, respectively). The following property of these functors is basic for us.

THEOREM 1.1 1) *Let (Γ, Λ) be an oriented graph and let $\beta \in \Gamma_0$ be a vertex that is $(+)$-accessible with respect to Λ. Let $V \in \mathcal{L}(\Gamma, \Lambda)$ be an indecomposable object. Then two cases are possible:*

a) *$V \approx L_\beta$ and $F_\beta^+V = 0$ (we recall that L_β is an irreducible object, defined by the condition $(L_\beta)_\gamma = 0$ for $\gamma \neq \beta$, $(L_\beta)_\beta = K$, $f_l = 0$ for all $l \in \Gamma_1$).*

b) *$F_\beta^+(V)$ is an indecomposable object, $F_\beta^-F_\beta^+(V) = V$, and the dimensions of the spaces $F_\beta^+(V)_\gamma$ can be calculated by the formula*

$$\begin{aligned} &\dim F_\beta^+(V)_\gamma = \dim V_\gamma \quad \textit{for } \gamma \neq \beta, \\ &\dim F_\beta^+(V)_\beta = -\dim V_\beta + \sum_{l \in \Gamma^\beta} \dim V_{\alpha(l)}. \end{aligned} \tag{1.1.1}$$

2) *If the vertex α is $(-)$-accessible with respect to Λ and if $V \in \mathcal{L}(\Gamma, \Lambda)$ is an indecomposable object, then two cases are possible:*

a) *$V \approx L_\alpha$, $F_\alpha^-(V) = 0$.*

b) $F_\alpha^-(V)$ *is an indecomposable object*, $F_\alpha^+F_\alpha^-(V) = V$,

$$\dim F_\alpha^-(V)_\gamma = \dim V_\gamma \text{ for } \gamma \neq \alpha, \tag{1.1.2}$$
$$\dim F_\alpha^-(V)_\alpha = -\dim V_\alpha + \sum_{l \in \Gamma^\alpha} \dim V_{\beta(l)}.$$

PROOF. If the vertex β is (+)-accessible with respect to Λ, then it is (−)-accessible with respect to $\sigma_\beta\Lambda$, and so the functor $F_\beta^-F_\beta^+$: $\mathcal{L}(\Gamma, \Lambda) \to \mathcal{L}(\Gamma, \Lambda)$ is defined. For each object $V \in \mathcal{L}(\Gamma, \Lambda)$ we construct a morphism i_V^β: $F_\beta^-F_\beta^+(V) \to V$ in the following way.

If $\gamma \neq \beta$, then $F_\beta^-F_\beta^+(V)_\gamma = V_\gamma$, and we put $(i_V^\beta)_\gamma$ = Id, the identity mapping.

For the definition of $(i_V^\beta)_\beta$ we note that in the sequence of mappings

$F_\beta^+(V)_\beta \xrightarrow{\tilde{h}} \bigoplus_{l \in \Gamma^\beta} V_{\alpha(l)} \xrightarrow{h} V_\beta$ (see definition 1.1) Ker h = Im $\tilde{h}$; we take for $(i_V^\beta)_\beta$ the natural mapping

$$F_\beta^-F_\beta^+(V)_\beta = \bigoplus_{l \in \Gamma^\beta} V_{\alpha(l)}/\mathrm{Im}\,\tilde{h} = \bigoplus_{l \in \Gamma^\beta} V_{\alpha(l)}/\mathrm{Ker}\,h \to V_\beta.$$

It is easy to verify that i_V^β is a morphism. Similarly, for each (−)-accessible vertex α we construct a morphism p_V^α: $V \to F_\alpha^+F_\alpha^-(V)$. Now we state the basic properties of the functors F_α^-, F_β^+ and the morphisms p_V^α, i_V^β.

LEMMA 1.1. 1) $F_\alpha^\pm(V_1 \oplus V_2) = F_\alpha^\pm(V_1) \oplus F_\alpha^\pm(V_2)$. 2) p_V^α *is an epimorphism and* i_V^β *is a monomorphism.* 3) *If* i_V^β *is an isomorphism, then the dimensions of the spaces* $F_\beta^+(V)_\gamma$ *can be calculated from* (1.1.1). *If* p_V^α *is an isomorphism, then the dimensions of the spaces* $F_\alpha^-(V)_\gamma$ *can be calculated from* (1.1.2). 4) *The object* Ker p_V^α *is concentrated at* α *(that is,* $(\mathrm{Ker}\, p_V^\alpha)_\gamma = 0$ *for* $\gamma \neq \alpha$*). The object* $V/\mathrm{Im}\ i_V^\beta$ *is concentrated at* β. 5) *If the object* V *has the form* F_α^+W *(*F_β^-W*, respectively), then* p_V^α *(*i_V^β*) is an isomorphism.* 6) *The object* V *is isomorphic to the direct sum of the objects* $F_\beta^-F_\beta^+(V)$ *and* $V/\mathrm{Im}\ i_V^\beta$ *(similarly,* $V \approx F_\alpha^+F_\alpha^-(V) \oplus \ker p_V^\alpha$*).*

PROOF. 1), 2), 3), 4) and 5) can be verified immediately. Let us prove 6).

We have to show that $V \approx F_\beta^-F_\beta^+(V) \oplus \tilde{V}$, where $\tilde{V} = V/\mathrm{Im}\ i_V^\beta$. The natural projection φ'_β: $V_\beta \to \tilde{V}_\beta$ has a section φ_β: $\tilde{V}_\beta \to V_\beta$ ($\varphi'_\beta \cdot \varphi_\beta$ = Id). If we put $\varphi_\gamma = 0$ for $\gamma \neq \beta$, we obtain a morphism φ: $\tilde{V} \to V$. It is clear that the morphisms φ: $\tilde{V} \to V$ and i_V^β: $F_\beta^-F_\beta^+(V) \to V$ give a decomposition of V into a direct sum. We can prove similarly that $V \approx F_\alpha^+F_\alpha^-(V) \oplus \mathrm{Ker}\ p_V^\alpha$.

We now prove Theorem 1.1. Let V be an indecomposable object of the category $\mathcal{L}(\Gamma, \Lambda)$, and β a (+)-accessible vertex with respect to Λ. Since

$V \approx F_\beta^- F_\beta^+ (V) \oplus V/\mathrm{Im}\, i_V^\beta$ and V is indecomposable, V coincides with one of the terms.

CASE I). $V = V/\mathrm{Im}\, i_V^\beta$. Then $V_\gamma = 0$ for $\gamma \neq \beta$ and, because V is indecomposable, $V \approx L_\beta$.

CASE II). $V = F_\beta^- F_\beta^+(V)$, that is, i_V^β is an isomorphism. Then (1.1.1) is satisfied by Lemma 1.1. We show that the object $W = F_\beta^+(V)$ is indecomposable. For suppose that $W = W_1 \oplus W_2$. Then $V = F_\beta^- (W_1) \oplus F_\beta^- (W_2)$ and so one of the terms (for example, $F_\beta^-(W_2)$) is 0. By 5) of Lemma 1.1, the morphism $p_V^\beta : W \to F_\beta^+ F_\beta^-(W)$ is an isomorphism, but $p_V^\beta (W_2) \subset F_\beta^+ F_\beta^-(W_2) = 0$, that is, $W_2 = 0$.

So we have shown that the object $F_\beta^+(V)$ is indecomposable. We can similarly prove 2) of Theorem 1.1.

We say that a sequence of vertices $\alpha_1, \alpha_2, \ldots, \alpha_k$ is (+)-accessible with respect to Λ if α_1 is (+)-accessible with respect to Λ, α_2 is (+)-accessible with respect to $\sigma_{\alpha_1}\Lambda$, α_3 is (+)-accessible with respect to $\sigma_{\alpha_2}\sigma_{\alpha_1}\Lambda$, and so on. We define a (−)-accessible sequence similarly.

COROLLARY 1.1. *Let* (Γ, Λ) *be an oriented graph and* $\alpha_1, \alpha_2, \ldots, \alpha_k$ *a (+)-accessible sequence.*

1) *For any* i $(1 \leqslant i \leqslant k)$, $F_{\alpha_1}^- \cdot \ldots \cdot F_{\alpha_{i-1}}^- (L_{\alpha_i})$ *is either* 0 *or an indecomposable object in* $\mathcal{L}(\Gamma, \Lambda)$ *(here* $L_{\alpha_i} \in \mathcal{L}(\Gamma, \sigma_{\alpha_{i-1}}\sigma_{\alpha_{i-2}} \ldots \sigma_{\alpha_1}\Lambda)$*).*[1]

2) *Let* $V \in \mathcal{L}(\Gamma, \Lambda)$ *be an indecomposable object, and*

$$F_{\alpha_k}^+ F_{\alpha_{k-1}}^+ \cdot \ldots \cdot F_{\alpha_1}^+ (V) = 0.$$

Then for some i

$$V \approx F_{\alpha_1}^- F_{\alpha_2}^- \cdot \ldots \cdot F_{\alpha_{i-1}}^- (L_{\alpha_i}).$$

We illustrate the application of the functors F_β^+ and F_α^- by the following theorem.

THEOREM 1.2. *Let* Γ *be a graph without cycles (in particular, without loops), and* Λ, Λ' *two orientations of it.*

1) *There exists a sequence of vertices* $\alpha_1, \ldots, \alpha_k$, *(+)-accessible with respect to* Λ, *such that* $\sigma_{\alpha_k}\sigma_{\alpha_{k-1}} \cdot \ldots \cdot \sigma_{\alpha_1}\Lambda = \Lambda'$.

2) *Let* $\mathcal{M}$, $\mathcal{M}'$ *be the sets of classes (to within isomorphism) of indecomposable objects in* $\mathcal{L}(\Gamma, \Lambda)$ *and* $\mathcal{L}(\Gamma, \Lambda')$, $\tilde{\mathcal{M}} \subset \mathcal{M}$ — *the set of classes of objects* $F_{\alpha_1}^- F_{\alpha_2}^- \cdot \ldots \cdot F_{\alpha_{i-1}}^- (L_{\alpha_i})$ $(1 \leqslant i \leqslant k)$, *and* $\tilde{\mathcal{M}}' \subset \mathcal{M}'$ *the set of classes of objects* $F_{\alpha_k}^+ \cdot \ldots \cdot F_{\alpha_{i+1}}^+ (L_{\alpha_i})$ $(1 \leqslant i \leqslant k)$. *Then the functor* $F_{\alpha_k}^+ \cdot \ldots \cdot F_{\alpha_1}^+$ *sets up a one-to-one correspondence between* $\mathcal{M} \setminus \tilde{\mathcal{M}}$ *and* $\mathcal{M}' \setminus \tilde{\mathcal{M}}'$.

---

[1] Where it cannot lead to misunderstanding, we denote by the same symbol L_α irreducible objects in all categories $\mathcal{L}(\Gamma, \Lambda)$, omitting the indication of the orientation Λ.

This theorem shows that, knowing the classification of indecomposable objects for Λ, we can easily carry it over to Λ'; in other words, problems that can be obtained from one another by reversing some of the arrows are equivalent in a certain sense.

Examples show that the same is true for graphs with cycles, but we are unable to prove it.

PROOF OF THEOREM 1.2. It is clear that 2) follows at once from 1) and Corollary 1.1. Let us prove 1).

It is sufficient to consider the case when the orientations Λ and Λ' differ in only one edge l. The graph $\Gamma \setminus l$ splits into two connected components. Let Γ' be the one that contains the vertex $\beta(l)$ ($\beta(l)$ is taken with the orientation of Λ). Let $\alpha_1, \ldots, \alpha_k$ be a numbering of the vertices of Γ' such that for any edge $l' \in \Gamma'_1$ the index of the vertex $\alpha(l')$ is greater than that of $\beta(l')$. (Such a numbering exists because Γ' is a graph without cycles.) It is easy to see that the sequence of vertices $\alpha_1, \ldots, \alpha_k$ is the one required (that is, it is (+)-accessible and $\sigma_{\alpha_k} \cdot \ldots \cdot \sigma_{\alpha_1} \Lambda = \Lambda'$). This proves Theorem 1.2.

It is often convenient to use a certain combination of functors $F_\alpha^\pm$ that takes the category $\mathcal{L}(\Gamma, \Lambda)$ into itself.

DEFINITION 1.2. Let (Γ, Λ) be an oriented graph without oriented cycles. We choose a numbering $\alpha_1, \ldots, \alpha_n$ of the vertices of Γ such that for any edge $l \in \Gamma_1$ the index of the vertex $\alpha(l)$ is greater than that of $\beta(l)$. We put $\Phi^+ = F_{\alpha_n}^+ \cdot \ldots \cdot F_{\alpha_2}^+ F_{\alpha_1}^+$, $\Phi^- = F_{\alpha_1}^- \cdot F_{\alpha_2}^- \cdot \ldots \cdot F_{\alpha_n}^-$. We call Φ^+ and Φ^- *Coxeter functors.*

LEMMA 1.2. 1) *The sequence* $\alpha_1, \ldots, \alpha_n$ *is (+)-accessible and* $\alpha_n, \ldots, \alpha_1$ *is (−)-accessible.* 2) *The functors* Φ^+ *and* Φ^- *take the category* $\mathcal{L}(\Gamma, \Lambda)$ *into itself.* 3) Φ^+ *and* Φ^- *do not depend on the freedom of choice in numbering the vertices.*

The proof of 1) and 2) is obvious. We prove 3) for Φ^+. We note firstly that if two different vertices $\gamma_1, \gamma_2 \in \Gamma_0$ are not joined by an edge and are (+)-accessible with respect to some orientation, then the functors $F_{\gamma_1}^+$ and $F_{\gamma_2}^+$ commute (that is, $F_{\gamma_2}^+ F_{\gamma_1}^+ = F_{\gamma_1}^+ F_{\gamma_2}^+$).

Let $\alpha_1, \ldots, \alpha_n$ and $\alpha'_1, \ldots \alpha'_n$ be two suitable numberings and let $\alpha_1 = \alpha'_m$. Then the vertices $\alpha'_1, \alpha'_2, \ldots, \alpha'_{m-1}$ are not joined to α_1 by an edge (if α_1 and α'_i $(i < m)$ are joined by an edge l, then $\alpha(l) = \alpha'_m = \alpha_1$ by virtue of the choice of the numbering of $\alpha'_1, \ldots, \alpha'_n$, but this contradicts the choice of the numbering of $\alpha_1, \ldots, \alpha_n$). Therefore $F_{\alpha'_m}^+ \cdot \ldots \cdot F_{\alpha'_1}^+ = F_{\alpha'_{m-1}}^+ \cdot \ldots \cdot F_{\alpha'_1}^+ F_{\alpha_1}^+$. Carrying out a similar argument with α_2, then with α_3, and so on, we prove that $F_{\alpha'_n}^+ \cdot \ldots \cdot F_{\alpha'_1}^+$ $F_{\alpha'_n}^+ \cdot \ldots \cdot F_{\alpha'_1}^+ = F_{\alpha_n}^+ \cdot \ldots \cdot F_{\alpha_1}^+$.

The proof is similar for the functor Φ^-.

Following [2] we can introduce the following definition.

DEFINITION 1.3. Let (Γ, Λ) be an oriented graph without oriented cycles. We say that an object $V \in \mathcal{L}(\Gamma, \Lambda)$ is (+)-(respectively, (−)-) irregular if $(\Phi^+)^k V = 0$ $((\Phi^-)^k V = 0)$ for some k. We say that an object V is regular if $V \approx (\Phi^-)^k (\Phi^+)^k V \approx (\Phi^+)^k (\Phi^-)^k V$ for all k.

NOTE 1. Using the morphisms p_V^α and i_V^β introduced in the proof of Theorem 1.1, we can construct a canonical epimorphism $p_V^k: V \to (\Phi^+)^k (\Phi^-)^k V$ and monomorphism $i_V^k: (\Phi^-)^k (\Phi^+)^k V \to V$. The object V is regular if and only if for all k these morphisms are isomorphisms.

NOTE 2. If an object V is annihilated by the functor $F_{\alpha_s}^+ \cdot \ldots \cdot F_{\alpha_1}^+$ ($\alpha_1, \ldots, \alpha_s$ is some (+)-accessible sequence), then this object is (+)-irregular. Moreover, the sequence $\alpha_1, \ldots, \alpha_s$ can be extended to $\alpha_1, \ldots, \alpha_s, \alpha_{s+1}, \ldots, \alpha_m$ so that $F_{\alpha_m}^+ \cdot \ldots \cdot F_{\alpha_{s+1}}^+ \cdot F_{\alpha_s}^+ \cdot \ldots \cdot F_{\alpha_1}^+ = (\Phi^+)^s$.

THEOREM 1.3. *Let (Γ, Λ) be an oriented graph without oriented cycles.* 1) *Each indecomposable object $V \in \mathcal{L}(\Gamma, \Lambda)$ is either regular or irregular.* 2) *Let $\alpha_1, \ldots, \alpha_n$ be a numbering of the vertices of Γ such that for any $l \in \Gamma_1$ the index of $\alpha(l)$ is greater than that of $\beta(l)$. Put* $V_i = F_{\alpha_1}^- F_{\alpha_2}^- \cdot \ldots \cdot F_{\alpha_{i-1}}^- (L_{\alpha_i}) \in \mathcal{L}(\Gamma, \Lambda)$, $\hat{V}_i = F_{\alpha_n}^+ \cdot \ldots \cdot F_{\alpha_{i+1}}^+ (L_{\alpha_i}) \in \mathcal{L}(\Gamma, \Lambda)$ *(here $1 \leqslant i \leqslant n$). Then $\Phi^+(V_i) = 0$ and any indecomposable object $V \in \mathcal{L}(\Gamma, \Lambda)$ for which $\Phi^+(V) = 0$ is isomorphic to one of the objects V_i. Similarly, $\Phi^-(\hat{V}_i) = 0$, and if V is indecomposable and $\Phi^-(V) = 0$, then $V \approx \hat{V}_i$ for some i.* 3) *Each (+)-(respectively, (−)-) irregular indecomposable object V has the form $(\Phi^-)^k V_i$ (respectively, $(\Phi^+)^k \hat{V}_i$) for some i, k.*

Theorem 1.3 follows immediately from Corollary 1.1.

With the help of this theorem it is possible, as was done in [2] for the classification of quadruples of subspaces, to distinguish "simple" (irregular) objects from more "complicated" (regular) objects; other methods are necessary for the investigation of regular objects.

§ 2. Graphs, Weyl groups and Coxeter transformations

In this section we define Weyl groups, roots, and Coxeter transformations, and we prove results that are needed subsequently. We mention two differences between our account and the conventional one.

a) We have only Dynkin diagrams with single arrows.

b) In the case of graphs with multiple edges we obtain a wider class of groups than, for example, in [7].

DEFINITION 2.1. Let Γ be a graph without loops.

(1) We denote by $\mathscr{E}_\Gamma$ the linear space over $\mathbf{Q}$ consisting of sequences $x = (x_\alpha)$ of rational numbers x_α $(\alpha \in \Gamma_0)$.

For each $\beta \in \Gamma_0$ we denote by $\bar{\beta}$ the vector in $\mathscr{E}_\Gamma$ such that $(\bar{\beta})_\alpha = 0$ for $\alpha \neq \beta$ and $(\bar{\beta})_\beta = 1$.

We call a vector $x = (x_\alpha)$ *integral* if $x_\alpha \in \mathbf{Z}$ for all $\alpha \in \Gamma_0$.

We call a vector $x = (x_\alpha)$ *positive* (written $x > 0$) if $x \neq 0$ and

$x_\alpha \geqslant 0$ for all $\alpha \in \Gamma_0$.

2) We denote by B the quadratic form on the space $\mathscr{E}_\Gamma$ defined by the formula $B(x) = \sum_{\alpha \in \Gamma_0} x_\alpha^2 - \sum_{l \in \Gamma_1} x_{\gamma_1(l)} \cdot x_{\gamma_2(l)}$, where $x = (x_\alpha)$, and $\gamma_1(l)$ and $\gamma_2(l)$ are the ends of the edge l. We denote by $\langle\,,\,\rangle$ the corresponding symmetric bilinear form.

3) For each $\beta \in \Gamma_0$ we denote by σ_β the linear transformation in $\mathscr{E}_\Gamma$ defined by the formula $(\sigma_\beta x)_\gamma = x_\gamma$ for $\gamma \neq \beta$, $(\sigma_\beta x)_\beta = -x_\beta + \sum_{l \in \Gamma^\beta} x_{\gamma(l)}$, where $\gamma(l)$ is the end-point of the edge l other than β.

We denote by W the semigroup of transformations of $\mathscr{E}_\Gamma$ generated by the σ_β $(\beta \in \Gamma_0)$.

LEMMA 2.1. 1) *If* $\alpha, \beta \in \Gamma_0$, $\alpha \neq \beta$, *then* $\langle \bar{\alpha}\, \bar{\alpha} \rangle = 1$ *and* $2\langle \bar{\alpha}\, \bar{\beta} \rangle$ *is the negative of the number of edges joining* α *and* β. 2) *Let* $\beta \in \Gamma_0$. *Then* $\sigma_\beta(x) = x - 2\langle \bar{\beta}, x \rangle \bar{\beta}$, $\sigma_\beta^2 = 1$. *In particular,* W *is a group.* 3) *The group* W *preserves the integral lattice in* $\mathscr{E}_\Gamma$ *and preserves the quadratic form* B. 4) *If the form* B *is positive definite* (*that is,* $B(x) > 0$ *for* $x \neq 0$), *then the group* W *is finite.*

PROOF. 1), 2) and 3) are verified immediately; 4) follows from 3).

For the proof of Gabriel's theorem the case where B is positive definite is interesting.

PROPOSITION 2.1. *The form* B *is positive definite for the graphs* A_n, D_n, E_6, E_7, E_8 *and only for them* (see [7], Chap. VI).

We give an outline of the proof of this proposition.

1. If Γ contains a subgraph of the form

(*)

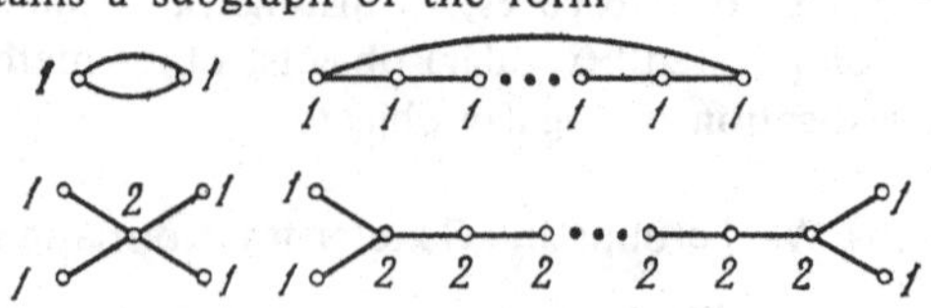

1 1   1 1 1 1 1 1

1 2 1 1 1   1 1 2 2 2 2 2 2 1

then the form B is not positive definite, because when we complete the numbers at the vertices in Fig. (*) by zeros, we obtain a vector $x \in \mathscr{E}_\Gamma$ for which $B(x) \leqslant 0$. Hence, if B is positive definite, then Γ has the form

(**)

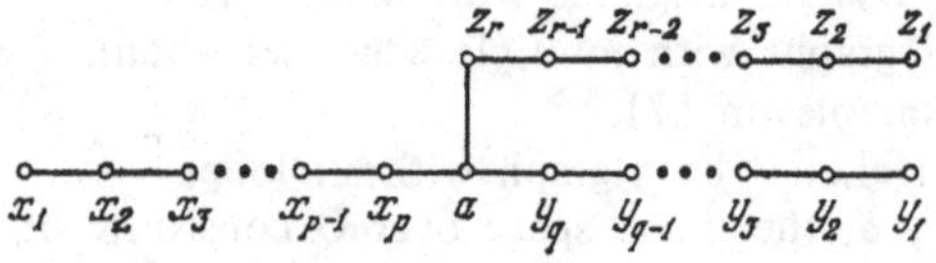

$z_r\ z_{r-1}\ z_{r-2}\ z_3\ z_2\ z_1$

$x_1\ x_2\ x_3\ x_{p-1}\ x_p\ a\ y_q\ y_{q-1}\ y_3\ y_2\ y_1$

where p, q, r are non-negative integers.

2 For each non-negative integer p we consider the quadratic form in $(p+1)$ variables $x_1, \ldots, x_{p+1}$

$$C_p(x_1, \ldots, x_{p+1}) = -x_1x_2 - x_2x_3 - \ldots - x_px_{p+1} + x_1^2 + \ldots + x_p^2 + \frac{p}{2(p+1)} x_{p+1}^2.$$

This form is non-negative definite, and the dimension of its null space is 1. Moreover, any vector $x \neq 0$ for which $C_p(x) = 0$ has all its coordinates non-zero.

To prove these facts it is sufficient to rewrite $C_p(x)$ in the form

$$C_p(x) = \sum_{i=1}^{p} \frac{i}{2(i+1)} \left(x_{i+1} - \frac{i+1}{i} x_i\right)^2.$$

3. We place the numbers $x_1, \ldots, x_p, y_1, \ldots, y_q, z_1, \ldots, z_r\ a$ at the vertices of Γ in accordance with Fig. (**). Then

$$B(x_i, y_i, z_i, a) = C_p(x_1, \ldots, x_p, a) + C_q(y_1, \ldots, y_q, a) + C_r(z_1, \ldots, z_r, a) + \left(1 - \frac{p}{2(p+1)} - \frac{q}{2(q+1)} - \frac{r}{2(r+1)}\right) a^2.$$

Hence it is clear that B is positive definite if and only if

$$\frac{p}{2(p+1)} + \frac{q}{2(q+1)} + \frac{r}{2(r+1)} < 1, \text{ that is, } \frac{1}{p+1} + \frac{1}{q+1} + \frac{1}{r+1} > 1.$$

4. We may suppose that $p \leqslant q \leqslant r$. We examine possible cases.

a) $p = 0$, q and r arbitrary. $A = \frac{1}{p+1} + \frac{1}{q+1} + \frac{1}{r+1} > 1$, that is, B is positive definite (series A_n).

b) $p = 1, q = 1, r$ arbitrary. $A > 1$ (series D_n),

c) $p = 1, q = 2, r = 2, 3, 4$. $A > 1$ (E_6, E_7, E_8),

d) $p = 1, q = 2, r \geqslant 5$. $A \leqslant 1$,

$p = 1, q = 3, r \geqslant 3$. $A \leqslant 1$,

$p \geqslant 2, q \geqslant 2, r \geqslant 2$. $A \leqslant 1$.

Thus B is positive definite for the graphs A_n, D_n, E_6, E_7, E_8 and only for them.

DEFINITION 2.2 A vector $x \in \mathscr{E}_\Gamma$ is called a *root* if for some $\beta \in \Gamma_0$, $w \in W$ we have $x = w\bar{\beta}$. The vectors $\bar{\beta}$ ($\beta \in \Gamma_0$) are called *simple roots*. A root x is called positive if $x > 0$ (see Definition 2.1).

LEMMA 2.2 1) *If x is a root, then x is an integral vector and $B(x) = 1$.* 2) *If x is a root, then $(-x)$ is a root.* 3) *If x is a root, then either $x > 0$ or $(-x) > 0$.*

PROOF. 1) follows from Lemma 2.1; 2) follows from the fact that $\sigma_\alpha(\bar{\alpha}) = -\bar{\alpha}$ for all $\alpha \in \Gamma_0$.

3) is needed only when B is positive definite and we prove it only in this case.

We can write the root x in the form $\sigma_{\alpha_1} \sigma_{\alpha_2} \cdot \ldots \cdot \sigma_{\alpha_k} \bar{\beta}$, where $\alpha_1, \ldots, \alpha_k, \beta \in \Gamma_0$. It is therefore sufficient to show that if $y > 0$ and $\alpha \in \Gamma_0$, then either $\sigma_\alpha y > 0$ or $y = \bar{\alpha}$ (and $-\sigma_\alpha y = +\bar{\alpha} > 0$).

Since $\|y\| = \|\bar{\alpha}\| = 1$, we have $|\langle\bar{\alpha}, y\rangle| \leqslant 1$. Moreover, $2\langle\bar{\alpha}, y\rangle \in \mathbf{Z}$. Hence $2\langle\bar{\alpha}, y\rangle$ takes one of the five values $2, 1, 0, -1, -2$.

a) $2\langle\bar{\alpha}, y\rangle = 2$. Then $\langle\bar{\alpha}, y\rangle = 1$, that is, $y = \bar{\alpha}$.

b) $2\langle\bar{\alpha}, y\rangle \leqslant 0$. Then $\sigma_\alpha(y) = y - 2\langle\bar{\alpha}, y\rangle\,\bar{\alpha} > 0$.

c) $2\langle\bar{\alpha}, y\rangle = 1$. Since $2\langle\bar{\alpha}, y\rangle = 2y_\alpha - \sum_{l \in \Gamma^\alpha} y_{\gamma(l)}$ ($\gamma(l)$ is the other end-point of the edge l), we have $y_\alpha > 0$, that is, $y_\alpha \geqslant 1$. Hence $\sigma_\alpha y = y - \bar{\alpha} > 0$.

This proves Lemma 2.2.

DEFINITION 2.3. Let Γ be a graph without loops, and let $\alpha_1, \ldots, \alpha_n$ be a numbering of its vertices. An element $c = \sigma_{\alpha_n} \cdot \ldots \cdot \sigma_{\alpha_1}$ (c depends on the choice of numbering) of the group W is called a *Coxeter transformation.*

LEMMA 2.3. *Suppose that the form B for the graph Γ is positive definite:*

1) *the transformation c in $\mathscr{E}_\Gamma$ has non non-zero invariant vectors;*

2) *if $x \in \mathscr{E}_\Gamma$, $x \neq 0$, then for some i the vector $c^i x$ is not positive.*

PROOF. 1) Suppose that $y \in \mathscr{E}_\Gamma$, $y \neq 0$ and $cy = y$. Since the transformations $\sigma_{\alpha_n}, \sigma_{\alpha_{n-1}}, \ldots, \sigma_{\alpha_2}$ do not change the coordinate corresponding to α_1 (that is, for any $z \in \mathscr{E}_\Gamma$ $(\sigma_{\alpha_i} z)_{\alpha_1} = z_{\alpha_1}$ for $i \neq 1$), we have $(\sigma_{\alpha_1} y)_{\alpha_1} = (cy)_{\alpha_1} = y_{\alpha_1}$. Hence $\sigma_{\alpha_1} y = y$ Similarly we can prove that $\sigma_{\alpha_2} y = y$, then $\sigma_{\alpha_3} y = y$, and so on.

For all $\alpha \in \Gamma_0$, $\sigma_\alpha y = y - 2\langle\bar{\alpha}, y\rangle\bar{\alpha} = y$, that is $\langle\bar{\alpha}, y\rangle = 0$. Since the vectors $\bar{\alpha}$ ($\alpha \in \Gamma_0$) form a basis of $\mathscr{E}_\Gamma$ and B is non-degenerate, $y = 0$.

2) Since W is a finite group, for some h we have $c^h = 1$. If all the vectors $x, cx, \ldots, c^{h-1}x$ are positive, then $y = x + cx + \ldots + c^{h-1}x$ is non-zero. Hence $cy = y$, which contradicts 1).

§ 3. Gabriel's theorem

Let (Γ, Λ) be an oriented graph. For each object $V \in \mathscr{L}(\Gamma, \Lambda)$ we regard the set of dimensions $\dim V_\alpha$ as a vector in $\mathscr{E}_\Gamma$ and denote it by $\dim V$.

THEOREM 3.1 (Gabriel [1]). 1) *If in $\mathscr{L}(\Gamma, \Lambda)$ there are only finitely many non-isomorphic indecomposable objects, then Γ coincides with one of the graphs A_n, D_n, E_6, E_7, E_8.*

2) *Let Γ be a graph of one of the types A_n, D_n, E_6, E_7, E_8, and Λ some orientation of it. Then in $\mathscr{L}(\Gamma, \Lambda)$ there are only finitely many non-isomprphic indecomposable objects. In addition, the mapping $V \mapsto \dim V$ sets up a one-to-one correspondence between classes of isomorphic indecomposable objects and positive roots in $\mathscr{E}_\Gamma$.*

We start with a proof due to Tits of the first part of the theorem.

TITS'S PROOF. Consider the objects $(V, f) \in \mathscr{L}(\Gamma, \Lambda)$ with a fixed dimension $\dim V = m = (m_\alpha)$.

If we fix a basis in each of the spaces V_α, then the object (V, f) is completely defined by the set of matrices A_l ($l \in \Gamma_1$), where A_l is the matrix of the mapping f_l: $V_{\alpha(l)} \to V_{\beta(l)}$. In each space V_α we change the basis by means of a non-singular $(m_\alpha \times m_\alpha)$ matrix g_α. Then the matrices A_l are replaced by the matrices

(*) $$A_l' = g_{\beta(l)}^{-1} A_l \, g_{\alpha(l)}.$$

Let A be the manifold of all sets of matrices A_l ($l \in \Gamma_1$) and G the group of all sets of non-singular matrices g_α ($\alpha \in \Gamma_0$). Then G acts on A according to (*); clearly, two objects of $\mathscr{L}(\Gamma, \Lambda)$ with given dimension m are isomorphic if and only if the sets of matrices $\{A_l\}$ corresponding to them lie in one orbit of G.

If in $\mathscr{L}(\Gamma, \Lambda)$ there are only finitely many indecomposable objects, then there are only finitely many non-isomorphic objects of dimension m. Therefore the manifold A splits into a finite number of orbits of G. It follows[1] that $\dim A \leqslant \dim G - 1$ (the -1 is explained by the fact that G has a 1-dimensional subgroup $G_0 = \{g(\lambda) | \lambda \in K^*\}$, $g(\lambda)_\alpha = \lambda \cdot 1_{V_\alpha}$, which acts on A identically). Clearly, $\dim G = \sum_{\alpha \in \Gamma_0} m_\alpha^2$, $\dim A = \sum_{l \in \Gamma_1} m_{\alpha(l)} m_{\beta(l)}$.

Therefore the condition $\dim A \leqslant \dim G - 1$ can be rewritten in the form[2] $B(m) > 0$ (if $m \neq 0$). In addition, it is easy to verify that $B((x_\alpha)) \geqslant B((|x_\alpha|))$ for all $x = (x_\alpha) \in \mathscr{E}_\Gamma$.

So we have shown that if in $\mathscr{L}(\Gamma, \Lambda)$ there are finitely many indecomposable objects, then the form B in $\mathscr{E}_\Gamma$ is positive definite.

As we have shown in Proposition 2.1, this holds only for the graphs A_n, D_n, E_6, E_7, E_8.

We now prove the second part of Gabriel's theorem.

LEMMA 3.1. *Suppose that* (Γ, Λ) *is an oriented graph,* $\beta \in \Gamma_0$ *a* (+)-*accessible vertex with respect to* Λ, *and* $V \in \mathscr{L}(\Gamma, \Lambda)$ *an indecomposable object. Then either* $F_\beta^+(V)$ *is an indecomposable object and* $\dim F_\beta^+(V) = \sigma_\beta(\dim V)$, *or* $V = L_\beta$, $F_\beta^+(V) = 0$, $\dim F_\beta^+(V) \neq \sigma_\beta(\dim V) < 0$. *A similar statement holds for a* (−)-*accessible vertex* α *and the functor* F_α^-.

This lemma is a reformulation of Theorem 1.1.

COROLLARY 3.1. *Suppose that the sequence of vertices* $\alpha_1, \ldots, \alpha_k$ *is* (+)-*accessible with respect to* Λ *and that* $V \in \mathscr{L}(\Gamma, \Lambda)$ *is an indecomposable object. Put* $V_j = F_{\alpha_j}^+ F_{\alpha_{j-1}}^+ \cdot \ldots \cdot F_{\alpha_1}^+ V$, $m_j = \sigma_{\alpha_j} \sigma_{\alpha_{j-1}} \cdot \ldots \cdot \sigma_{\alpha_1}(\dim V)$ $(0 \leqslant j \leqslant k)$. *Let i be the last index such that* $m_j > 0$ *for* $j \leqslant i$. *Then the* V_j *are indecomposable objects for* $j \leqslant i$, *and* $V = F_{\alpha_1}^- \cdot \ldots \cdot F_{\alpha_j}^- V_j$. *If* $i < k$, *then* $V_{i+1} = V_{i+2} = \ldots = V_k = 0$, $V_i = L_{\alpha_{i+1}}$, $V = F_{\alpha_1}^- \cdot \ldots \cdot F_{\alpha_i}^- (L_{\alpha_{i+1}})$. *Similar statements are true when* (+) *is replaced by* (−).

We now show that in the case of a graph Γ of type A_n, D_n, E_6, E_7 or E_8 (that is, B is positive definite), indecomposable objects correspond to positive roots.

a) Let $V \in \mathscr{L}(\Gamma, \Lambda)$ be an indecomposable object.

---

[1] This argument is suitable only for an infinite field K. If $K = \mathbf{F}_q$ is a finite field, we must use the fact that the number of non-isomorphic objects of dimension m increases no faster than a polynomial in m, and the number of orbits of G on the manifold A is not less than $C \cdot q^{\dim A - (\dim G - 1)}$.

[2] We can clearly restrict ourselves to graphs without loops.

We choose a numbering $\alpha_1, \alpha_2, \ldots, \alpha_n$ of the vertices of Γ such that for any edge $l \in \Gamma_1$ the vertex $\alpha(l)$ has an index greater than that of $\beta(l)$. Let $c = \sigma_{\alpha_n} \cdot \ldots \cdot \sigma_{\alpha_1}$ be the corresponding Coxeter transformation.

By Lemma 2.3, for some k the vector $c^k(\dim V) \in \mathscr{E}_\Gamma$ is not positive.

If we consider the (+)-accessible sequence $\beta_1, \beta_2, \ldots, \beta_{nk} = (\alpha_1, \ldots, \alpha_n, \alpha_1, \ldots, \alpha_n, \ldots, \alpha_1, \ldots, \alpha_n)$ (k times), then we have $\sigma_{\beta_{nk}} \cdot \ldots \cdot \sigma_{\beta_1}(\dim V) = c^k(\dim V) \not> 0$. From Corollary 3.1 it follows that there is an index $i < kn$ (depending only on $\dim V$) such that $V = F^-_{\beta_1} \cdot F^-_{\beta_2} \cdot \ldots \cdot F^-_{\beta_i}(L_{\beta_{i+1}})$, $\dim V = \sigma_{\beta_1} \cdot \ldots \cdot \sigma_{\beta_i}(\bar{\beta}_{i+1})$. It follows that $\dim V$ is a positive root and V is determined by the vector $\dim V$.

b) Let x be a positive root.

By Lemma 2.3, $c^k x \not> 0$ for some k. Consider the (+)-accessible sequence $\beta_1, \beta_2, \ldots, \beta_{nk} = (\alpha_1, \ldots, \alpha_n, \ldots, \alpha_1, \ldots, \alpha_n)$ (k times). Then $\sigma_{\beta_{nk}} \cdot \ldots \cdot \sigma_{\beta_1}(x) = c^k(x) \not> 0$. Let i be the last index for which $\sigma_{\beta_i}\sigma_{\beta_{i-1}} \cdot \ldots \cdot \sigma_{\beta_1}(x) > 0$. It is obvious from the proof of 3) in Lemma 2.2 that $\sigma_{\beta_i} \cdot \ldots \cdot \sigma_{\beta_1}(x) = \bar{\beta}_{i+1}$.

It follows that Corollary 3.1 that $V = F^-_{\beta_1} F^-_{\beta_2} \cdot \ldots \cdot F^-_{\beta_i}(L_{\beta_{i+1}}) \in \mathscr{L}(\Gamma, \Lambda)$ is an an indecomposable object and $\dim V = \sigma_{\beta_1} \cdot \ldots \cdot \sigma_{\beta_i}(\bar{\beta}_{i+1}) = x$.

This concludes the proof of Gabriel's theorem.

NOTE 1. When B is positive definite, the set of roots coincides with the set of integral vectors $x \in \mathscr{E}_\Gamma$ for which $B(x) = 1$ (this is easy to see from Lemma 2.3 and the proof of Lemma 2.2).

NOTE 2. It is interesting to consider categories $\mathscr{L}(\Gamma, \Lambda)$, for which the canonical form of an object of dimension m depends on fewer than $C \cdot |m|^2$ parameters (here $|m| = \Sigma\, |m_\alpha|$, $\alpha \in \Gamma_0$). From the proof it is obvious that for this it is necessary that B should be non-negative definite.

As in Proposition 2.1 we can show that B is non-negative definite for the graphs A_n, D_n, E_6, E_7, E_8 and $\hat{A}_0$, $\hat{A}_n$, $\hat{D}_n$, $\hat{E}_6$, $\hat{E}_7$, $\hat{E}_8$, where

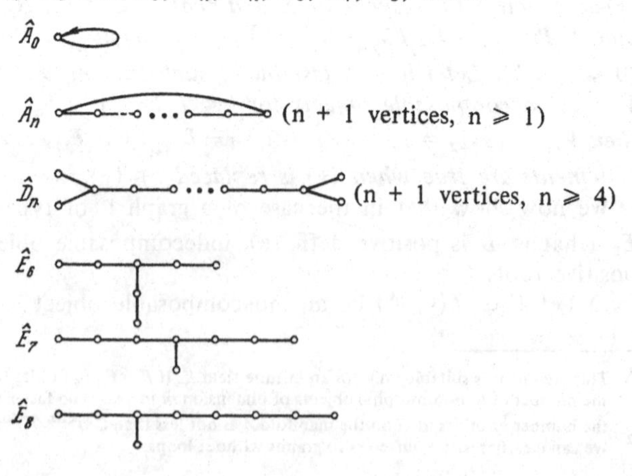

(the graphs $\hat{A}_n$, $\hat{D}_n$, $\hat{E}_6$, $\hat{E}_7$, $\hat{E}_8$ are extensions of the Dynkin diagrams (see [7])).

In a recent article Nazarova has given a classification of indecomposable objects for these graphs. In addition, she has shown there that such a classification for the remaining graphs would contain a classification of pairs of non-commuting operators (that is, in a certain sense it is impossible to give such a classification).

§ 4. Some open questions

Let Γ be a finite connected graph without loops and Λ an orientation of it.

CONJECTURES. 1) Suppose that $x \in \mathscr{E}_\Gamma$ is an integral vector, $x > 0$, $B(x) > 0$ and x is not a root. Then any object $V \in \mathscr{L}(\Gamma, \Lambda)$ for which $\dim V = x$ is decomposable.

2) If x is a positive root, then there is exactly one (to within isomorphism) indecomposable object $V \in \mathscr{L}(\Gamma, \Lambda)$, for which $\dim V = x$.

3) If V is an indecomposable object in $\mathscr{L}(\Gamma, \Lambda)$ and $B(\dim V) \leqslant 0$, then there are infinitely many non-isomorphic indecomposable objects $V' \in \mathscr{L}(\Gamma, \Lambda)$ with $\dim V' = \dim V$ (we suppose that K is an infinite field).

4) If Λ and Λ' are two orientations of Γ and $V \in \mathscr{L}(\Gamma, \Lambda')$ is an indecomposable object, then there is an indecomposable object $V' \in \mathscr{L}(\Gamma, \Lambda')$ such that $\dim V' = \dim V$.

We illustrate this conjecture by the example of the graph (Γ, Λ)

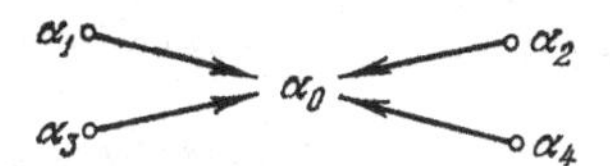

(quadruple of subspaces).

For each $x \in \mathscr{E}_\Gamma$ we put $\rho(x) = -2\langle\bar{\alpha}_0, x\rangle$ (if $x = (x_0, x_1, x_2, x_3, x_4)$, then $\rho(x) = x_1 + x_2 + x_3 + x_4 - 2x_0$).

In [2] all the indecomposable objects in the category $\mathscr{L}(\Gamma, \Lambda)$ are described. They are of the following types.

1. Irregular indecomposable objects (see the end of § 1). Such objects are in one-to-one correspondence with positive roots x for which $\rho(x) \neq 0$.

2. Regular indecomposable objects V for which $B(\dim V) \neq 0$. These objects are in one-to-one correspondence with positive roots x for which $\rho(x) = 0$.

3) Regular objects V for which $B(\dim V) = 0$. In this case $\dim V$ has the form $\dim V = (2n, n, n, n, n)$, $\rho(\dim V) = 0$. Indecomposable objects with fixed dimension $m = (2n, n, n, n, n)$ depend on one parameter. If $m \in \mathscr{E}_\Gamma$ is an integral vector such that $m > 0$ and $B(m) = 0$, then it has the form $m = (2n, n, n, n, n)$ $(n > 0)$ and there are indecomposable objects V for which $\dim V = m$.

If f is a linear transformation in n-dimensional space consisting of one

Jordan block then the quadruple of subspaces corresponding to it (see the Introduction) is a quadruple of the third type.

References

[1] P. Gabriel, Unzerlegbare Darstellungen I, Manuscripta Math. **6** (1972), 71–103.

[2] I. M. Gelfand and V. A. Ponomarev, Problems of linear algebra and classification of quadruples of subspaces in a finite-dimensional vector space, Colloquia Mathematica Societatis Ianos Bolyai, 5, Hilbert space operators, Tihany (Hungary), 1970, 163–237 (in English). (For a brief account, see Dokl. Akad. Nauk SSSR **197** (1971), 762–765 = Soviet Math. Doklady **12** (1971), 535–539.)

[3] L. A. Nazarova and A. V. Roiter, Representations of partially ordered sets, in the collection "Investigations in the theory of representations", Izdat. Nauka, Leningrad 1972, 5–31.

[4] I. M Gelfand, The cohomology of infinite-dimensional Lie algebras. Actes Congrès Internat. Math. Nice 1970, vol. 1. (1970), 95–111 (in English).

[5] I. M. Gelfand and V. A. Ponomarev, Indecomposable representations of the Lorentz group, Uspekhi Mat. Nauk **23**: 2 (1968), 3–60, MR **37** # 5325. = Russian Math. Surveys **23**: 2 (1968), 1–58.

[6] C. W. Curtis and I. Reiner, Representation theory of finite groups and associative algebras, Interscience, New York–London 1962, MR **26** # 2519. Translation: *Teoriya predstavlenii konechnykh grupp i assotsiativnykh algebr,* Izdat. Nauka, Moscow 1969.

[7] N. Bourbaki, Éléments de mathématique, XXVI, Groupes et algèbres de Lie, Hermann & Co., Paris 1960, MR **24** # A2641. Translation: *Gruppy i algebry Li,* Izdat. Mir, Moscow 1972.

Received by the Editors, 18 December 1972.

Translated by E. J. F. Primrose.

Dedicated to the memory of
Ivan Georgievich Petrovskii

FREE MODULAR LATTICES AND THEIR REPRESENTATIONS

I. M. Gel'fand and V. A. Ponomarev

Let L be a modular lattice, and V a finite-dimensional vector space over a field k. A representation of L in V is a morphism from L into the lattice $\mathscr{L}(V)$ of all subspaces of V. In this paper we study representations of finitely generated free modular lattices D^r.

An element a of a lattice L is called perfect if for every indecomposable representation $\rho: L \to \mathscr{L}(k^n)$ the subspace $\rho(a)$ of $V = k^n$ is such that $\rho(a) = V$ or $\rho(a) = 0$. We construct and study certain important sublattices of D^r, called "cubicles." All elements of the cubicles are perfect.

There are indecomposable representations connected with the cubicles. It will be shown that almost all these representations, except the elementary ones, have the important property of complete irreducibility; here a representation ρ of L is called completely irreducible if the sublattice $\rho(L) \subset \mathscr{L}(k^n)$ is isomorphic to the lattice $P(Q, n-1)$ of linear submanifolds of projective space over the field Q of rational numbers.

Contents

§1. Definitions and statement of results

1.1. Lattices. A *lattice* L is a set with two operations: intersection and sum. If $a, b \in L$, we denote their intersection[1] by ab and their sum by

[1] The intersection of elements a and b of L is often denoted $a \cap b$. We use the notation ab to avoid clumsy formulae. For $a + b$ the notation $a \cup b$ or $a \vee b$, is also used fairly frequently.

We denote the sum of the elements $a_1, \ldots, a_n$ by $\sum_{i=1}^{n} a_i$, and their intersection $a_1 \cap a_2 \cap \ldots \cap a_n$ by $\bigcap_{i=1}^{n} a_i$.

$a + b$. Each of these operations is commutative and associative. Moreover, for any $a, b \in L$ we have the identities

$$aa = a, \quad a + a = a, \quad a(a + b) = a, \quad a + ab = a.$$

An order relation is defined in a lattice L by

$$(a \subseteq b) \Leftrightarrow (ab = a).$$

If a, b, c are arbitrary elements of a lattice L, then it is easy to show that $a(b + c) \supseteq ab + ac$.

A lattice L is called *distributive* if for any $a, b, c \in L$

$$a(b + c) = ab + ac \text{ and } a + bc = (a + b)(a + c).$$

A lattice L is called *modular* (or *Dedekind*) if for any $a, b, c \in L$ such that $a \subseteq b$,

$$b(a + c) = a + bc.$$

This relation is called *Dedekind's axiom*.

EXAMPLE 1. Let V be a finite-dimensional vector space over a field k, and $V = k^n$. The set of all subspaces of V is a lattice in which EF is the intersection of the subspaces E and F, and $E + F$ is their sum, that is, $(E + F) = \{x + y \mid x \in E, y \in F\}$. We denote this lattice by $\mathcal{L}(V)$ or $\mathcal{L}(k^n)$. It is well-known that for $n > 1$ the lattice $\mathcal{L}(k^n)$ is modular, but not distributive.

Let $\mathbf{P}$ be the projective space generated by $V \cong k^n$. Then the lattice $\mathcal{L}(k^n)$ is isomorphic to the lattice of linear submanifolds of $\mathbf{P}$. We denote the latter lattice by $\mathbf{P}(k, n - 1)$ or $\mathbf{P}(k, V)$, and call it the *projective geometry over* k.

EXAMPLE 2. Let M be an arbitrary module over a commutative ring A. Then the set of all submodules of M is a modular lattice under the operations of intersection and sum.

1.2. Basic definitions. Let L be a modular lattice, and V a finite-dimensional vector space over a field k. A representation of L in V is a *morphism from L into the lattice $\mathcal{L}(V)$*. Thus, a representation $\rho: L \to \mathcal{L}(V)$ associates with each element $x \in L$ a subspace $\rho(x) \subseteq V$ such that for all $x, y \in L$

$$\rho(xy) = \rho(x)\rho(y) \quad \text{and} \quad \rho(x + y) = \rho(x) + \rho(y).$$

Let ρ_1 and ρ_2 be representations of a lattice L in spaces V_1 and V_2, respectively. We set $\rho(x) = \rho_1(x) \oplus \rho_2(x)$ for every $x \in X$, where $\rho_1(x) \oplus \rho_2(x)$ is the subspace of $V_1 \oplus V_2$ consisting of all pairs (ξ, η) such that $\xi \in \rho_1(x)$ and $\eta \in \rho_2(x)$. It is not hard to show that this defines a representation ρ in the space $V = V_1 \oplus V_2$. This representation is called the *direct sum of* ρ_1 and ρ_2 and is denoted by $\rho = \rho_1 \oplus \rho_2$.

A representation ρ is decomposable if it is isomorphic to the direct sum $\rho_1 \oplus \rho_2$ of two non-zero representations ρ_1 and ρ_2. It is easy to see that a representation ρ in a space V is decomposable if and only if there exist subspaces U_1 and U_2 such that $U_1U_2 = 0$ and $U_1 + U_2 = V$, and that $\rho(a) = U_1\rho(a) + U_2\rho(a)$ for every $a \in L$.

DEFINITION. An element a of a modular lattice L is called *perfect* if for every field k and every representation $\rho: L \to \mathcal{L}(V) = \mathcal{L}(k^n)$ the subspace $\rho(a) \subseteq V$ has the following property: there is a subspace U complementary to $\rho(a)$ (that is, $U\rho(a) = 0$ and $U + \rho(a) = V$) such that the subspaces U and $\rho(a)$ define a decomposition of ρ into the direct sum of subrepresentations, that is, $\rho(x) = U\rho(x) + \rho(a)\rho(x)$ for every $x \in L$.

It is easy to check that this definition is equivalent to the following: an element $a \in L$ is called *perfect* if, for every indecomposable representation $\rho: L \to \mathcal{L}(k^n)$ with $V = k^n$, either $\rho(a) = V$ or $\rho(a) = 0$.

Two elements a and b of a modular lattice L are called *linearly equivalent* if $\rho(a) = \rho(b)$ in every representation $\rho: L \to \mathcal{L}(k^n)$ for any k and n. In this case we write $a \cong b$. It can be shown that if L is the free modular lattice with 4 generators, then in L there are unequal, but linearly equivalent, elements.

Such examples are of interest to us in connection with the following problem. A modular lattice L is called *linear* if for any $x, y \in L$ and every representation $\rho: L \to \mathcal{L}(k^n)$ we have $x = y$ if and only if $\rho(x) = \rho(y)$. This leads to the following question: can a linear lattice be characterized by adding to the axioms for a modular lattice finitely many identities?

1.3. Cubicles in the lattice D^r. In this paper we study representations of the free modular lattice D^r with r generators $e_1, \ldots, e_r$.

The key idea in this paper is the construction of an important sublattice B of D^r whose elements are all perfect. For each integer $l \geqslant 1$ we construct sublattices $B^+(l)$ and $B^-(l)$, each consisting of 2^r perfect elements. We call $B^+(l)$ the *l-th upper cubicle* and $B^-(l)$ the *l-th lower cubicle*.

It is quite simple to define the upper cubicle $B^+(1)$. We set $h_t(1) = \sum_{j \neq t} e_j$. The sublattice of D^r generated by the elements $h_1(1), \ldots, h_r(1)$ is then the upper cubicle $B^+(1)$. We shall prove that 1) $B^+(1)$ is a Boolean algebra with 2^r elements (see §3), and 2) every element $x \in B^+(1)$ is perfect (see §4). The lattice $B^+(1)$ is, thus, isomorphic to the lattice of vertices of an r-dimensional cube with the natural ordering. The element $h_i(1)$ corresponds to the point $(1, \ldots, 1, 0, 1, \ldots, 1)$ with 0 in the i-th place.

Now we construct the lattice $B^+(l)$ with $l > 1$. The elements of the cubicle $B^+(l)$ are "constructed" from certain important polynomials[1] $e_{i_1 \cdots i_l}$, which are of independent interest. We proceed to define these polynomials.

Let $l \geqslant 1$ and $I = \{1, \ldots, r\}$. We denote by $A(r, l)$ the set whose elements are sequences of integers $\alpha = (i_1, \ldots, i_l)$ with $i_\lambda \in I$ such that

[1] The elements of D^r are also called lattice polynomials, or simply polynomials.

$i_1 \neq i_2, i_2 \neq i_3, \ldots, i_{l-1} \neq i_l$. In particular, $A(r, 1) = I$. For fixed $\alpha \in A(r, l)$ we construct a set $\Gamma(\alpha)$ consisting of elements $\beta \in A(r, l-1)$ in the following way:

$$\Gamma(\alpha) = \{\beta = (k_1, \ldots, k_{l-1}) \in A(r, l-1) \mid k_1 \notin \{i_1, i_2\}, k_2 \notin \{i_2, i_3\}, \ldots \\ \ldots, k_{l-1} \notin \{i_{l-1}, i_l\}\}.$$

Note that $k_1 \neq k_2, k_2 \neq k_3, \ldots, k_{l-2} \neq k_{l-1}$, because $\beta \in A(r, l-1)$. With each $\alpha \in A(r, l)$ we now associate an element $e_\alpha \in D^r$ by the following rule. Let $l = 1$ and $\alpha = (i_1)$. We set $e_\alpha = e_{i_1}$. For $l = 2$ and $\alpha = (i_1, i_2)$ with $i_1 \neq i_2$ we set

$$e_\alpha = e_{i_1 i_2} = e_{i_1} \sum_{\beta \in \Gamma(\alpha)} e_\beta = e_{i_1} \sum_{j \neq i_1, i_2} e_j.$$

In general, for arbitrary l and $\alpha \in A(r, l)$ we set by induction

$$e_\alpha = e_{i_1 \ldots i_l} = e_{i_1} \sum_{\beta \in \Gamma(\alpha)} e_\beta. \tag{1.1}$$

Now we introduce the elements $h_t(l)$, from which we construct the l-th cubicle, just as $B^+(1)$ was constructed from the elements $h_t(1)$. We denote by $A_t(r, l)$ the subset of $A(r, l)$ consisting of all $\alpha = (i_1, \ldots, i_{l-1}, t)$ whose last index is fixed and equal to t. We set

$$e_t(l) = \sum_{\alpha \in A_t(r, l)} e_\alpha, \tag{1.2}$$

$$h_t(l) = \sum_{j \neq t} e_j(l). \tag{1.3}$$

The sublattice of D^r generated by $h_1(l), \ldots, h_r(l)$ is called the *l-th upper cubicle* $B^+(l)$.

It is fairly elementary to prove (see §3) that for every $l \geqslant 1$ the sublattice $B^+(l)$ is a Boolean algebra. It is vastly more complicated to prove that the number of elements in $B^+(l)$ is equal to 2^r and that every element $x \in B^+(l)$ is perfect (§§4–7). We shall also prove (§3) that the elements of any cubicles $B^+(l)$ and $B^+(m)$ can be ordered in the following way: for every $x_l \in B^+(l)$ and every $y_m \in B^+(m)$, if $l < m$, then $x_l \supset x_m$. It follows from this that the collection of elements of all of the cubicles $B^+(l)$ is also a lattice, which we denote by B^+.

We denote by $B^-(l)$ the sublattice of D^r dual[1)] to $B^+(l)$. The sublattice $B^-(l)$ is called the *l-th lower cubicle*. Apparently, the following is true.

[1)] A lattice polynomial $g(x_1, \ldots, x_r)$ in the variables $x_1, \ldots, x_r$ is dual to a polynomial $f(x_1, \ldots, x_r)$ if g is obtained from f by changing the operation of intersection into addition and addition into intersection. For example, the polynomials $x_1(x_2 + x_3 + \ldots + x_r)$ and $x_1 + x_2 x_3 \ldots x_r$ are dual to each other. Let a_1 and a_2 be elements of D^r, and $a_i = f_i(e_1, \ldots, e_r)$ for $i = 1, 2$ be lattice polynomials. The elements a_1 and a_2 are dual if $f_2 = g_1(e_1, \ldots, e_r)$, where $g_1(e_1, \ldots, e_r)$ is the polynomial dual to $f_1(e_1, \ldots, e_r)$. We say that $B^-(l)$ is dual to $B^+(l)$ if $B^-(l)$ is the set of all elements dual to the elements of $B^+(l)$.

PROPOSITION. *Let $x^+ \in B^+(l)$ and $y^- \in B^-(m)$. Then $y^- \subset x^+$ for every l and m.*

We are able to prove this proposition only up to linear equivalence, that is, for every representation $\rho: D^r \to \mathcal{L}(k^n)$ and any $x^+ \in B^+(l)$ and $y^- \in B^-(m)$ we have $\rho(y^-) \subseteq \rho(x^+)$.

MAIN THEOREM. *The elements of the sublattices $B^+(l)$ and $B^-(l)$ are perfect.*

CONJECTURE. *Let a be a perfect element in D^r. Then there exists an $l \geqslant 1$ such that either $a \in B^+(l)$ or $a \in B^-(l)$.*

REMARK 1. The elements $e_\alpha = e_{i_1 \ldots i_l}$ used to construct the perfect elements in $B^+(l)$ are of considerable independent interest. Below, in §1.4, we construct completely irreducible representations $\rho_{t,l}$ $(t = 1, \ldots, r;\ l = 1, 2, \ldots)$. For these representations the images $\rho_{t,l}(e_\alpha)$ of the elements $e_\alpha \in D^r$ are one-dimensional subspaces of $V_{t,l}$. These $\rho_{t,l}$, with $t \in \{1, \ldots, r\}$, are called *completely irreducible representations of the first kind.*

REMARK 2. We shall also construct completely irreducible representations $\rho_{0,l}$. of the second kind. In these representations the elements e_α, $\alpha \in A_t(r, l)$, are replaced by elements $f_{i_1 \ldots i_l 0} \in D^r$, where $i_\nu \in I$ and $i_\nu \neq i_{\nu+1}$. The elements $f_{i_1 \ldots i_l 0}$ are "constructed" in the following manner. The set $\Gamma(\alpha) = \Gamma(i_1, \ldots, i_{l-1}, 0) \subseteq A(r, l-1)$ consists in this case of all elements $\beta = (k_1, \ldots, k_{l-1}) \in A(r, l-1)$ such that $k_1 \notin \{i_1, i_2\}$, $k_2 \notin \{i_2, i_3\}, \ldots, k_{l-2} \notin \{i_{l-2}, i_{l-1}\}$, $k_{l-1} \notin \{i_{l-1}\}$. We note that all k_λ in β are different from zero and therefore e_β is defined by (1.1). We set

$f_\alpha = f_{i_1 \ldots i_{l-1} 0} = e_{i_1} \sum_{\beta \in \Gamma(\alpha)} e_\beta$. For example $f_{i_1 0} = e_{i_1} \sum_{t \neq i_1} e_t$.

By analogy to the elements $e_t(l)$, $t \in I$, we set $f_0(l) = \sum_\alpha f_\alpha$, where $\alpha = (i_1, \ldots, i_{l-1}, 0)$, $i_\nu \in I$, and the summation is over all such α.

We shall prove later that $f_0(l)$ is linearly equivalent to the smallest element of $B^+(l-1)$, that is, $f_0(l) \simeq \bigcap_{i=1}^r h_i(l-1)$. Apparently, in D^r we have $f_0(l) = \bigcap_{i=1}^r h_i(l-1)$. for every $l \geqslant 2$. However, we can prove this only for $l = 2$.

1.4. Representations of the first and second kind. We denote by $\rho(L)$ the sublattice of $\mathcal{L}(k^n)$ consisting of all elements $\rho(a)$, $a \in L$.

DEFINITION. A representation ρ of a modular lattice L in a space V over a field k of characteristic zero is called *completely irreducible* if $\rho(L) \cong \mathbf{P}(\mathbf{Q}, m)$, where $\mathbf{Q}$ is the field of rational numbers and $m = (\dim_k V) - 1$.

We first construct the representations $\rho_{t,l}$ of the first kind in the spaces $V_{t,l}$, $t \in \{1, \ldots, r\}$, $l = 1, 2, \ldots$ Clearly, a representation ρ of D^r in a space V is completely determined by the subspaces

$\rho(e_j)$ $(j = 1, \ldots, r)$ of V. For brevity we set $\rho_{t,l}(e_j) = E_{j,t} \subseteq V_{t,l}$.

Let $l = 1$. Then $\rho_{t,1}$, $t \in \{1, \ldots, r\}$, is the representation in the one-dimensional space $V \cong k'$ such that $\rho_{t,1}(e_j) = 0$ if $j \neq t$ and $\rho_{t,1}(e_t) = V$.

Now let $l > 1$. We denote by $W_{t,l}$ the linear vector space over k with the basis $\{\eta_\alpha\}$, where $\alpha = (i_1, \ldots, i_{l-1}, t)$ ranges over the whole set $A_t(r, l)$ $(\alpha \in A_t(r, l) \Leftrightarrow \alpha = (i_1, \ldots, i_{l-1}, t)$ and t is fixed).

We denote by $Z_{t,l}$ the subspace of $W_{t,l}$ spanned by all possible vectors $g_{\alpha,k} = \sum_{i_k}' \eta_{i_1 \ldots i_k \ldots i_{l-1} t}$, where $1 \leqslant k \leqslant l-1$ and $\sum'$ is summation over those $\alpha = (i_1, \ldots, i_k, \ldots, i_{l-1}, t)$ in which $i_1, \ldots, i_{k-1}, i_{k+1}, \ldots, i_{l-1}$ are fixed. Next, we set $V_{t,l} = W_{t,l}/Z_{t,l}$. The images of the vectors η_α in the factor space $V_{t,l}$ are denoted by ξ_α. Thus, $V_{t,l}$ is the vector space over k spanned by the vectors ξ_α for which

$$\sum_{i_k}' \xi_{i_1 \ldots i_k \ldots i_{l-1} t} = 0 \text{ for every } k.$$

By $E_{j,t}$ we denote the subspace of $V_{t,l}$ spanned by all vectors ξ_α such that $\alpha = (i_1, i_2, \ldots, i_{l-1}, t) = (j, i_2, \ldots, i_{l-1}, t)$ (where the index $i_1 = j$ is fixed). We define a representation $\rho_{t,l}$ in $V_{t,l}$ by setting $\rho_{t,l}(e_j) = E_{j,t}$.

Now we define the representations $\rho_{0,l}$ of the second kind $(l = 1, 2, \ldots)$.

For $l = 1$, $\rho_{0,l}$ is the representation in the one-dimensional space $V \cong k'$ such that $\rho_{0,1}(e_i) = 0$ for all $i = 1, \ldots, r$.

For $l > 1$ we define a set $A_0(r, l)$ in the following way:

$$A_0(r, l) = \{\alpha = (i_1, \ldots, i_{l-1}, 0) \mid i_\lambda \in I = \{1, \ldots, r\},\ i_1 \neq i_2,\ i_2 \neq i_3, \ldots, i_{l-2} \neq i_{l-1}\}.$$

Clearly, $A_0(r, l) \cong A(r, l-1)$. The representation $\rho_{0,l}$ is constructed on $A_0(r, l)$ in the same way as $\rho_{t,l}$ is constructed on $A_t(r, l)$. Namely, we denote by $W_{0,l}$ the vector space over k with the basis $\{\eta_\alpha\}$, where $\alpha \in A_0(r, l)$. Further, let $Z_{0,l}$ be the subspace of $W_{0,l}$ spanned by all possible vectors $g_{\alpha,k} = \sum_{i_k} \eta_{i_1 \ldots i_k \ldots i_{l-1} 0}$. We denote by $V_{0,l}$ the factor space $W_{0,l}/Z_{0,l}$ and by ξ_α the image of η_α under the canonical map $W_{0,l} \to V_{0,l}$. The subspace of $V_{0,l}$ spanned by all vectors ξ_α with $\alpha = (i_1, i_2, \ldots, i_{l-1}, 0) = (j, i_2, \ldots, i_{l-1}, 0)$ is denoted by $E_{j,0}$. We define a representation $\rho_{0,l}$ in $V_{0,l}$ by setting $\rho_{0,l}(e_j) = E_{j,0}$.

REMARK. Our definition of the representations $\rho_{t,l}$ of D^r makes sense for all $r \geqslant 1$. It is known [2] that the lattices D^1, D^2, D^3 are finite, and each of these lattices has only finitely many non-isomorphic indecomposable representations. It can be shown that the number of such representations is 2, 4, and 9, respectively, and that they coincide with representations $\rho_{t,l}$, $l \leqslant r$. It can be shown by direct computation that for $r \in \{1, 2, 3\}$ and $l > r$ the space $V_{t,l}$ is equal to 0, hence, the corresponding representation $\rho_{t,l} = 0$.

The second main result of this paper is the following theorem.

THEOREM 1.1. (I) *For every* $t \in \{0, 1, \ldots, r\}$ *and all* $r, l \geqslant 1$ *the representation* $\rho_{t,l}\colon D^r \to \mathcal{L}(k, V_{t,l})$ *is indecomposable.*

(II) *If the characteristic of* k *is* 0, *if* $r > 3$, $l > 1$, *and if* $(t, r, l) \notin \{(1, 4, 2), (2, 4, 2), (3, 4, 2), (4, 4, 2)\}$, *then the representation* $\rho_{t,l}\colon D^r \to \mathcal{L}(k, V_{t,l})$ *is completely irreducible.*

Other properties of the representations $\rho_{t,l}$ are described by the following propositions.

PROPOSITION 1.1. *Let* $\rho_{t,l}$ *be a representation of the first kind* ($t \in \{1, \ldots, r\}$) *in the space* $V_{t,l}$. *We denote by* $v_{t,l}$ *the element* $\bigcap_{i \neq t} h_i(l)$ *of* $B^+(l)$.

(I) *For every element* $x \in B^+ = \bigcup_s B^+(s)$ *such that* $x \supseteq v_{t,l}$ *we have* $\rho_{t,l}(x) = V_{t,l}$. *In particular, this equality holds for all* $x \in B^+(m)$, $m < l$.

(II) *For every element* $y \in B^+$ *such that* $y \subseteq h_t(l)$ *we have* $\rho_{t,l}(y) = 0$. *In particular, this equality holds for all* $y \in B^+(n)$, $n > l$.

(III) *For every* $\alpha = (i_1, \ldots, i_{l-1}, t) \in A_t(r, l)$ *we have* $\rho_{t,l}(e_\alpha) = k(\xi_\alpha)$, *where* $k(\xi_\alpha)$ *is the one-dimensional subspace of* $V_{t,l}$ *spanned by the vector* ξ_α.

PROPOSITION 1.2. *Let* $\rho_{0,l}$ *be a representation of the second kind in the space* $V_{0,l}$. *We denote by* $v_{\theta,l-1}$ *the element* $\bigcap_i h_i(l-1)$ *of* $B^+(l-1)$.

(I) *For every element* $x \in B^+$ *such that* $x \supseteq v_{\theta,l-1}$ *we have* $\rho_{0,l}(x) = V_{0,l}$. *This means that* $\rho_{0,l}(x) = V_{0,l}$ *for* $x \in B^+(m)$, $m \leqslant l-1$.

(II) *For every element* $y \in B^+$ *such that* $y \subset v_{\theta,l-1}$ *we have* $\rho_{0,l}(y) = 0$. *This means that* $\rho_{0,l}(y) = 0$ *for every* $y \in B^+(n)$, $n \geqslant l$.

We shall now briefly describe the representations $\rho_{\bar{t},l}$ associated with the lower cubicles $B^-(l)$.

DEFINITION. Let ρ be a representation of a modular lattice L in a space V over a field k. We denote by V^* the space dual to V. A representation ρ^* in V^* is called *dual* to ρ if $\rho^*(x) = (\rho(x))^\perp$ for every $x \in L$, where $(\rho(x))^\perp$ is the subspace of functionals in V^* that vanish on $\rho(x)$.

We set $\rho_{\bar{t},l} = (\rho_{t,l})^*$. Thus, $\rho_{\bar{t},l}$ is the representation in $V^*_{t,l}$ such that $\rho_{\bar{t},l}(e_i) = (\rho_{t,l}(e_i))^\perp$ for all $i = 1, \ldots, r$.

We do not describe in detail the properties of the representations $\rho_{\bar{t},l}$, because they are dual to those of Theorem 1.1 and Propositions 1.1 and 1.2.

1.5. As we have already mentioned, the $\rho_{t,l}$ describe all the indecomposable representations of the lattices D^1, D^2, D^3. For the lattices D^r with $r \geqslant 4$ this is not the case. We describe below how to split off from an arbitrary representation ρ of D^r, $r \geqslant 4$, indecomposable representations $\rho_{t,l}$ and $\rho_{\bar{t},l}$. By $v^+_{\theta,l} = v_{\theta,l}$ we denote the smallest element of the cubicle $B^+(l)$. It follows easily from the definition of $B^+(l)$ that $v_{\theta,l} = \bigcap_{i=1}^r h_i(l)$, where $h_i(l)$ is defined by (1.3). It can be shown (see §3) that $v^+_{\theta,1} \supset v^+_{\theta,2} \supset \cdots \supset v^+_{\theta,l} \supset \cdots$. Dual to the element $v^+_{\theta,l} \in D^r$ is $v^-_{\bar{I},l}$, which is the largest element in $B^-(l)$. Then $v^-_{\bar{I},l} = \sum_{i=1}^{r} h^-_{\bar{i}}(l)$, where $h^-_i(l)$ is polynomial dual to $h_i(l)$. The elements $v^-_{\bar{I},l}$ also form a chain $v^-_{\bar{I},1} \subset v^-_{\bar{I},2} \subset \cdots \subset v^-_{\bar{I},l} \subset \cdots$.

Let ρ be any representation of D^r. We write $\rho(v^+_{\theta,l}) = V^+_{\theta,l}$ and $\rho(v^-_{\bar{I},l}) = V^-_{\bar{I},l}$. We define $V^+_{\theta,\infty} = \bigcap_{l=1}^{\infty} V^+_{\theta,l}$ and $V^-_{\bar{I},\infty} = \bigcup_{l=1}^{\infty} V^-_{\bar{I},l}$.

We shall prove (§6) that $V^-_{\bar{I},\infty} \subseteq V_{\theta,\infty}$. Thus, the subspaces $V^-_{\bar{I},l}$ and $V^+_{\theta,l}$ form a chain

$$V^-_{\bar{I},1} \subseteq V^-_{\bar{I},2} \subseteq \ldots \subseteq V^-_{\bar{I},\infty} \subseteq V^+_{\theta,\infty} \subseteq \ldots \subseteq V^+_{\theta,2} \subseteq V^+_{\theta,1}.$$

THEOREM 1.2. *Let ρ be a representation of D^r, $r \geqslant 4$, in a space V, and let $0 \subset V^-_{\bar{I},\infty} \subset V^+_{\theta,\infty} \subset V$, that is, all the terms of this chain are distinct. Then ρ is decomposable into a direct sum $\rho = \rho^- \oplus \rho_\lambda \oplus \rho^+$, where $\rho^- = \rho\,|_{V^-_{\bar{I},\infty}}$ is the restriction of ρ to the subspace $V^-_{\bar{I},\infty}$ and $(\rho^- \oplus \rho_\lambda) = \rho\,|_{V^+_{\theta,\infty}}$.*

If the representations ρ^- and ρ^+ are decomposable, then $\rho^+ \cong \bigoplus_{t,l} \rho_{t,l}$ and $\rho^- \cong \bigoplus_{t,l} \rho_{\bar{t},l}$, where $t \in \{0, 1, \ldots, r\}$ $(l = 1, 2, \ldots)$. If also the representation ρ_λ decomposes into a direct sum $\rho_\lambda = \bigoplus_{\lambda,s} \tau_{\lambda,s}$ of indecomposable representations $\tau_{\lambda,s}$, then among the $\tau_{\lambda,s}$ there are no representations isomorphic to $\rho_{t,l}$ and $\rho_{\bar{t},l}$.

We shall prove this theorem in §6. For the lattice D^4 the representations ρ_λ were studied in [5], where they were called regular. The indecomposable regular representations of D^4 are completely classified. For a specification of these indecomposable representations one needs not only discrete invariants (of the type of a dimension), but also continuous parameters (analogous to the eigenvalues of a linear transformation).

Very little is known about the regular representations of the lattices D^r $(r \geqslant 5)$. It is clear (see [5], [7]) only that the classification problem (up to similarity) of an arbitrary set of linear transformations $A_1, \ldots, A_n$ $(n \geqslant 2)$, $A_i : V \to V$, reduces to a special case of the classification problem of indecomposable regular representations of D^r $(r \geqslant 5)$.

§2. The category of representations

In this section, which is of an auxiliary role, we describe some elementary properties of the category of representations. Also, we introduce the notion of an admissible subspace, and we formulate a simple criterion for decomposability of representations.

2.1. The category $\mathcal{R}(L, k)$. Let ρ_1 and ρ_2 be representations of a modular lattice L in finite-dimensional spaces V_1 and V_2 over one and the same field

k. A morphism $\tilde{u}: \rho_1 \to \rho_2$ is a linear mapping $u: V_1 \to V_2$ such that $u\rho_1(x) \subseteq \rho_2(x)$ for every $x \in L$, where $u\rho_1(x)$ is the image of the subspace $\rho_1(x)$. When there is no ambiguity, we denote a morphism $\tilde{u}: \rho_1 \to \rho_2$ by $u: \rho_1 \to \rho_2$.

We denote by $\mathrm{Hom}(\rho_1, \rho_2)$ the set of all morphism from ρ_1 to ρ_2. Note that $\mathrm{Hom}(\rho_1, \rho_2)$ is a vector space over k.

It is not hard to verify that we now have a category $\mathcal{R}(L, k)$, that of finite-dimensional representations of L over k.

REMARK. Let $\tilde{u}: \rho_1 \to \rho_2$ be a morphism in $\mathcal{R}(L, k)$ and $u: V_1 \to V_2$ the corresponding linear transformation. It is not true that the set of all subspaces $u\rho_1(x), x \in L$, defines a representation of L in V_2. If $x, y \in L$, then $u\rho_1(x+y) = u(\rho_1(x) + \rho_1(y)) = u\rho_1(x) + u\rho_1(y)$. However, in general, $u\rho_1(xy) = u(\rho_1(x)\rho_1(y)) \neq (u\rho_1(x))(u\rho_1(y))$. We can only assert that $u\rho_1(xy) \subseteq (u\rho_1(x))(u\rho_1(y))$.

For any two objects ρ_1 and ρ_2 in $\mathcal{R}(L, k)$ there is the direct sum $\rho_1 \oplus \rho_2$. Namely, let $\rho_1, \rho_2 \in \mathcal{R}(L, k)$ be representations in spaces V_1 and V_2. We set $V = V_1 \oplus V_2$. For every $x \in L$ we define $\rho(x) = \rho_1(x) \oplus \rho_2(x) \subseteq V_1 \oplus V_2$. It can be shown that this defines a representation ρ in $V_1 \oplus V_2$.

It is not hard to check that $\mathcal{R}(L, k)$ is an additive category.

We now return to the category $\mathcal{R}(D^r, k)$ of representations of a free modular lattice D^r. For brevity, we denote $\mathcal{R}(D^r, k)$ by $\mathcal{R}$.

It is easy to show $\mathcal{R}$ is additive, and that every morphism $\tilde{u}: \rho_1 \to \rho_2$ has a kernel and a cokernel,[1)] that is, the category $\mathcal{R}$ is pre-Abelian. However, it is not Abelian, because the canonical mapping[2)] $\mathrm{Coim}\,\tilde{u} \to \mathrm{Im}\,\tilde{u}$ is not an isomorphism for an arbitrary morphism $\tilde{u}$.

EXAMPLE. Let $\mathcal{R} = \mathcal{R}(D^3, k)$. We define representations ρ_1 and ρ_2 in spaces V_1 and V_2 in the following way:

$$V_1 \cong V_2 \cong k^1,$$

$$\rho_1(e_1) = \rho_1(e_2) = V_1, \quad \rho_1(e_3) = 0, \quad \rho_2(e_1) = \rho_2(e_2) = \rho_2(e_3) = V_2.$$

Next let $u: V_1 \to V_2$ be any isomorphism (in the category of linear spaces), and let $\tilde{u}: \rho_1 \to \rho_2$ be the morphism corresponding to the mapping u. It is not hard to check that $\mathrm{Ker}\,\tilde{u} = 0$ and $\mathrm{Coker}\,\tilde{u} = 0$. Consequently, in the canonical decomposition $\rho_1 \to \mathrm{Coim}\,\tilde{u} \to \mathrm{Im}\,\tilde{u} \to \rho_2$ we have

---

1) We recall that a *kernel* of a morphism $\tilde{u}: \rho_1 \to \rho_2$ is a subobject $\tilde{\mu}: \rho' \to \rho_1$ of ρ_1 such that for every τ of R there is an exact sequence of vector spaces
$0 \to \mathrm{Hom}(\tau, \rho') \to \mathrm{Hom}(\tau, \rho_1) \to \mathrm{Hom}(\tau, \rho_2)$. In other words, $\tilde{\mu}$ is a monomorphism such that if $\tilde{u}\,\tilde{w} = 0$ with $\tilde{w} \in \mathrm{Hom}(\tau, \rho_1)$, then there exists a morphism $\tilde{w}' \in \mathrm{Hom}(\tau, \rho')$ such that $\tilde{w} = \tilde{\mu}\,\tilde{w}'$.
The cokernel of a morphism $\tilde{u}: \rho_1 \to \rho_2$ is a factor object $\tilde{\pi}: \rho_2 \to \rho''$ such that for every object τ of R there is an exact sequence of vector spaces $\mathrm{Hom}(\rho_1, \tau) \leftarrow \mathrm{Hom}(\rho_2, \tau) \leftarrow \mathrm{Hom}(\rho'', \tau) \leftarrow 0$.

2) We recall that $\mathrm{Im}\,\tilde{u}$ is the kernel of the cokernel of $\tilde{u}$, and $\mathrm{Coim}\,\tilde{u}$ the cokernel of the kernel of $\tilde{u}$.

$\rho_1 \cong \operatorname{Coim} \tilde{u}$ and $\rho_2 \cong \operatorname{Im} \tilde{u}$. Therefore, the mapping $\operatorname{Coim} \tilde{u} \to \operatorname{Im} \tilde{u}$ is not an isomorphism.

2.2. Decomposable representations and admissible subspaces.

DEFINITION. Let ρ be a representation of a modular lattice L in a linear space V. A subspace U of V is *admissible* relative to ρ if for any $x, y \in L$ one of the following conditions is satisfied:

(I) $$U(\rho(x) + \rho(y)) = U\rho(x) + U\rho(y);$$

(II) $$U + \rho(x)\rho(y) = (U + \rho(x))(U + \rho(y)).$$

The equivalence of (I) and (II) follows from a more general statement.

LEMMA 2.1. *Let L be a modular lattice and let $x, y, z \in L$. The sublattice generated by x, y, z is distributive if one of the following conditions is satisfied*: 1) $z(x + y) = zx + zy$ *and* 2) $z + xy = (z + x)(z + y)$.

A proof of this assertion can be found, for example, in Birkhoff [2]. Three elements x, y, z of a modular lattice L satisfying the conditions of Lemma 2.1 are called a distributive triple. It follows from Lemma 2.1 that the following equalities are equivalent:

$z(x + y) = zx + zy$, $x(y + z) = xy + xz$, $y(x + z) = yx + yz$, $z + xy = (z + x)(z + y)$, $x + yz = (x + y)(x + z)$, $y + xz = (y + x)(y + z)$.

PROPOSITION 2.1. *Let ρ be a representation of a lattice L in a space V. Let U be a subspace of V and let $U'' = V/U$ the factor space. Let $\theta : V \to U''$ be the canonical mapping. Then the following conditions are equivalent:*

1°. *The subspace U is admissible relative to ρ.*

2°. *The correspondence $x \to U\rho(x)$ defines a representation in U.*

3°. *The correspondence $x \mapsto \theta\rho(x)$ defines a representation in U''.*

PROOF. Let us show, say, that 2° follows from 1°. Let U be an admissible subspace. Then $U\rho(x + y) = U(\rho(x) + \rho(y)) = U\rho(x) + U\rho(y)$. Moreover, $U\rho(xy) = U\rho(x)\rho(y) = (U\rho(x))(U\rho(y))$. Consequently, the rule $x \mapsto U\rho(x)$ defines a representation in U. This representation is called *admissible*; it is also called the *restriction of* ρ to U and is denoted by ρ^U or $\rho|_U$. The proofs of the remaining parts of the proposition are elementary.

The representation in V/U, where U is an admissible subspace, is called an admissible factor representation.

We say that a representation $\rho \in \mathcal{R}(L, k)$ is decomposable if it is isomorphic to a direct sum $\bigoplus_{i=1}^{n} \rho_i$, $n \geqslant 2$, of representations $\rho_i \neq 0$.

PROPOSITION 2.2. *A representation $\rho \in \mathcal{R}(L, k)$ in V is decomposable if and only if there exist non-zero subspaces $U_1, \ldots, U_n$ such that $V = U_1 \oplus \cdots \oplus U_n$, and if for every $x \in L$*

(2.1) $$\rho(x) = \sum_{i=1}^{n} U_i\rho(x).$$

PROOF. The necessity of (2.1) is clear. To prove the sufficiency we show that every subspace U_j is admissible, that is
$U_j(\rho(x) + \rho(y)) = U_j\rho(x) + U_j\rho(y)$ for any $x, y \in L$.

By (2.1) we have $\rho(x) + \rho(y) = \sum_{i=1}^{n} U_i\rho(x) + \sum_{i=1}^{n} U_i\rho(y)$. Using Dedekind's axiom, we find

$$U_j(\rho(x) + \rho(y)) = U_j(\sum_{i=1}^{n}(U_i\rho(x) + U_i\rho(y))) =$$
$$= U_j(U_j\rho(x) + U_j\rho(y) + \sum_{i\neq j}(U_i\rho(x) + U_i\rho(y))) =$$
$$= U_j\rho(x) + U_j\rho(y) + U_j\sum_{i\neq j}(U_i\rho(x) + U_i\rho(y)).$$

Note that $U_j(\sum_{i\neq j} U_i\rho(x) + U_i\rho(y)) \subseteq U_j\sum_{i\neq j} U_i = 0$. Consequently, $U_j(\rho(x) + \rho(y)) = U_j\rho(x) + U_j\rho(y)$.

This proves that every subspace U_j is admissible and means that the correspondence $x \mapsto U_j\rho(x)$ defines a subrepresentation ρ^{U_j} in U_j. It is easy to check that $\rho = \bigoplus_{i=1}^{n} \rho^{U_i}$, so that ρ is decomposable.

We now assume that $V = U_1 \oplus \cdots \oplus U_n$, and that each of the subspaces U_i is admissible relative to ρ. The following example shows that we cannot, in general, assert that ρ is equal to the sum of its restrictions ρ^{U_i}.

EXAMPLE. Let D^2 be the free modular lattice with two generators. (Note that D^2 consists of four elements $e_1, e_2, e_1e_2, e_1 + e_2$.) Let V denote the 8-dimensional vector space over a field k with the basis $\xi_1, \ldots, \xi_8$. We define a representation ρ of D^2 in V in the following way. We set

$$\rho(e_1) = k\xi_1 + k\xi_5 + k(\xi_2 + \xi_6), \quad \rho(e_2) = k\xi_3 + k\xi_7 + k(\xi_4 + \xi_8),$$

where $k\xi_i$ and $k(\xi_j + \xi_l)$ are the one-dimensional subspaces spanned by the vectors ξ_i and $\xi_j + \xi_l$, respectively. Let $U_1 = \sum_{i=1}^{4} k\xi_i$ and $U_2 = \sum_{i=5}^{8} k\xi_i$. It is easy to check that each of the subspaces U_i is admissible relative to ρ, but that $\rho \neq \rho^{U_1} + \rho^{U_2}$.

In conclusion of this section we state a simple criterion for the decomposability of a representation ρ of D^r.

PROPOSITION 2.3. *Let ρ be a representation of D^r in a space V. Then ρ is decomposable into a direct sum $\rho = \bigoplus_{i=1}^{n} \rho_i$ of representations ρ_i if and only if there exist non-zero subspaces $U_1, \ldots, U_n$ with $V = \bigoplus_{i=1}^{n} U_i$ such that $\rho(e_t) = \sum_{i=1}^{n} \rho(e_t)U_i$ for every $t \in \{1, \ldots, r\}$, where the e_t are generators of D^r.*

§3. Construction and elementary properties of the cubicles $B^+(l)$ and $B^-(l)$

3.1. Definition of the cubicles $B^+(l)$ and $B^-(l)$. The sublattices $B^+(l)$ and $B^-(l)$ were defined in §1. To make the text independent, we repeat here the definition and introduce some notation.

The sublattice of D^r generated by the elements $h_t(1) = \sum_{i \neq t} e_i$ is, by definition, the upper cubicle $B^+(1)$.

We define the sublattices $B^+(l)$ for $l \geqslant 2$. First we construct important polynomials $e_{i_1 \ldots i_l}$. Let $I = \{1, \ldots, r\}$. We set $A(r, 1) = I$, and let $A(r, l)$ be the subset of all sequences $(i_1, \ldots, i_l)$, $i_\lambda \in I$, with $i_1 \neq i_2, i_2 \neq i_3, \ldots$, that is, $A(r, l) = \{\alpha = (i_1, \ldots, i_l) | i \in I$ and $\forall \lambda i_\lambda \neq i_{\lambda+1}\}$. From an element $\alpha = (i_1, \ldots, i_l)$, $l \geqslant 2$, we construct a set $\Gamma(\alpha) \subset A(r, l-1)$ in the following way:

$$\Gamma(\alpha) = \{\beta = (k_1, \ldots, k_{l-1}) | \beta \in A(r, l-1);\ k_1 \notin \{i_1, i_2\},\ k_2 \notin \{i_2, i_3\}, \ldots \\ \ldots, k_{l-1} \notin \{i_{l-1}, i_l\}\}.$$

Note that from $\beta = (k_1, \ldots, k_{l-1}) \in A(r, l-1)$ it follows that $k_1 \neq k_2, k_2 \neq k_3, \ldots, k_{l-2} \neq k_{l-1}$. The polynomials $e_\alpha = e_{i_1 \ldots i_l}$ are defined by induction on l:

if $\alpha = (i_1) \in A(r, 1)$, then $e_\alpha = e_{(i_1)} = e_{i_1}$;

if $\alpha = (i_1, i_2) \in A(r, 2)$, then $e_\alpha = e_{i_1 i_2} = e_{i_1} \sum_{\beta \in \Gamma(\alpha)} e_\beta$;

if $\alpha = (i_1, \ldots, i_l) \in A(r, l)$, then $e_\alpha = e_{i_1 \ldots i_l} = e_{i_1} \sum_{\beta \in \Gamma(\alpha)} e_\beta$.

EXAMPLE. Let $\alpha = (i_1, i_2)$; then $\Gamma(\alpha) = \{\beta = (k_1) \mid k_1 \notin \{i_1, i_2\}\}$, that is,

$$e_\alpha = e_{i_1 i_2} = e_{i_1} \sum_{\beta \in \Gamma(\alpha)} e_\beta = e_{i_1} \sum_{j \neq i_1 i_2} e_j.$$

Now we define generators $h_t(l)$ $(t = 1, \ldots, r)$ of $B^+(l)$. We set $A_t(r, l) = \{\alpha = (i_1, \ldots, i_{l-1}, t) \mid \alpha \in A(r, l),\ t \text{ is fixed }\}$,

$$e_t(l) = \sum_{\alpha \in A_t(r, l)} e_\alpha, \quad h_t(l) = \sum_{i \neq t} e_i(l).$$

The sublattice of D^r generated by $h_1(l), \ldots, h_r(l)$ is denoted by $B^+(l)$ and is called the l-th upper cubicle. The sublattice dual to $B^+(l)$ is denoted by $B^-(l)$ and is called the l-th lower cubicle.

3.2. A structural lemma and its consequences. In this section we prove that $B^+(l)$ and $B^-(l)$ are Boolean algebras.

LEMMA 3.1. *Let L be an arbitrary modular lattice, and $\{e_1, \ldots, e_r\}$ a finite set of elements of L. Then the sublattice B generated by the elements $h_j = \sum_{i \neq j} e_i$ $(j = 1, \ldots, r)$ is a Boolean algebra.*

PROOF. Let C be a non-empty subset of $I = \{1, \ldots, r\}$. We claim that the following identity holds in L:

$$\bigcap_{i\in C} h_i = \sum_{i\in C} e_i h_i + \sum_{k\in I-C} e_k. \tag{3.1}$$

If C consists of a single element, $C = \{j\}$, then (3.1) takes the form

$$h_j = e_j h_j + \sum_{k\neq j} e_k. \tag{3.2}$$

By definition, $\sum_{k\neq j} e_k = h_j$. Thus, in the case $C = \{j\}$ we must prove that $h_j = e_j h_j + h_j$, which is obviously true.

Suppose that (3.1) is proved for every subset C of m elements ($m < r$). We show that then (3.1) holds for every subset C_1 containing C and consisting of $m + 1$ elements. Suppose, for example, that $C_1 = C \cup \{s\}$, where $s \notin C$. Then

$$\bigcap_{j\in C_1} h_j = h_s\Big(\bigcap_{j\in C} h_j\Big) = \Big(\sum_{t\neq s} e_t\Big)\Big(\sum_{i\in C} e_i h_i + \sum_{k\in I-C} e_k\Big) = \tag{3.3}$$
$$= \Big(\sum_{t\neq s} e_t\Big)\Big(\sum_{i\in C} e_i h_i + \sum_{k\in I-C_1} e_k + e_s\Big).$$

It follows from $s \notin C$ that $\sum_{t\neq s} e_t \supseteq \sum_{i\in C} e_i \supseteq \sum_{i\in C} e_i h_i$ and $\sum_{t\neq s} e_t \supseteq \sum_{k\in I-C_1} e_k$. Consequently, by Dedekind's axiom,

$$\bigcap_{j\in C_1} h_j = \sum_{i\in C} e_i h_i + \sum_{k\in I-C_1} e_k + \Big(\sum_{t\neq s} e_t\Big) e_s = \tag{3.4}$$
$$= \sum_{i\in C} e_i h_i + \sum_{k\in I-C_1} e_k + e_s h_s = \sum_{i\in C_1} e_i h_i + \sum_{k\in I-C_1} e_k.$$

We have denoted by B the sublattice of L generated by the elements $h_j = \sum_{t\neq j} e_t$. We claim that every element $v \in B$ can be written in the form

$$v = \sum_{i\in a} e_i + \sum_{j\in a'} e_j h_j,$$

where $\varnothing \subseteq a$ and $a' = I - a$.

Note that in the case $v = h_j$ we have proved in (3.2) that $h_j = \sum_{i\in I-\{j\}} e_i + e_j h_j$. Now let v_1 and v_2 be two elements of L such that

$$v_q = \sum_{i\in a_q} e_i + \sum_{j\in a'_q} e_j h_j \qquad (q = 1, 2).$$

It easily follows from the identity $e_i h + e_i = e_i$ that

$$v_1 + v_2 = \sum_{i\in a_1\cup a_2} e_i + \sum_{j\in (a_1\cup a_2)'} e_j h_j.$$

Applying (3.1), we can write in the case $a \neq I$

$$\sum_{i\in a} e_i + \sum_{j\in a'} e_j h_j = \bigcap_{j\in a'} h_j.$$

In accordance with this identity, in case $a_q \neq I$,

$$v_1 v_2 = (\bigcap_{i\in a_1'} h_i)(\bigcap_{i\in a_2'} h_i) = \bigcap_{i\in a_1'\cup a_2'} h_i.$$

Since $a_1' \cup a_2' = (a_1 \cap a_2)'$, we have

$$v_1 v_2 = \bigcap_{i\in(a_1\cap a_2)'} h_i = \sum_{i\in a_1\cap a_2} e_i + \sum_{j\in(a_1\cap a_2)'} e_j h_j.$$

It is not hard to verify that the formula

$$v_1 v_2 = \sum_{i\in a_1\cap a_2} e_i + \sum_{j\in(a_1\cap a_2)'} e_j h_j$$

remains true when $a_1 = I$ or $a_2 = I$.

It follows from the relations we have proved that every element $v \in B$ can be written in the form $v = \sum_{i\in a} e_i + \sum_{j\in a'} e_j h_j$. We denote such an element by v_a.

We denote by $\mathscr{B}(I)$ the set of all subsets of $I = \{1, \ldots, r\}$. It is well known that $\mathscr{B}(I)$ is a Boolean algebra with 2^r elements. We have proved that $v_a + v_b = v_{a\cup b}$ and $v_a v_b = v_{a\cap b}$ for any $a, b \in \mathscr{B}(I)$. This shows that the correspondence $a \to v_a$ is a morphism of $\mathscr{B}(I)$ onto B.

It is not hard to prove from this that B is a Boolean algebra, and the number of elements of B is 2^m, where $m \leqslant r$.

COROLLARY 3.1. *Each sublattice $B^+(l)$, $B^-(l)$ ($l = 1, 2, \ldots$) of D^r is a Boolean algebra.*

COROLLARY 3.2. *Every element v of $B^+(l)$ can be written in the form*

$$v = v_{a,l} = \sum_{i\in a} e_i(l) + \sum_{j\in a'} e_j(l) h_j(l),$$

where a is an arbitrary subset of I ($\varnothing \subseteq a \subseteq I$). If $a \neq I$,

$$v_{a,l} = \bigcap_{j\in a'} h_j(l).$$

The minimal element of $B^+(l)$ is $v_{\emptyset,l} = \sum_{i\in I} e_i(l)h_i(l) = \bigcap_{i\in I} h_i(l)$, and the maximal element is $v_{I,l} = \sum_{i\in I} e_i(l)$.

We shall prove later that each sublattice $B^+(l)$ and $B^-(l)$ consists of 2^r elements, so that $B^+(l) \cong \mathscr{B}(I)$ and $B^-(l) \cong \mathscr{B}(I)$.

Occasionally we denote an arbitrary element of $B^+(l)$ and $B^-(l)$ by $v^+(l)$ and $v^-(l)$, respectively.

3.3. Ordering of the sublattices $B^+(l)$ and $B^-(l)$. PROPOSITION 3.1. *Let $v^+(l) \in B^+(l)$ and $v^+(m) \in B^+(m)$. If $l < m$, then $v^+(l) \supseteq v^+(m)$. Similarly, if $l < m$, then $v^-(l) \subseteq v^-(m)$, where $v^-(i) \in B^-(i)$.*

The proof of this proposition rests on two lemmas.

LEMMA 3.2. *Let $\alpha = (i_1, \ldots, i_l) \in A(r, l)$ for $l \geqslant 2$. We write*

$\pi(\alpha) = (i_1, \ldots, i_{l-1})$. *Then the elements* e_α *and* $e_{\pi(\alpha)}$ *of* D^r *can be ordered as follows*: $e_{\pi(\alpha)} \supseteq e_\alpha$.

The proof is by induction on l. Let $l = 2$, that is, $\alpha = (i_1, i_2)$. Then $\pi(\alpha) = (i_1)$ and $e_\alpha = e_{i_1 i_2} = e_{i_1} \sum_{j \neq i_1 i_2} e_j$ and $e_{\pi(\alpha)} = e_{i_1}$. Clearly, $e_\alpha \subseteq e_{\pi(\alpha)}$.

Suppose that the lemma has been proved for every $\alpha \in A(r, \lambda)$ with $\lambda < l$. We prove it for $\alpha = (i_1, \ldots, i_l)$. By definition,

$$e_\alpha = e_{i_1 \ldots i_l} = e_{i_1} \sum_{\beta \in \Gamma(\alpha)} e_\beta,$$

where

$$\Gamma(\alpha) = \{\beta = (k_1, \ldots, k_{l-1}) \in A(r, l-1) \mid k_1 \notin \{i_1, i_2\}, \ k_2 \notin \{i_2, i_3\}, \ \ldots, \ k_{l-1} \notin \{i_{l-1}, i_l\}\}.$$

Similarly,

$$e_{\pi(\alpha)} = e_{i_1 \ldots i_{l-1}} = e_{i_1} \sum_{\beta' \in \Gamma(\pi(\alpha))} e_{\beta'},$$

where

$$\Gamma(\pi(\alpha)) = \{\beta' = (m_1, \ldots, m_{l-2}) \in A(r, l-2) \mid m_1 \notin \{i_1, i_2\}, \ m_2 \notin \{i_2, i_3\}, \ldots m_{l-2} \notin \{i_{l-2}, i_{l-1}\}\}.$$

Clearly, for any $\beta = (k_1, \ldots, k_{l-1}) \in \Gamma(\alpha)$ we can find an element $\beta' \in \Gamma(\pi(\alpha))$ such that $\beta' = \pi(\beta) = (k_1, \ldots, k_{l-2})$. By induction on such β and $\beta' = \pi(\beta)$ we have $e_\beta \subseteq e_{\pi(\beta)}$. Consequently $\sum_{\beta \in \Gamma(\alpha)} e_\beta \subseteq \sum_{\beta' \in \Gamma(\pi(\alpha))} e_{\beta'}$, hence $e_\alpha \subseteq e_{\pi(\alpha)}$.

LEMMA 3.3. *Let* $h_t(l-1) \in B^+(l-1)$ *for* $l > 1$ *and* $t \in I$. *Then* $e_\alpha \subseteq h_t(l-1)$ *for every* $\alpha \in A(r, l)$.

PROOF. By definition,

$$h_t(l-1) = \sum_{j \neq t} e_j(l-1) = \sum_{j \neq t} \sum_{\beta \in A_j(r, l-1)} e_\beta.$$

We consider first the case when $\alpha = (i_1, \ldots, i_{l-1}, i_l)$ with $i_{l-1} \neq t$. Then $\pi(\alpha) = (i_1, \ldots, i_{l-1}) \in A_{i_{l-1}}(r, l-1)$, and so

$$e_{\pi(\alpha)} \subseteq \sum_{j \neq t} \sum_{\beta \in A_j(r, l-1)} e_\beta = h_t(l-1).$$

By the preceding lemma, $e_\alpha \subseteq e_{\pi(\alpha)}$. Consequently $e_\alpha \subseteq h_t(l-1)$.

Now we consider the case $\alpha = (i_1, \ldots, i_{l-1}, i_l)$, where $i_{l-1} = t$. By definition, $e_\alpha = e_{i_1} \sum_{\beta \in \Gamma(\alpha)} e_\beta$, and $k_{l-1} \notin \{i_{l-1}, i_l\}$ for every $\beta = (k_1, \ldots, k_{l-1}) \in \Gamma(\alpha)$. In this case, $i_{l-1} = t$. Therefore $k_{l-1} \neq t$. Consequently, $\Gamma(\alpha) \subset \sum_{j \neq t} A_j(r, l-1)$, and so

$$\sum_{\beta \in \Gamma(\alpha)} e_\beta \subseteq \sum_{j \neq t} \sum_{\beta \in A_j(r, l)} e_\beta = h_t(l-1).$$

A fortiori $e_\alpha = e_{i_1} \sum_{\beta\in\Gamma(\alpha)} e_\beta \subseteq h_t(l-1)$.

PROOF OF PROPOSITION 3.1. It was proved in Corollary 3.2 that the maximal element of $B^+(l)$ is $v_{I,l} = \sum_{i\in I} e_i(l) = \sum_{\alpha\in A(r,l)} e_\alpha$, and the minimal element of $B^+(l-1)$ is $v_{\theta,l-1} = \bigcap_{t\in I} h_t(l-1)$. It is clear from Lemma 3.3 that for any $\alpha \in A(r, l)$,

$$\bigcap_{t\in I} h_t(l-1) \supseteq e_\alpha.$$

Therefore,

$$v_{\theta,l-1} = \bigcap_{t\in I} h_t(l-1) \supseteq \sum_{\alpha\in A(r,l)} e_\alpha = v_{I,l}.$$

Now if $v^+(l-1)$ and $v^-(l-1)$ are arbitrary elements of $B^+(l-1)$ and $B^+(l)$, respectively, then $v^+(l-1) \supseteq v_{\theta,l-1} \supseteq v_{I,l} \supseteq v^+(l)$.

The corresponding statement for the cubicles $B^-(l)$ and $B^-(m)$ is obtained by duality.

We denote by B^+ the subset of D^r that is the union of the $B^+(l)$, $l = 1, 2, \ldots$. Similarly, $B^- = \bigcup_{l=1}^{\infty} B^-(l)$.

COROLLARY 3.3. *B^+ and B^- are sublattices of D^r.*

3.4. Fundamental properties of cubicles. We have, thus, proved the following theorem.

THEOREM 3.1. (1) *Every sublattice $B^+(l)$ and $B^-(l)$ ($l = 1, 2, \ldots$) is a Boolean algebra. Any element $v_{a,l} \in B^+(l)$ can be written in the following form:*

$$v_{a,l} = \sum_{i\in a} e_i(l) + \sum_{j\in a'} e_j(l)\, h_j(l),$$

where a is an arbitrary subset of $I = \{1, \ldots, r\}$, and $a' = I - a$. If $a \neq I$, then $v_{a,l} = \bigcap_{j\in a} h_j(l)$.

(2) *Let $v^+(l) \in B^+(l)$ and $v^+(m) \in B^+(m)$. If $l < m$, then*

$$v^+(l) \supseteq v^+(m). \tag{3.5}$$

Similarly, if $l < m$, then

$$v^-(l) \subseteq v^-(m), \tag{3.6}$$

where $v^-(j) \in B^-(j)$.

REMARK. Later we shall prove the stronger statement:

$$(l<m) \Rightarrow \begin{cases} v^+(l) \supset v^+(m), \\ v^-(l) \subset v^-(m), \end{cases}$$

where $v^+(j) \in B^+(j)$ and $v^-(j) \in B^-(j)$.

CONJECTURE. *Let $v^+(l) \in B^+(l)$ and $v^-(m) \in B^-(m)$. Then $v^-(m) \subset v^+(l)$ for all l and m.*

3.5. In this subsection we introduce another definition of the elements

$e_\alpha \in D^r$, which appears rather clumsy, but leads to shorter formulae than before. First, however, we examine an example.

Let $r = 5$, and let D^5 be generated by the five elements $e_1, \ldots, e_5$. Let $\alpha = (2, 1, 5)$, that is $\alpha \in A(5, 3)$. Then it is not difficult to compute that $\Gamma(\alpha) = \{(3, 2), (3, 4), (4, 2), (4, 3), (5, 2), (5, 3), (5, 4)\}$ hence

$$e_\alpha = e_{2,1,5} = e_2 \sum_{\beta \in \Gamma(\alpha)} e_\beta = e_2 (e_{3,2} + e_{3,4} + e_{4,2} + e_{4,3} + e_{5,2} + e_{5,3} + e_{5,4}).$$

If in this formula we substitute $e_{i,j} = e_i \sum_{t \neq j, i} e_t$, then we obtain the final formula for the polynomial $e_{2,1,5}$. However, we do not write it down because of its extreme length.

It can be shown that in D^r

$$e_{2,1,5} = e_2(e_3 + e_5 + e_{4,2} + e_{4,3}) = \\ = e_2(e_3 + e_5 + e_4(e_1 + e_3 + e_5) + e_4(e_1 + e_2 + e_5)),$$

and also

$$e_{2,1,5} = e_2(e_4 + e_5 + e_{3,2} + e_{3,4}) = \\ = e_2(e_4 + e_5 + e_3(e_1 + e_4 + e_5) + e_3(e_1 + e_2 + e_5)).$$

It is clear from this example that our method for writing the formulae for the elements e_α, $\alpha \in A(5, 3)$, as $e_\alpha = e_{i_1 i_2 i_3} = e_{i_1} \sum_{\beta \in \Gamma(\alpha)} e_\beta$ is not very economical.

Let $\alpha = (i_1, \ldots, i_l) \in A(r, l)$ with $l \geqslant 3$. We describe an inductive method of constructing polynomials e'_α from an element $\alpha \in A(r, l)$. For every $\alpha \in A(r, l)$ with $l \geqslant 3$ we construct an entire family of polynomials $\{e'_\alpha\}$. We associate with $\alpha = (i_1, \ldots, i_l)$, $l \geqslant 3$, a fixed sequence $(k_1, \ldots, k_{l-2})$ of numbers $k_i \in I$ such that $k_1 \neq k_2$, $k_2 \neq k_3, \ldots, k_{l-3} \neq k_{l-2}$, and $k_1 \notin \{i_1, i_2, i_3\}$, $k_2 \notin \{i_2, i_3, i_4\}, \ldots, k_{l-2} \notin \{i_{l-2}, i_{l-1}, i_l\}$. We set

$$H_1 = \{\delta = (t_1) \mid t_1 \in I - \{k_1, i_1, i_2\}\},$$
$$H_2 = \{\delta = (k_1, t_2) \mid t_2 \in I - \{k_1, k_2, i_2, i_3\}\},$$
$$\cdots$$
$$H_\lambda = \{\delta = (k_1, \ldots, k_{\lambda-1}, t_\lambda) \mid t_\lambda \in I - \{k_{\lambda-1}, k_\lambda, i_\lambda, i_{\lambda+1}\}\},$$
$$\cdots$$
$$H_{l-1} = \{\delta = (k_1, \ldots, k_{l-2}, t_{l-1}) \mid t_{l-1} \in I - \{k_{l-2}, i_{l-1}, i_l\}\}.$$

We denote by H the disjoint union of $H_1, \ldots, H_{l-1}$. We set

$$e'_\alpha = e'_{i_1 \ldots i_l} = e_{i_1} \sum_{\delta \in H} e'_\delta,$$

where δ ranges over the whole of H, that is, $H_1, \ldots, H_{l-1}$. This definition is ambiguous. It depends on the choice of the numbers $k_1, \ldots, k_{l-2}$, and, of course, on the choice of e'_δ. For example, $e'_{i_1 i_2 i_3}$, depends on the choice of k_1, that is, there are $r - 3$ (or $r - 2$ if $i_1 = i_3$) polynomials $e'_{i_1 i_2 i_3}$. The polynomial $e'_{i_1 i_2 i_3 i_4}$ depends on the sequence (k_1, k_2) and on the choice of e'_δ, where $\delta = (k_1, k_2, t)$, and so on.

We propose the following conjecture.

CONJECTURE. *For every $\alpha \in A(r, l)$ with $l \geqslant 3$, and every e'_α*

$$e'_\alpha = e_\alpha .$$

We are able to prove this only for $l = 3$. For $l > 3$ we can prove only the weaker assertion that, for every α and every e'_α, the elements e_α and e'_α are linearly equivalent.

§4. Representations of the first upper cubicle

By definition, the cubicle $B^+(l)$ is the sublattice of D^r generated by the elements $h_1(1), \ldots, h_r(1)$, where $h_t(1) = \sum_{j \neq t} e_j$. In this section we prove that all elements of $B^+(1)$ are perfect.

4.1. Atomic representations and their connection with representations of $B^+(l)$. We define the most trivial among the indecomposable representations of D^r – the atomic representations $\rho_{j,1}$ for $j \in \{0, 1, \ldots, r\}$.

a) The representation $\rho_{0,1}$ in the one-dimensional space $V \cong k^1$ is defined by $\rho_{0,1}(e_i) = 0$ for all $i \in \{1, \ldots, r\}$. It follows that $\rho_{0,1}(x) = 0$ for every $x \in D^r$.

b) The representation $\rho_{t,1}$ for $t \in \{1, \ldots, r\}$ in $V \cong k^1$ is defined by $\rho_{t,1}(e_j) = 0$ for $t = j$ and $\rho_{t,1}(e_t) = V$.

Presently we describe the connection between the atomic representation $\rho_{t,1}$ and those of $B^+(1)$. In §3 we provide that $B^+(1)$ is a Boolean algebra with the minimal element $v_{\theta,1} = \bigcap_t h_t(1)$. Now we prove that $v_{\theta,1}$ is perfect, that is, the restriction of ρ to $\rho(v_{\theta,1})$ is a direct summand of ρ. We denote $\rho(v_{\theta,1})$ by $V_{\theta,1}$.

PROPOSITION 4.1. *Let ρ be a representation of D^r in a space V over k, and let $v_{\theta,1}$ be the minimal element of $B^+(1)$. Then ρ decomposes into a direct sum*

$$\rho = \Big(\bigoplus_{j \in I \cup \{0\}} \tilde{\rho}_{j,1} \Big) \oplus \tau_{\theta,1},$$

where $\tau_{\theta,1}$ is the restriction of ρ to the subspace $V_{\theta,1} = \rho(v_{\theta,1})$, and each $\tilde{\rho}_{j,1}$ is a multiple of the atomic representation $\rho_{j,1}$, that is, $\tilde{\rho}_{j,1} = \underbrace{\rho_{j,1} \oplus \ldots \oplus \rho_{j,1}}_{m_j}$, where $m_j \geqslant 0$.

PROOF. We indicate how to choose subspaces U_j such that $V = (\bigoplus_{j \in I^0} U_j) \oplus V_{\theta,1}$, where $I^0 = I \cup \{0\} = \{0, 1, \ldots, r\}$ and ρ decomposes into a direct sum relative to these subspaces.

We claim that the U_j can be chosen as subspaces satisfying the following relations:

$$(4.1) \qquad U_0 \sum_{i=1}^{r} \rho(e_i) = 0, \quad U_0 + \sum_{i=1}^{r} \rho(e_i) = V,$$

and for any $j \neq 0$, $j \in I$,

$$(4.2) \qquad U_j\rho(h_j) = 0, \quad U_j + \rho(e_jh_j) = \rho(e_j),$$

where $h_j = h_j(1) = \sum_{t \neq j} e_t$.

Step 1. We recall that any element of $B^+(1)$ can be written in the form

$$(4.3) \qquad v_{a,1} = \sum_{i \in a} e_i + \sum_{j \in a'} e_jh_j,$$

where a is a subset of $I = \{1, \ldots, r\}$ and $a' = I - a$.

We write $V_{a,1} = \rho(v_{a,1})$ and claim that if in V subspaces U_j, $j \in \{0, 1, \ldots, r\}$, are chosen to satisfy (4.1) and (4.2), then for any $a \subseteq I$

$$(4.4) \qquad V_{a,1} = \sum_{j \in a} U_j + V_{\theta,1}.$$

We prove (4.4) first for the case of one-element subsets $a = \{t\}$, $t \in \{1, \ldots, r\}$, that is, that

$$(4.5) \qquad V_{\{t\},1} = U_t + V_{\theta,1}.$$

Let $\rho(e_i) = E_i$ and $\rho(h_i) = H_i$. It follows from (4.3) that $V_{\{t\},1} = E_t + \sum_{i \neq t} E_iH_i$ and $V_{\theta,1} = \sum_{i=1}^{r} E_iH_i$. Then we find that

$$U_t + V_{\theta,1} = U_t + \sum_{i=1}^{r} E_iH_i = U_t + E_tH_t + \sum_{i \neq t} E_iH_i.$$

By construction (see (4.2)), $U_t + E_tH_t = E_t$, consequently,

$$U_t + V_{\theta,1} = E_t + \sum_{i \neq t} E_iH_i = V_{\{t\},1}.$$

This proves (4.5).

In §3 we have proved that $B^+(1)$ is a Boolean algebra, and that $v_{a\cup b,1} = v_{a,1} + v_{b,1}$ for any subsets $a, b \subseteq I$; in particular, $v_{a,1} = \sum_{t \in a} v_{\{t\},1}$. It follows that every subspace $\rho(v_{a,1}) = V_{a,1}$ can be represented as a sum $V_{a,1} = \sum_{t \in a} V_{\{t\},1}$. Putting $V_{\{t\},1} = U_t + V_{\theta,1}$ in this formula, we obtain

$$V_{a,1} = \sum_{t \in a} (U_t + V_{\theta,1}) = \sum_{t \in a} U_t + V_{\theta,1}.$$

This proves (4.4).

Step 2. We show that our chosen subspaces U_j are such that $V \cong V_{\theta,1} \oplus U_0 \oplus U_1 \oplus \ldots \oplus U_r$. We write $a_t = \{t+1, \ldots, r\}$

and $W_t = V_{a_t,1} = \rho(v_{a_t})$. Note that (see (4.3))
$W_t = \sum_{i=1}^{t} E_iH_i + \sum_{j=t+1}^{r} E_j$. It follows easily from the relations $E_iH_i \subseteq E_i$ that the subspaces W_t, $t \in \{0, 1, \ldots, r\}$ form a chain

$$W_r \subseteq W_{r-1} \subseteq \ldots \subseteq W_2 \subseteq W_1 \subseteq W_0 \subseteq V,$$

where $W_r = \sum_{i=1}^{r} E_iH_i = V_{\theta,1}$, $W_0 = \sum_{i=1}^{r} E_i = V_{I,1}$.

We claim that the U_j, $j \in \{0, 1, \ldots, r\}$, subject to (4.1) and (4.2), are connected with the W_j by the following equations:

$$W_0 + U_0 = V, \quad W_0U_0 = 0, \tag{4.6}$$

and for every $t \in \{1, \ldots, r)$

$$U_t + W_t = W_{t-1}. \tag{4.7}$$

$$U_tW_t = 0. \tag{4.8}$$

Note that (4.6) is the same as (4.1), because $W_0 = \sum_{i=1}^{r} E_i = \sum_{i=1}^{r} \rho(e_i)$. We have proved earlier (see 4.4) that $V_{a,1} = \sum_{i \in a} U_i + V_{\theta,1}$ for every $a \subseteq I$.

Consequently, $W_t = V_{a_t,1} = \sum_{j=t+1}^{r} U_j + V_{\theta,1} = \sum_{j=t+1}^{r} U_j + W_r$. Now (4.7) evidently follows from this equation. Note that

$W_t = \sum_{i=1}^{t} E_iH_i + \sum_{j=t+1}^{r} E_j \subseteq E_tH_t + \sum_{j \neq t} E_j = E_tH_t + H_t = H_t$. From this, using (4.2) ($U_tH_t = 0$), we obtain $W_tU_t \subseteq H_tU_t = 0$, that is, $U_tW_t = 0$. This proves (4.8).

It follows easily from (4.6)–(4.8) that
$V = \sum_{t=0}^{r} U_t + W_r = \sum_{t=0}^{r} U_t + V_{\theta,1}$ and that this sum is direct.

Step 3. We claim that for every $i \in \{1, \ldots, r\}$

$$E_i = \sum_{j=0}^{r} E_iU_j + E_iV_{\theta,1}. \tag{4.9}$$

To prove this we show first that $E_iU_j = 0$ for $i \neq j$, and $E_iU_i = U_i$. For by construction, $U_0 \sum_{i=1}^{r} E_i = 0$ and $U_jH_j = U_j \sum_{i \neq j} E_i = 0$, $U_i + E_iH_i = E_i$. Consequently, for every $i \neq 0$ we have: 1) $U_0E_i = 0$, 2) $U_jE_i = 0$ for $j \neq 0$, $j \neq i$, and 3) $E_iU_i = U_i$.

Thus, we can rewrite the right-hand side of (4.9) in the following form:

$$\sum_{j=0}^{r} E_iU_j + E_iV_{\theta,1} = E_iU_i + E_iV_{\theta,1} = U_i + E_iV_{\theta,1}. \tag{4.10}$$

Let us find $E_iV_{\theta,1}$. By definition, $V_{\theta,1} = \rho(v_{\theta,1}) = \rho(\bigcap_{j=1}^{r} h_j) = \bigcap_{j=1}^{r} H_j$, where

$H_j = \sum_{t \neq j} E_t$. Hence $E_i H_j|_{i \neq j} = E_i$, and therefore

$$E_i V_{\theta,1} = E_i (\bigcap_{j=1}^{r} H_j) = \bigcap_{j=1}^{r} E_i H_j = E_i H_i.$$

We have to show that $E_i = U_i + V_{\theta,1} E_i = U_i + E_i H_i$, which is true by construction (4.2). This proves (4.9).

Step 4. We combine the results of steps 2 and 3. We have proved that $V = \sum_{i=0}^{r} U_i + V_{\theta,1}$ and that this sum is direct. Further, we have proved that every subspace E_i ($E_i = \rho(e_i)$) is representable as a sum $E_i = \sum_{j=0}^{r} E_i U_j + E_i V_{\theta,1}$. Consequently, by Proposition 2.3, ρ splits into the direct sum $\rho = \bigoplus_{j=0}^{r} \tilde{\rho}_{j,1} \oplus \tau_{\theta,1}$, where $\tilde{\rho}_{j,1} = \rho^{U_j}$ and $\tau_{\theta,1} = \rho^{V_{\theta,1}}$.

Step 5. We claim that $\tilde{\rho}_{j,1}$ is a multiple of the atomic representation $\rho_{j,1}$, that is, $\tilde{\rho}_{j,1} = \underbrace{\rho_{j,1} \oplus \ldots \oplus \rho_{j,1}}_{m_j}$ with $m_j \geqslant 0$.

First we study the representation $\tilde{\rho}_{0,1}$. It follows from Proposition 2.3 that the subspace U_0 is admissible, therefore, $\tilde{\rho}_{0,1}(e_i) = U_0 \rho(e_i) = U_0 E_i$. We have just proved in b) that $U_0 E_i = 0$ for every i. Thus, the subrepresentation $\tilde{\rho}_{0,1}$ in U_0 is such that $\tilde{\rho}_{0,1}(e_i) = 0$ for every i. If $\dim U_0 = m_0 > 0$, then $\tilde{\rho}_{0,1}$ is different from zero, and, clearly, splits into the direct sum $\tilde{\rho}_{0,1} = \overbrace{\rho_{0,1} \oplus \ldots \oplus \rho_{0,1}}^{m_0}$ of atomic representations $\rho_{0,1}$.

Similarly, for the $\tilde{\rho}_{j,1}$ with $j \neq 0$, we obtain

$$\tilde{\rho}_{j,1}(e_i) = U_j \rho(e_i) = U_j E_i = \begin{cases} 0 & \text{if } i \neq j, \\ U_j & \text{if } i = j. \end{cases}$$

If $\dim U_j = m_j > 0$, then it is easy to see that $\tilde{\rho}_{j,1}$ splits into a direct sum of atomic representations $\tilde{\rho}_{j,1} = \rho_{j,1} \oplus \ldots \oplus \rho_{j,1}$.

PROPOSITION 4.2. *Let $v_{a,1}$ be an arbitrary element of $B^+(1)$ corresponding to the subset $a \in I$. Let ρ be a representation of D^r in V such that the subspace $V_{a,1} = \rho(v_{a,1})$ is different from zero and from V. Then ρ splits into a direct sum*

$$\rho = (\bigoplus_{j \in a' \cup \{0\}} \tilde{\rho}_{j,1}) \oplus \tau_{a,1},$$

where $a' = I - a$; each of the representations $\tilde{\rho}_{j,1}$ is a multiple of an atomic one: $\tilde{\rho}_{j,1} = \overbrace{\rho_{j,1} \oplus \ldots \oplus \rho_{j,1}}^{m_j}$ with $m_j \geqslant 0$: Also, $\tau_{a,1}$ is the restriction of ρ to $V_{a,1}$. The representation $\tau_{a,1}$ is such that if it splits into a direct sum of indecomposable representations $\tau_{a,1} = \sum \tau_j$, then *none of the atomic representations $\rho_{i,1}$ with $i \in a' \cup \{0\}$ occurs in this sum.*

PROOF. In Proposition 4.1 we have shown how to choose subspaces U_t, $t \in \{0, 1, \ldots, r\}$, such that ρ splits into a direct sum $\rho = (\bigoplus_{j \in I \cup \{0\}} \tilde{\rho}_{j,1}) \oplus \tau_{\theta,1}$ relative to $U_0, U_1, \ldots, U_r$, and $V_{\theta,1}$. Here $\tilde{\rho}_{j,1} = \rho|_{U_j}$ is a multiple of the atomic representation $\rho_{j,1}$, and $\tau_\theta = \rho|_{V_{\theta,1}}$, where $V_{\theta,1} = \rho(v_{\theta,1})$, is the image of the minimal element $v_{\theta,1} \in B^+(1)$. The subspaces U_j are such that $V_{a1} = \sum_{j \in a} U_j + V_{\theta,1}$. Consequently,

$$\rho = (\bigoplus_{j \in a' \cup \{0\}} \tilde{\rho}_{j,1}) \oplus (\bigoplus_{k \in a} \tilde{\rho}_{k,1}) \oplus \tau_{\theta,1} = (\bigoplus_{j \in a' \cup \{0\}} \tilde{\rho}_{j,1}) \oplus \tau_{a,1},$$

where $\tau_{a,1} = (\bigoplus_{k \in a} \rho_{k,1}) \oplus \tau_{\theta,1}$ is the restriction of ρ to $V_{a,1}$.

We now claim that $\tau_{a,1}$ does not contain as direct summands the atomic representations $\rho_{j,1}$, $j \in a' \cup \{0\}$. We assume the contrary. For example, let $\tau_{a,1} = \rho_{j,1} \oplus \tau'_{a,1}$, where $j \in a'$, and let $V_{a,1} = W_j + V'_{a,1}$ be the corresponding decomposition of $V_{a,1}$ into a direct sum of subspaces $W_j \neq 0$ and $V'_{a,1} \neq V_{a,1}$. Let us find out what the subspace $\tau_{a,1}(h_j)$ is, where $h_j = \sum_{t \neq j} e_t$. By Proposition 2.3, $V_{a,1}$ is admissible, therefore $\tau_{a,1}(h_j) = V_{a,1}\rho(h_j) = \rho(v_{a,1})\rho(h_j) = \rho(v_{a,1}h_j)$. We recall that $v_{a,1} = \bigcap_{i \in a'} h_i$ and, by assumption $j \in a'$. Consequently, $h_j v_{a,1} = h_j \cap \bigcap_{i \in a'} h_i = \bigcap_{i \in a'} h_i = v_{a,1}$. Finally, $\tau_{a,1}(h_j) = V_{a,1}$. On the other hand, we have assumed that $\tau_{a,1} \cong \rho_{j,1} + \tau'_{a,1}$. Therefore, $\tau_{a,1}(h_j) = \rho_{j,1}(h_j) + \tau'_{a,1}(h_j)$. Since the representation $\rho_{j,1}$ is atomic, we see that $\rho_{j,1}(e_t) = 0$ if $j \neq t$. Therefore, $\rho_{j,1}(h_j) = \rho_{j,1} \sum_{t \neq j} e_t = \sum_{t \neq j} \rho_{j,1}(e_t) = 0$. Consequently, $\tau_{a,1}(h_j) = 0 + \tau'_{a,1}(h_j) \subseteq V'_{a,1} \neq V_{a,1}$. So we have obtained a contradiction. It can be proved similarly that $\tau_{a,1}$ does not contain the atomic representations $\rho_{0,1}$.

COROLLARY 4.1. *Every element $v_{a,1} \in B^+(1)$ is perfect.*

We have already mentioned that every subspace $V_{a,1} = \rho(v_{a,1})$ is admissible. But we can prove the more general assertion that every element $v_{a,1} \in B^+(1)$ is neutral[1] in D^r.

COROLLARY 4.2. *The Boolean algebra $B^+(1)$ consists of 2^r elements.*

PROOF. We choose a representation ρ which is a direct sum $\rho_{1,1} \oplus \ldots \oplus \rho_{r,1}$ of the atomic representations $\rho_{t,1}$ for $t \in \{1, \ldots, r\}$. Then it is easy to see that the representation space V of ρ is r-dimensional: $V \cong k^r$. Here all the subspaces $\rho(e_i)$ are one-dimensional, and $\sum_{i=1}^{r} \rho(e_i) = V$.

---

[1] An element a of a modular lattice L is called *neutral* if $a(x + y) = ax + ay$ for all $x, y \in L$.

It is easy to see that the sublattice $\rho(B^+(1))$ generated by the $\rho(e_i)$ is a Boolean algebra with 2^r elements, that is, $\rho(B^+(1)) \cong \mathscr{B}(I)$, where $\mathscr{B}(I)$ is the Boolean algebra of subsets of $I = \{1, \ldots, r\}$. Earlier we have proved that $B^+(1)$ is a factor algebra of $\mathscr{B}(I)$. Now we conclude that $B^+(1) \cong \mathscr{B}(I)$.

In the following sections we show that every element in the cubicle $B^+(l)$ is perfect. However, to prove this we need more complicated techniques (the functors Φ^+ and Φ^-).

4.2. Representations of the first lower cubicle. The results of this subsection are dual to those of 4.1. The first lower cubicle $B^-(1)$ is the sublattice of D^r generated by the r elements $h_t^- = \bigcap_{i \neq t} e_i$ $(t = 1, \ldots, r)$.

In addition, we write $h_0^- = \bigcap_{i \in I} e_i$. Clearly, $h_0^- \subset h_i^-$ for every $i \neq 0$ and $h_0^- = h_i^- h_j^-$ for all $i, j \in I$, $i \neq j$. Thus, h_0^- is the minimal element of $B^-(1)$. Then any element $v_{a,1} \in B^-(1)$ can be written in the form

$$v_{\bar{a},1}^- = h_0^- + \sum_{t \in a} h_t^-,$$

where $a \subseteq I$. Since $h_0^- = v_{\emptyset,1}^-$ is the minimal element of $B^-(1)$, we see that if $a \neq \emptyset$, then

$$v_{\bar{a},1}^- = \sum_{t \in a} h_t^-.$$

Let $\rho_{\bar{t},1}^-$ be dual to $\rho_{t,1}$. Namely, $\rho_{\bar{t},1}^-$ is the representation in $V \cong k^1$ such that $\rho_{\bar{0},1}^-(e_i) = V$, $\rho_{\bar{t},1}^-(e_i) = V$ for all i, $t \neq i$, $t \neq 0$ and $\rho_{\bar{t},1}^-(e_t) = 0$. The representation $\rho_{\bar{t},1}^-$ is called Φ^--atomic, and $\rho_{t,1} = \rho_{t,1}^+$ is called Φ^+-atomic.

The dual to Proposition 4.2 is the following assertion.

PROPOSITION 4.3. *Let ρ be a representation of D^r in V such that the subspace $V_{\bar{a},1}^- = \rho(v_{\bar{a},1}^-)$ is different from zero and V. Then ρ splits into a direct sum $\rho = \rho' \oplus \tau_{a,1}$, where $\tau_{a,1} = \rho\,|_{\tilde{V}_{a,1}}$. Here $\tau_{a,1}$ can be represented as a sum $\tau_{a,1} = \bigoplus_{j \in a \cup \{0\}} \tilde{\rho}_{\bar{j},1}^-$, where each $\tilde{\rho}_{\bar{j},1}^-$ is a multiple of the atomic representation $\rho_{\bar{j},1}^-$. If ρ' splits into a direct sum, then the atomic representations $\rho_{\bar{j},1}^-$, $j \in a$ do not occur in this sum.*

We have also proved that each element $v_{a,1}^- \in B^-(1)$ is perfect.

§5. The functors Φ^+ and Φ^-

In this section we define functors Φ^+ and Φ^- from $\mathscr{R}(D^r, k)$ into $\mathscr{R}(D^r, k)$, and use them to study representations of the cubicles $B^+(l)$, $l > 1$. In essence, these functors were first defined in [5]. A modification of them, under the name "Coxeter functors", was used effectively in [1]. They also play a decisive role in the proofs of the main results of this paper. However, we still do not understand their connection with the lattice operations sufficiently well.

5.1. Definition of the functors and their fundamental properties. Let ρ be a representation of D^r in V. We define a space V^1 and a representation ρ^1 in V^1 in the following way:

$$V^1=\{(\xi_1, \ldots, \xi_r) \mid \xi_i \in \rho(e_i),\ \sum_{i=1}^{r} \xi_i = 0\},$$

$$\rho^1(e_i)=\{(\xi_1, \ldots, \xi_{i-1}, 0, \xi_{i+1}, \ldots, \xi_r) \in V^1\},$$

where e_i $(i = 1, \ldots, r)$ are generators of D^r.

In other words, let $R = \bigoplus_{i=1}^{r} \rho(e_i)$ and let $\nabla\colon R \to V$ be the linear transformation defined by $\nabla(\xi_1, \ldots, \xi_r) = \sum_{i=1}^{r} \xi_i$. Then $V^1 = \operatorname{Ker} \nabla$. We set $G'_i = \{(\xi_1, \ldots, \xi_{i-1}, 0, \xi_{i+1}, \ldots, \xi_r) \in R\}$, that is, $G'_i = \bigoplus_{j\neq i} \rho(e_j)$. Then $\rho^1(e_i) = V^1 G'_i$.

Thus, from the representation $\rho \in \mathcal{R}$ we have constructed another representation $\rho^1 \in \mathcal{R}$. It is not hard to check that the correspondence $\rho \mapsto \rho^1$ is functorial. We denote this functor by Φ^+.

A representation ρ^{-1} is constructed dually. We set $Q = \bigoplus_{i=1}^{r} (V/\rho(e_i))$. We denote by $\beta_i\colon V \to V/\rho(e_i)$ the natural map, and by $\mu\colon V \to Q$ the map defined by $\mu\xi = (\beta_1 \xi, \ldots, \beta_r \xi)$. Further, we set $V^{-1} = \operatorname{Coker} \mu = Q/\operatorname{Im} \mu$. We define a representation ρ^{-1} in V^{-1} in the following way. We write $Q_i = \{(0, \ldots, 0, \beta_i \xi, 0, \ldots, 0) \mid \xi \in V\}$. Let $\theta\colon Q \to V^{-1} = \operatorname{Coker} \mu$ be the natural map. We set

$$\rho^{-1}(e_i) \overset{\text{def}}{=} \theta Q_i,$$

where θQ_i is the image of the subspace Q_i $(\theta Q_i \subseteq V^{-1})$.

It is not hard to check that the correspondence $\rho \mapsto \rho^{-1}$ is functorial. We denote this functor by Φ^- $(\Phi^-\colon \mathcal{R} \to \mathcal{R})$ and we write $\rho^{-1} = \Phi^-\rho$.

We describe first some simple properties of the functors Φ^+ and Φ^-.

PROPOSITION 5.1. *Let ρ and $\rho^1 = \Phi^+\rho$ and $\rho^{-1} = \Phi^-\rho$ be representations of the lattice D^r in the spaces V, V^1, and V^{-1}, respectively.*

(I) *If ρ is such that $\sum_{i=1}^{r} \rho(e_i) = V$, then*

$$\dim V^1 = \sum_{i=1}^{r} \dim \rho(e_i) - \dim V.$$

(II) *If ρ is such that $\sum_{j\neq i} \rho(e_j) = V$, then*

$$\dim \rho^1(e_i) = \sum_{j\neq i} \rho(e_j) - \dim V.$$

(III) *If* ρ *is such that* $\bigcap_{i=1}^{r} \rho(e_i) = 0$, *then*

$$\dim V^{-1} = \sum_{i=1}^{r} \dim V/\rho(e_i) - \dim V.$$

(IV) *If* ρ *is such that* $\bigcap_{j \neq i} \rho(e_i) = 0$, *then*

$$\dim \rho^{-1}(e_i) = \dim V/\rho(e_i).$$

PROPOSITION 5.2. *Let* ρ *be a representation of* D^r.

(I). $\Phi^+(\rho^*) \cong (\Phi^-\rho)^*$, *where* ρ^* *is the dual to* ρ.

(II) *If* $\rho = \bigoplus_{i=1}^{n} \rho_i$, *then* $\Phi^+(\rho) \cong \bigoplus_{i=1}^{n} (\Phi^+\rho_i)$ *and* $\Phi^-\rho \cong \bigoplus_{i=1}^{n} (\Phi^-\rho_i)$.

The proof of these propositions is not difficult and reduces to a direct verification.

Next we prove a number of important properties of $\Phi^+\Phi^-$.

(I) There is a natural way of defining a monomorphism i: $\Phi^-\Phi^+\rho \to \rho$ such that $\Phi^-\Phi^+\rho$ is isomorphic to the subrepresentation $\rho\,|_{V_{\theta,1}}$ – the restriction of ρ to $V_{\theta,1} = \rho(v_{\theta,1})$, where $v_{\theta,1}$ is the minimal element of $B^+(1)$.

(II) There is a natural way of defining an epimorphism p: $\rho \to \Phi^+\Phi^-\rho$ such that $\Phi^+\Phi^-\rho$ is isomorphic to the factor representation $\rho/\tau_{\bar{I},1}$, where $\tau_{I,1}$ is the restriction of ρ to $V_{\bar{I},1} = \rho(v_{\bar{I},1})$, where $v_{\bar{I},1}$ is the maximal element of $B^-(1)$.

If ρ is indecomposable, then so are, as a rule, $\Phi^+(\rho)$ and $\Phi^-(\rho)$. More accurately, we have the following assertions.

(III) If ρ is indecomposable and $\Phi^+\rho \neq 0$, then $\Phi^+(\rho)$ is also indecomposable; in this case the monomorphism i: $\Phi^-\Phi^+\rho \to \rho$ defined above is an isomorphism. If ρ is indecomposable and $\Phi^+\rho = 0$, then ρ is Φ^+-atomic: $\rho \cong \rho_{j,1}$ for some $j \in \{0, 1, \ldots, r\}$.

(IV) If ρ is indecomposable and $\Phi^-\rho \neq 0$, then $\Phi^-(\rho)$ is also indecomposable, and the epimorphism p: $\rho \to \Phi^+\Phi^-\rho$ is an isomorphism. If ρ is indecomposable and $\Phi^-\rho = 0$, then ρ is Φ^--atomic: $\rho \cong \rho_{\bar{j},1}$ for some $j \in \{0, 1, \ldots, r\}$.

We prove (I) and (III) below in the framework of the more general Proposition 5.4. (II) and (IV) are dual to (I) and (III).

5.2. The elementary maps φ_i. We denote by φ_i the linear map φ_i: $V^1 \to V$ from the representation space V^1 of $\rho^1 = \Phi^+\rho$ to the representation space V of ρ that is defined by the formula

$$\varphi_i(\xi_1, \ldots, \xi_{i-1}, \xi_i, \xi_{i+1}, \ldots, \xi_r) = \xi_i. \tag{5.1}$$

We call these maps elementary. It follows at once from the definition of ρ^1 that for any $i \in \{1, \ldots, r\}$

(5.2) $$\rho^1(e_i) = \operatorname{Ker} \varphi_i,$$

(5.3) $$\operatorname{Im} \varphi_i = \rho(e_i) \sum_{t \neq i} \rho(e_t) = \rho(e_i h_i),$$

where $h_i = \sum_{t \neq i} e_t$. Note that φ_i does not define a morphism from ρ^1 to ρ.

The φ_i have another property, which could also serve as their definition. First some notation. We note by μ the embedding $\mu: V^1 \hookrightarrow R = \bigoplus_{i=1}^{r} \rho(e_i)$.

It follows from the definition $V^1 = \operatorname{Ker} \nabla$ that the following sequence of vector spaces is exact: $V \overset{\nabla}{\leftarrow} R \overset{\mu}{\leftarrow} V^1 \leftarrow 0$. We set $G_i = \{(0, \ldots, 0, \xi_i, 0, \ldots, 0) \mid \xi_i \in \rho(e_i)\}$. Then $R = \sum_{i=1}^{r} G_i$ and this sum is direct. We denote by π_i the projection $\pi_i: R \to R$ with kernel $G_i' = \sum_{t \neq i} G_t$ and image G_i.

PROPOSITION 5.3. (I) $\varphi_i = \nabla \pi_i \mu$ *and* (II) $\rho^1(e_i) = \operatorname{Ker} \varphi_i = \operatorname{Ker} \pi_i \mu$.

The proof is elementary.

We now describe an important "construction." Later (§§6, 7) we shall prove that this construction "builds up" from $V^1_{a,l} = \rho^1(v_{a,l})$, $v_{a,l} \in B^+(l)$, the subspace $V_{a,l+1} = \rho(v_{a,l+1})$, $v_{a,l+1} \in B^+(l+1)$, where a is some subset of $I = \{1, \ldots, r\}$.

CONSTRUCTION 5.1. Let ρ be a representation in V, τ^1 a representation of $\rho^1 = \Phi^+\rho$, and U^1 be the representation space of τ^1.

We set $U = \sum_{i=1}^{r} \varphi_i U^1$ $(U \subseteq V)$. Then a representation τ in U is given by $\tau(e_i) = \varphi_i U^1$.

PROPOSITION 5.4. *Let τ^1 be a direct summand of $\rho^1 = \Phi^+\rho$. Then the representation τ defined from τ^1 by the construction 5.1 has the following properties*:

(I) $\tau \cong \Phi^-\tau^1$;

(II) $\tau^1 \cong \Phi^+\tau$;

(III) τ *is a direct summand of* ρ.

Before proving Proposition 5.4, we illustrate it by one of its consequences. In §3 we have introduced the element $v_{\theta,1} = \bigcap_{i=1}^{r} h_i = \sum_{i=1}^{r} e_i h_i$, where $h_i = \sum_{j \neq i} e_j$. This $v_{\theta,1}$ is the minimal element of $B^+(1)$. We denote by $\tau_{\theta,1}$ the restriction of ρ to $V_{\theta,1} = \rho(v_{\theta,1})$.

COROLLARY 5.1. $\tau_{\theta,1} \cong \Phi^-\Phi^+\rho$.

PROOF OF THE COROLLARY. We have proved in Proposition 4.1 that $\tau_{\theta,1}$ is a direct summand of ρ. Therefore,

$$\tau_{\theta,1}(e_i) = V_{\theta,1}\rho(e_i) = \rho(\bigcap_j h_j)\rho(e_i) = \rho(\bigcap_j h_j e_i).$$

Since $h_j = \sum_{t \neq j} e_t$, we see that $e_i h_j = e_i$ if $i \neq j$, and hence $\tau_{\theta,1}(e_i) = \rho(e_i h_i)$.

In Proposition 5.4 we set $U^1 = V^1$. Then $U = \sum_{i=1}^{r} \varphi_i V^1$ and the representation τ in U is given by $\tau(e_i) = \varphi_i V^1$. By Proposition 5.4, $\tau \cong \Phi^- \rho^1 = \Phi^- \Phi^+ \rho$.

We claim that $\tau = \tau_{\theta,1}$. It follows easily from the definition of the elementary map φ_i that $\varphi_i V^1 = \rho(e_i h_i)$. Therefore,

$$U = \sum_{i=1}^{r} \varphi_i V^1 = \sum_{i=1}^{r} \rho(e_i h_i) = \rho\left(\sum_{i=1}^{r} e_i h_i\right) = \rho(v_{\theta,1}) = V_{\theta,1}.$$

Moreover, by definition $\tau(e_i) = \varphi_i V^1 = \rho(e_i h_i)$. Thus $\tau_{\theta,1} = \tau \cong \Phi^- \Phi^+ \rho$, and the Corollary is proved.

PROOF OF PROPOSITION 5.4. Suppose that $\rho^1 = \Phi^+ \rho$ splits into a direct sum $\rho^1 = \tau_2^1 \oplus \tau_2^1$. We denote by U_1^1 and U_2^1 the representation spaces of τ_1^1 and τ_2^1. Then $V^1 = U_1^1 + U_2^1$, $U_1^1 U_2^1 = 0$, and for every $i \in \{1, \ldots, r\}$

$$\rho^1(e_i) = U_1^1 \rho^1(e_i) + U_2^1 \rho^1(e_i). \tag{5.4}$$

Next we set $R_j = \sum_{i=1}^{r} \pi_i \mu U_j^1$ $(j = 1, 2)$. Note that $R_j \subseteq R = \bigoplus_{i=1}^{r} \rho(e_i)$. We list some properties of R_1 and R_2.

a) Firstly, $R_1 R_2 = 0$. We recall (see Proposition 5.3(II)) that $\rho^1(e_i) = \operatorname{Ker} \pi_i \mu$. Consequently, we can rewrite (5.4) as:

$$\operatorname{Ker} \pi_i \mu = U_1^1 \operatorname{Ker} \pi_i \mu + U_2^1 \operatorname{Ker} \pi_i \mu. \tag{5.4'}$$

By construction, the subspaces U_1^1 and U_2^1 do not intersect. Using this fact and (5.4′), we deduce easily that for any $i \in I$ the subspaces $\pi_i \mu U_1^1$ and $\pi_i \mu U_2^1$ also do not intersect:

$$(\pi_i \mu U_1^1)(\pi_i \mu U_2^1) = 0. \tag{5.5}$$

Note that $\pi_i \mu U_j^1 \subseteq \operatorname{Im} \pi_i = G_i$. Also, $R = \sum_{i=1}^{r} G_i$, and this sum is direct. From these properties, the definition $R_j = \sum_{i=1}^{r} \pi_i \mu U_j^1$, and (5.5), it is not hard to show that $R_1 R_2 = 0$.

b) Next we claim that $R_j(\mu V^1) = \mu U_j^1$ $(j = 1, 2)$. Since $\mu: V^1 \to R$ is an embedding, we set $\mu V^1 = V^1$ and $\mu U_j^1 = U_j^1$. By definition of $\pi_i \mu$, $\pi_i \mu(\xi_1, \ldots, \xi_i, \ldots, \xi_r) = (0, \ldots, 0, \xi_i, 0, \ldots, 0)$. Consequently, $\sum_{i=1}^{r} \pi_i \mu U_j^1 \supseteq U_j^1$, in other words, $U_j^1 \subseteq R_j$. Using this relation, the equation $V^1 = U_1^1 + U_2^1$, and applying Dedekind's axiom, we obtain

$$R_1 V^1 = R_1(U_1^1 + U_2^1) = U_1^1 + R_1 U_2^1.$$

Since $U_2^1 \subseteq R_2$, we have $R_1 U_2^1 \subseteq R_1 R_2 = 0$, hence $R_1 U_2^1 = 0$. Consequently,

$$R_1 V^1 = U_1^1. \tag{5.6}$$

The equation $R_2 V^1 = U_2^1$ is proved similarly.

c) We denote by τ_j the representation in $U_j = \sum_{i=1}^{r} \varphi_i U_j^1$ $(j = 1, 2)$ given by $\tau_j(e_i) = \varphi_i U_j^1$. We now claim that $\tau_j \cong \Phi^- \tau_j^1$.

We note that $\nabla R_j = \nabla\left(\sum_{i=1}^{r} \pi_i \mu U_j^1\right) = \sum_{i=1}^{r} \nabla \pi_i \mu U_j^1 = \sum_{i=1}^{r} \varphi_i U_j^1 = U_j$.

It follows from this equality that we can define the restriction of $\nabla \colon R \to V$ to $R_j \subseteq R$ and $U_j \subseteq V$. We denote this restriction by $\nabla_j \colon R_j \to U_j$. Since $\nabla R_j = U_j$, it is an epimorphism. Also, $\operatorname{Ker} \nabla_j = R_j \operatorname{Ker} \nabla = R_j V^1$. We have shown (see (5.6)) that $R_j V^1 = U_j^1$. Consequently, $\operatorname{Ker} \nabla_j = U_j^1$. We denote by μ_j the embedding $U_j^1 \subset\to R_j$. Clearly, μ_j is the restriction of μ to U_j^1 and R_j. Thus, the following sequence of vector spaces is exact:

$$0 \leftarrow U_j \overset{\nabla_j}{\leftarrow} R_j \overset{\mu_j}{\leftarrow} U_j^1 \leftarrow 0.$$

By definition, the representation τ_j^1 in U_j^1 is a direct summand of ρ^1 hence $\tau_j^1(e_i) = U_j^1 \rho^1(e_i)$. According to Proposition 5.2, $\rho^1(e_i) = \operatorname{Ker} \pi_i \mu$. Therefore $\tau_j^1(e_i) = U_j^1 (\operatorname{Ker} \pi_i \mu)$. It clearly follows that $\pi_i \mu U_j^1 \cong U_j^1 / \tau_j^1(e_i)$. By definition, $R_j = \sum_i \pi_i \mu U_j^1$, and as we know $\pi_i \mu U_j^1 \subseteq \pi_i R = G_i$. Thus the sum $\sum_i \pi_i \mu U_j^1$ is direct, hence $R_j \cong \bigoplus_{i=1}^{r} U_j^1 / \tau_j^1(e_i)$.

We denote by $(\pi_i \mu)_j \colon U_j^1 \to R_j$ the restriction of $\pi_i \mu$ to U_j^1 and R_j. By definition of π_i, $\sum_i \pi_i = 1$. Therefore $\sum_i \pi_i \mu = \left(\sum_i \pi_i\right) \mu = \mu$. In particular, it follows that $\mu_j = \sum_i (\pi_i \mu)_j$.

So we have shown that (I) $R_j \overset{\text{def}}{=} \sum_i \pi_i \mu U_j^1 \cong \bigoplus_{i=1}^{r} U_j^1 / \tau_j^1(e_i)$; (II) $\mu_j = \sum_{i=1}^{r} (\pi_i \mu)_j$; (III) $U_j = \nabla_j R_j \cong R_j / \operatorname{Ker} \nabla_j = R_j / \operatorname{Im} \mu_j = \operatorname{Coker} \mu_j$; (IV) the representation τ_j in U_j is such that $\tau_j(e_i) = \varphi_i U_j^1 \overset{\text{def}}{=} \nabla \pi_i \mu U_j^1 = \nabla(\pi_i \mu U_j^1) = \nabla_j(\pi_i \mu U_j^1)$. By comparing these properties with the definition of Φ^-, it is not difficult to convince ourselves that $\tau_j \cong \Phi^- \tau_j^1$. This proves (I) of Proposition 5.4.

(II) We outline first a proof of the isomorphism $\tau_j^1 \cong \Phi^+ \tau_j$. It is easy to verify that $\nabla_j \colon (\pi_i \mu U_j^1) \to U_j$ is a monomorphism, consequently, $\pi_i \mu U_j^1 \cong \nabla(\pi_i \mu U_j^1) = \tau_j(e_i)$. Therefore, $R_j \overset{\text{def}}{=} \sum_i \pi_i \mu U_j^1 \cong \bigoplus_{i=1}^{r} \tau_j(e_i)$. We omit the rest of the proof.

(III) We claim that each τ_j $(j = 1, 2)$ is a direct summand of ρ. To prove this, we show that $\tau_1 \oplus \tau_2 = \tau_{\theta,1}$, where $\tau_{\theta,1} = \rho \mid_{V_{\theta,1}}$ and

$V_{\theta,1} = \rho(v_{\theta,1})$. Note that in the proof of Corollary 5.1 we have used only Proposition 5.4, (I). Thus, we may now assume that Corollary 5.1 has been proved (that is, that $\tau_{\theta,1} \cong \Phi^-\Phi^+\rho$).

By assumption, $V^1 = U_1^1 + U_2^1$, and this sum is direct. In Corollary 5.1 we have proved that $\sum_{i=1}^{r} \varphi_i V^1 = V_{\theta,1}$. Therefore, $V_{\theta,1} = \sum_{i=1}^{r} \varphi_i(U_1^1 + U_2^1) = \sum_{i=1}^{r} \varphi_i U_1^1 + \sum_{i=1}^{r} \varphi_i U_2^1 = U_1 + U_2$.

We claim that U_1 and U_2 do not intersect. In (I) we have shown that $U_j = \nabla R_j$. Consequently, $U_1 + U_2 = \nabla(R_1 + R_2)$. There we have also shown that R_1 and R_2 do not intersect, and that

$\operatorname{Ker} \nabla \overset{\text{def}}{=} V^1 = U_1^1 + U_2^1 = V^1 R_1 + V^1 R_2$. Thus, we can write : $\operatorname{Ker} \nabla = R_1 \operatorname{Ker} \nabla + R_2 \operatorname{Ker} \nabla$. From this and from $R_1 R_2 = 0$, it follows easily that $(\nabla R_1)(\nabla R_2) = 0$, that is, $U_1 U_2 = 0$.

Thus, $V_{\theta,1} \cong U_1 \oplus U_2$. Consequently, $\tau_{\theta,1} = \rho \mid_{V_{\theta,1}}$ splits into the direct sum $\tau_{\theta,1} = \tau_1 + \tau_2$, where $\tau_j \cong \Phi^- \tau_j^1$. Now $\tau_{\theta,1}$ is a direct summand of ρ: $\rho = \rho(1) \oplus \tau_{\theta,1}$. Thus, $\rho = \rho(1) \oplus \tau_1 \oplus \tau_2$, and Proposition 5.4 is proved.

COROLLARY 5.2. *Let $\rho^1 = \Phi^+\rho$ be decomposable:*

$\rho^1 = \bigoplus_{j=1}^{n} \tau_j^1$, $n \geqslant 2$, $\tau_j^1 \neq 0$. *Then ρ splits into the direct sum* $\rho = \rho(1) \oplus (\bigoplus_{j=1}^{n} \tau_j)$, *where τ_j is constructed from τ_j^1 as in* 5.1. *Moreover,*

a) $\tau_j \cong \Phi^- \tau_j^1$;

b) $\bigoplus_{j=1}^{n} \tau_j \cong \Phi^-\Phi^+\rho \cong \tau_{\theta,1}$, where $\tau_{\theta 1} = \rho \mid_{V_{\theta,1}}$;

c) $\Phi^+\rho(1) = 0$ *and* $\rho(1) = \bigoplus_{t=0}^{r} k_t \rho_{t,1}^+$, *where $\rho_{t,1}^+$ is an atomic representation.*

PROOF. It follows from the proof of Proposition 5.4 that

$\rho = \rho(1) \oplus (\bigoplus_{j=1}^{n} \tau_j)$, where $\tau_j \cong \Phi^- \tau_j^1$ and $\bigoplus_{j=1}^{n} \tau_j = \tau_{\theta,1} \cong \Phi^-\Phi^+\rho$. By Proposition 4.1, $\rho(1)$ is the direct sum of atomic representations

$\rho_{t,1}^+$, $t \in \{0, 1, \ldots, r\}$, that is, $\rho = \bigoplus_{t=1}^{r} k_t \rho_{t,1}^+$. It is not hard to check that $\Phi^+\rho_{t,1}^+ = 0$ for every atomic representation $\rho_{t,1}^+$. Therefore, $\Phi^+\rho(1) = 0$.

PROPOSITION 5.5 (I) $\Phi^-\Phi^+\rho \cong \rho \Leftrightarrow$ *for all j* $: \sum_{i \neq j} \rho(e_i) = V$, *where V is the representation space of* ρ.

(II) $\Phi^+\Phi^-\rho \cong \rho \Leftrightarrow \bigcap_{i \neq j} \rho(e_i) = 0$ *for all j.*

PROOF. (I). By Corollary 5.1, $\Phi^-\Phi^+\rho \cong \rho$ is equivalent to $\tau_{\theta,1} = \rho$. By definition, $\tau_{\theta,1} = \rho \mid_{V_{\theta,1}}$, where $V_{\theta,1} = \rho(v_{\theta,1}) = \rho(\bigcap_j h_j) = \bigcap_j \rho(h_j)$.

Thus, $\Phi^-\Phi^+\rho \cong \rho \Leftrightarrow V = \bigcap_i \rho(h_j)$. Clearly, $V = \bigcap_j \rho(h_j) \Leftrightarrow V = \rho(h_j) = \rho\left(\sum_{i \neq j} e_i\right)$.

for all j.

(II) This assertion is dual to (I).

PROPOSITION 5.6. *Let ρ be an indecomposable representation.*

(I) *If* $\Phi^+\rho \neq 0$ *then* $\rho^1 = \Phi^+\rho$ *is also indecomposable, and* $\rho \cong \Phi^-\rho^1 = \Phi^-\Phi^+\rho$;

(II) *If* $\Phi^+\rho = 0$, *then* ρ *is* Φ^+*-atomic*: $\rho \cong \rho^+_{t,1}$, $t \in \{0, 1, \ldots, r\}$;

(III) *If* $\Phi^-\rho \neq 0$, *then* $\rho^{-1} = \Phi^-\rho$ *is also indecomposable, and* $\rho \cong \Phi^+\Phi^-\rho$;

(IV) *If* $\Phi^-\rho = 0$, *then* $\rho \cong \rho^-_{t,1}$, $t \in \{0, 1, \ldots, r\}$.

PROOF. (I) and (II) clearly follow from Corollary 5.1; (III) and (IV) are dual to (I) and (II).

5.3. The maps φ^l_i. From a given representation ρ we can define not only the representations ρ^{-1} and ρ^1, but also an infinite series of representations $\ldots \rho^{-l}, \rho^{-l+1}, \ldots, \rho^{-1}, \rho^1, \ldots, \rho^{l-1}, \rho^l, \ldots$, where $\rho^l \cong \Phi^+\rho^{l-1}$ and $\rho^{-l} \cong \Phi^-\rho^{-l+1}$. To avoid loss of generality, we occasionally denote ρ by ρ^0, and the representation space of ρ^0 by V^0. Thus, if $l \geqslant 1$, then

$$V^l = \{(\xi_1^{l-1}, \ldots, \xi_r^{l-1}) \mid \xi_i^{l-1} \in \rho^{l-1}(e_i), \sum_{i=1}^{r} \xi_i^{l-1} = 0\}$$

and for any $i \in I$,

$$\rho^l(e_i) = \{(\xi_1^{l-1}, \ldots, \xi_{i-1}^{l-1}, 0, \xi_{i+1}^{l-1}, \ldots, \xi_r^{l-1}) \in V^l\}.$$

We define the elementary map φ^l_i: $V^l \to V^{l-1}$, $l \geqslant 1$, by setting $\varphi^l_i(\xi_1^{l-1}, \ldots, \xi_i^{l-1}, \ldots, \xi_r^{l-1}) = \xi_i^{l-1}$. Sometimes we denote φ^l_i by φ_{i_l}. It follows from this definition that

$$\text{Ker } \varphi^l_i = \rho^l(e_i), \tag{5.7}$$

$$\text{Im } \varphi^l_i = \rho^{l-1}(e_i h_i) \subseteq \rho^{l-1}(e_i). \tag{5.8}$$

Next, we set $\varphi(i_k, \ldots, i_l) = \varphi_{i_k} \circ \varphi_{i_{k+1}} \circ \ldots \circ \varphi_{i_l}$, $1 \leqslant k \leqslant l$. Thus, $\varphi(i_k, \ldots, i_l)$: $V^l \to V^{k-1}$. It follows from (5.7) and (5.8) that if $i_j = i_{j+1}$, then $\varphi(i_k, \ldots, i_j, i_{j+1}, \ldots, i_l) = 0$.

We now describe a generalization of the Construction 5.1.

CONSTRUCTION 5.2. Let $\rho^l = (\Phi^+)^l\rho^0$, and let V^0 and V^l be the representation spaces of ρ^0 and ρ^l. Let τ^l be a subrepresentation of ρ^l, and U^l the representation space of $\tau^l (U^l \subseteq V^l)$. We set $U^0 = \sum_{i_1, \ldots, i_l} \varphi(i_1, \ldots, i_l)U^l$, where the summation is over all sequences $i_1, \ldots, i_l$ with $i_j \in I = \{1, \ldots, r\}$. We define a representation τ^0 in U^0 by setting $\tau^0(e_j) = \varphi^1_j \sum_{i_2, \ldots, i_l} \varphi(i_2, \ldots, i_l)U^l$. Thus, τ^0 is a subrepresentation of ρ^0.

PROPOSITION 5.7. *Suppose that τ^l is a direct summand of*

$\rho^l = (\Phi^+)^l \rho^0$. *Then the subrepresentation* $\tau^0 \subseteq \rho^0$, *which we construct from* τ^l *by the method described above, has the following properties*:

(I) $\tau^0 \cong (\Phi^-)^l \tau^l$;

(II) $\tau^l \cong (\Phi^+)^l \tau^0$;

(III) τ^0 *is a direct summand of the representation* ρ^0.

REMARK. We shall prove later (in §6) that the Construction 5.2 has another property which is, perhaps, for us the most essential. Let $\tau^l = \rho^l|_{V_{a,1}}$, where $V_{a,1} = \rho(v_{a,1})$ and $v_{a,1} \in B^+(1)$. Then τ^0, which is constructed from τ^l, is such that $\tau^0 = \rho^0|_{V_{a,l+1}}$, where $V_{a,l+1} = \rho(v_{a,l+1})$ and $v_{a,l+1} \in B^+(l+1)$. Here $v_{a,1}$ and $v_{a,l+1}$ are the elements of $B^+(1)$ and $B^+(l+1)$ corresponding to the same subset $a \subseteq I$.

PROOF. We introduce the following notation:

$$\left.\begin{aligned} U^\lambda &= \sum_{i_{\lambda+1},\ldots,i_l} \varphi(i_{\lambda+1},\ldots,i_l)\,U^l, \\ F_j^\lambda &= \varphi_j^{\lambda+1} \sum_{i_{\lambda+2},\ldots,i_l} \varphi(i_{\lambda+2},\ldots,i_l)\,U^l \end{aligned}\right\} 0 \leqslant \lambda \leqslant l-1.$$

In the space $U^\lambda \subseteq V^\lambda$ we define a representation τ^λ by setting $\tau^\lambda(e_j) = F_j^\lambda$.

We claim that for all λ with $0 \leqslant \lambda \leqslant l-1$

$$\tau^\lambda \cong \Phi^- \tau^{\lambda+1} \tag{5.9}$$

and that τ^λ is a direct summand of $\rho^\lambda = (\Phi^+)^\lambda \rho^0$.

For $\lambda = l-1$, this assertion follows from Proposition 5.4, if we replace ρ by ρ^{l-1} and use the fact that $\rho^l = \Phi^+ \rho^{l-1}$.

Suppose now that (5.7) has been proved for $\lambda = k+1, \ldots, l-1$. We prove it for $\lambda = k$. By definition $\rho^{k+1} = \Phi^+\rho^k$, and we take it as proved that τ^{k+1} is a direct summand of ρ^{k+1}. We denote by $\bar\tau^k$ the representation in $\bar U_k = \sum_{i=1}^r \varphi_i^{k+1} U^{k+1}$ for which $\bar\tau^k(e_i) = \varphi_i^{k+1} U^{k+1}$. Then according to Proposition 5.4 $\bar\tau^k$ is a direct summand of ρ^k and $\bar\tau^k \cong \Phi^- \tau^{k+1}$. We claim that $\bar\tau^k = \tau^k$. Indeed, we take it as proved that

$U^{k+1} = \sum_{i_{k+2},\ldots,i_l} \varphi(i_{k+2},\ldots,i_l)U^l$. Therefore,

$\varphi_i^{k+1} U^{k+1} = \varphi_i^{k+1} \sum_{i_{k+2},\ldots,i_l} \varphi(i_{k+2},\ldots,i_l)U^l = F_i^k$. Consequently,

$$\begin{aligned} \bar U_k &= \sum_{i=1}^r \varphi_i^{k+1} U^{k+1} = \sum_{i=1}^r \varphi_i^{k+1} \sum_{i_{k+2},\ldots,i_l} \varphi(i_{k+2},\ldots,i_l)\,U^l = \\ &= \sum_i \sum_{i_{k+2},\ldots,i_l} \varphi_i^{k+1} \varphi(i_{k+2},\ldots,i_l)\,U^l = \sum_{i_{k+1},\ldots,i_l} \varphi(i_{k+1}, i_{k+2},\ldots,i_l)\,U^l = U^k. \end{aligned}$$

Thus, $\tau^k \cong \Phi^-\tau^{k+1}$, and τ^k is a direct summand of ρ^k. Also, by Proposition 5.4, $\tau^{k+1} \cong \Phi^+\tau^k$. By induction we conclude that these assertions are true for any $k \geqslant 0$. Consequently,
$\tau^0 \cong \Phi^-\tau^1 \cong \Phi^-(\Phi^-\tau^2) \cong \ldots \cong (\Phi^-)^l\tau^l$ and, similarly, $\tau^l \cong (\Phi^+)^l\tau^0$.

We write $\overline{V}_{\theta,l} = \sum_{i_1,\ldots,i_l} \varphi(i_1, \ldots, i_l)V^l$ By $\tau_{\theta,l}$ we denote the representation in $\overline{V}_{\theta,l}$ for which

$$\tau_{0,l}(e_i) = \varphi_i^1 \sum_{i_2,\ldots,i_l} \varphi(i_2, \ldots, i_l)V^l.$$

COROLLARY 5.3. *$\tau_{\theta,l}$ is a direct summand of ρ^0, and $\tau_{\theta,l} \cong (\Phi^-)^l(\Phi^+)^l\rho$.*

COROLLARY 5.4. *If $\rho^l = (\Phi^+)^l\rho^0$ splits into a direct sum*
$\rho^l = \bigoplus_{j=1}^{n} \tau_j^l$, $n \geqslant 2$, $\tau_j^l \neq 0$, then ρ^0 is also decomposable:
$\rho^0 = \rho(l) \oplus (\bigoplus_{j=1}^{n} \tau_j)$, where τ_j is constructed from τ_j^l by the Construction 5.2. Here $\tau_j \cong (\Phi^-)^l\tau_j^l$, and $(\Phi^+)^l\rho(l) = 0$.

The proof of this Proposition is, essentially, a combination of the arguments in Proposition 5.6 and Corollaries 5.2 and 5.3.

§6. Proof of the theorem on perfect elements

6.1. In this section we prove the main theorem: that all elements of $B^+(l)$ and $B^-(l)$ are perfect. We assume that the following assertion has been proved: let ρ be an arbitrary representation of D^r, $r \geqslant 4$, $\rho^l = (\Phi^+)^l\rho$, and let V and V^l be the representation spaces of ρ and ρ^l, respectively. Then

$$(6.1) \qquad \rho(e_{i_1\ldots i_l t}) = \varphi(i_1, \ldots, i_l)\rho^l(e_t),$$

$$(6.2) \qquad \rho(f_{i_1\ldots i_l 0}) = \varphi(i_1, \ldots, i_l)V^l.$$

The proofs of these assertions, which are in §7, are the central and most complicated part of this paper.

THE REPRESENTATIONS $\rho_{t,l}^+$ AND $\rho_{t,l}^-$. We define representations $\rho_{t,l}^+$ and $\rho_{t,l}^-$ ($t \in \{0, 1, \ldots, r\}$, $l = 1, 2, \ldots$) by means of the atomic representations $\rho_{t,1}$ and $\rho_{t,1}^-$ that were constructed in §4. We set, by definition,

$$\rho_{t,1}^+ = \rho_{t,1}, \quad \rho_{t,l}^+ = (\Phi^-)^{l-1}\rho_{t,1}^+, \quad \rho_{t,l}^- = (\Phi^+)^{l-1}\rho_{t,1}^-.$$

We shall prove later (§8) that $\rho_{t,l}^+ \cong \rho_{t,l}$, where $\rho_{t,l}$ are the representations defined in §1. The functorial definition we have given just now is more convenient in those cases when we need not go deeply into the "inner structure" of $\rho_{t,l}$.

Let ρ be a representation of D^r in V. We write
$\dim \rho = (n; m^1, \ldots, m^r)$, where $n = \dim V$ and $m^i = \dim \rho(e_i)$. For

example, the atomic representations $\rho^+_{t,1}$ and $\rho^-_{t,1}$ have $\dim \rho^+_{0,1} = (1;\ 0, \ldots, 0)$ and, for $t \neq 0$, $\dim \rho^+_{t,1} = (1;\ 0, \ldots, 0, 1, 0, \ldots, 0)$, where $m^t = 1$; $\dim \rho^-_{0,1} = (1;\ 1, \ldots, 1)$ and, for $t \neq 0$ $\dim \rho^-_{t,1} = (1;\ 1, \ldots, 1, 0, 1, \ldots, 1)$, where $m^t = 0$.

PROPOSITION 6.1. (I) $\rho^-_{t,l} \cong (\rho^+_{t,l})^*$, *where* $*$ *denotes the conjugate representation.*

(II) *The representations* $\rho^+_{t,l}$ *and* $\rho^-_{t,l}$ *of* D^r, $r \geqslant 4$, *are indecomposable for all* $t \in \{0, 1, \ldots, r\}$ *and* $l = 1, 2, \ldots$.

PROPOSITION 6.2. *We set* $\dim \rho^+_{t,l} = (n_{t,l};\ m^1_{t,l}, \ldots, m^r_{t,l})$.

(I) *The sequence* $n_{t,l}$ *satisfies the recurrence relations*

$$n_{t,l} = (r-2)n_{t,l-1} - n_{t,l-2}, \qquad l \geqslant 3, \tag{6.3}$$

with the initial conditions $n_{t,1} = 1$, $n_{t,2} = r - 2$ *for* $t \neq 0$ *and* $n_{0,1} = 1$, $n_{0,2} = r - 1$. *Hence it is clear, in particular, that, for* $t \neq 0$, $n_{t,l}$ *does not depend on* t.

(II) *There exist intergers* m_l *and* $m_{0,l}$ *such that*

$$\begin{cases} m^i_{t,l} = \begin{cases} m_l & \text{if } i \neq t \neq 0, \\ m_l + (-1)^{l+1} & \text{if } i = t \neq 0, \end{cases} \\ m^i_{0,l} = m_{0,l}. \end{cases} \tag{6.4}$$

These numbers satisfy the relations

$$\begin{cases} m_1 = m_{0,1} = 0, \\ rm_l = n_l + n_{l-1} + (-1)^l, \\ rm_{0,l} = n_{0,l} + n_{0,l-1}. \end{cases} \tag{6.5}$$

Here n_l *stands for* $n_{t,l}$ *with* $t \neq 0$.

We prove first Propositions 6.1 (II) and 6.2 by induction on l.

Let $t \neq 0$, so that $\rho^+_{t,l}$ is a representation of the first kind. For $l = 1$ we have already proved that $\dim \rho^+_{t,1} = (1;\ 0, \ldots, 0, 1, 0, \ldots, 0)$ and that $\rho^+_{t,1}$ is indecomposable.

By definition $\rho^+_{t,2} = \Phi^-\rho^+_{t,1}$. It is not hard to verify that $\dim \rho^+_{t,2} = (r-2;\ 1, \ldots, 1, 0, 1, \ldots, 1)$, where $m^t_{t,2} = 0$. It was shown in Proposition 5.6 that if ρ is indecomposable and $\Phi^-\rho \neq 0$, then $\Phi^-\rho$ is also indecomposable. Consequently, $\rho^+_{t,2}$ is indecomposable. We have shown that $n_{t,1} = 1$ and $n_{t,2} = r - 2$ for $t \neq 0$. It is also easy to check that (6.4) and (6.5) hold for the numbers $m^i_{t,1}$ and $m^i_{t,2}$.

Suppose that Propositions 6.1 (II) and 6.2 have been proved for all $k \leqslant l$, so that the $\rho^+_{t,k}$, $t \neq 0$ are indecomposable, and that the numbers $n_{t,k} = n_k$ $(t \neq 0)$ satisfy the relations

$$\begin{cases} n_1 = 1, \quad n_2 = r-2, \\ n_k = (r-2)n_{k-1} - n_{k-2} \qquad (k = 3, \ldots, l). \end{cases} \tag{6.6}$$

We now prove these assertions for $k = l + 1$, where $l \geqslant 2$.

a) We show first that $\rho^+_{t,l+1}$, $l \geqslant 2$, is indecomposable. Note that from (6.6) for $r \geqslant 4$ it follows that

$$n_k > 1 \text{ for } k = 2, 3, \dots, l \tag{6.7}$$

We claim that $\rho^+_{t,l+1} = \Phi^-\rho^+_{t,l}$ $(t \neq 0)$ is different from zero. Assume the contrary, that $\Phi^-\rho^+_{t,l} = 0$. Since $\rho^+_{t,l}$ is indecomposable, $\rho^+_{t,l}$ is atomic (see Proposition 5.6): $\rho^+_{t,l} \cong \rho^-_{s,1}$. By definition, $\rho^-_{s,1}$ is a representation in a one-dimensional space, $\dim V^+_{t,l} = 1$. Since $\dim V^+_{t,l} = n_l$, we have reached a contradiction to the fact that $n_l > 1$ for $l \geqslant 2$ (see (6.7)).

Thus, $\rho^{+\prime}_{t,l+1} = \Phi^-\rho^+_{t,l} \neq 0$. Hence, by Proposition 5.6, $\rho^+_{t,l+1}$ and $\rho^+_{t,l}$ are indecomposable, and $\rho^+_{t,l} \cong \Phi^+\rho^+_{t,l+1}$, that is, $\rho^+_{t,l} \cong \Phi^+\Phi^-\rho^+_{t,l}$.

b) We now prove (6.3)–(6.5) in the case $l + 1$. In Proposition 5.5 we have proved that if $\Phi^+\Phi^-\rho \cong \rho$, then $\rho(\bigcap_{j\neq i} e_j) = 0$ for every i. Hence it follows elementarily (see Proposition 5.1) that

$\Phi^-\rho(e_i) = \dim V - \dim \rho(e_i)$ and $\dim V^{-1} = \sum_{i=1}^{r} \dim \Phi^-\rho(e_i) - \dim V$. Applying this to $\rho^+_{t,l}$ and $\rho^+_{t,l+1} = \Phi^-\rho_{t,l}$, we obtain

$$m^i_{t,l+1} = n_{t,l} - m^i_{t,l}, \tag{6.8}$$

$$n_{t,l+1} = \sum_{i=1}^{r} m^i_{t,l+1} - n_{t,l}. \tag{6.9}$$

Since by the inductive hypothesis $n_{t,l} = n_l$ and $m^i_{t,l} = m_l$ if $i \neq t$ and $m^t_{t,l} = m_l + (-1)^{l+1}$, we see that

$$m^i_{t,l+1} = \begin{cases} n_l - m_l & \text{if } i \neq t, \\ n_l - m_l + (-1)^{l+2} & \text{if } i = t. \end{cases} \tag{6.10}$$

Thus, we have proved (6.4) in the case $l + 1$.

Substituting (6.10) in (6.9), we find $n_{t,l+1} = (r - 1)n_l - rm_l + (-1)^l$. By the inductive hypothesis we have $rm_l = n_l + n_{l-1} + (-1)^l$. Consequently, $n_{t,l+1} = (r - 2)n_l - n_{l-1}$. This proves (6.3) in the case $l + 1$.

We set, by definition, $m_{l+1} = n_l - m_l$. Then from (6.9), (6.10), and (6.3) just proved for $l + 1$ it follows that $rm_{l+1} = n_l + n_{l-1} + (-1)^{l+1}$. Thus, we have proved (6.5) in the case $l + 1$.

Combining the results of a) and b), we see that we have proved Propositions 6.1 (II) and (6.2) for $t \neq 0$. The proofs for $t = 0$ are similar.

Now we prove that $\rho^+_{t,l}$ and $\rho^-_{t,l}$ are conjugate. Now $\rho^+_{t,1}$ and $\rho^-_{t,1}$ are conjugate by definition. Suppose that $(\rho^-_{t,l}) \cong (\rho^+_{t,l})^*$. Then $(\rho^-_{t,l+1})^* = (\Phi^+\rho^-_{t,l})^* \cong$ (by Proposition 5.2) $\cong \Phi^-(\rho^-_{t,l})^* \cong \Phi^-\rho^+_{t,l} = \rho^+_{t,l+1}$.

6.2. The lattices D^1, D^2, D^3 are known to be finite. Therefore, each of these lattices has only finitely many non-isomorphic indecomposable representations, namely 2, 4 and 9, respectively. Each of these representations can be written either in the form $\rho^+_{t,\,l}$, $t \in \{0, 1, \ldots, r\}$, $l \in \{1, \ldots, r\}$, or in the form $\rho^-_{t,\,l}$.

For example, the 9 different indecomposable representations of D^3 are the following: $\rho^+_{t,\,1}$, $\rho^+_{t,\,2}$, where $t \in \{0, 1, 2, 3\}$ and $\rho^+_{0,\,3}$. Each of these representations can also be written in the form $\rho^-_{t,\,k}$. Namely, if $t \neq 0$, then $\rho^+_{t,\,1} \cong \rho^-_{t,\,2}$ and $\rho^+_{t,\,2} \cong \rho^-_{t,\,1}$ and $\rho^+_{0,\,1} \cong \rho^-_{0,\,3}$, $\rho^+_{0,\,2} \cong \rho^-_{0,\,2}$, $\rho^+_{0,\,3} \cong \rho^-_{0,\,1}$. The "dimensions" $\dim \rho_{t,\,l}$ of these representations are given in the following table:

$$(1;\ 0,\ 0,\ 0),\quad (1;\ 1,\ 0,\ 0),\quad (1;\ 0,\ 1,\ 0),\quad (1;\ 0,\ 0,\ 1),$$
$$(2;\ 1,\ 1,\ 1),\quad (1;\ 0,\ 1,\ 1),\quad (1;\ 1,\ 0,\ 1),\quad (1;\ 1,\ 1,\ 0).$$
$$(1;\ 1,\ 1,\ 1).$$

Thus, among the indecomposable representations of D^3, four are Φ^+-atomic, and four are Φ^--atomic, so that the spaces of these eight representations are one-dimensional. Only the representation $\rho^+_{0,\,2} \cong \rho^-_{0,\,2}$ is such that $\dim \rho^+_{0,\,2} = (2;\ 1,\ 1,\ 1)$.

6.3. With each representation ρ of D^r in V there is associated in a natural way an oriented graph Γ:

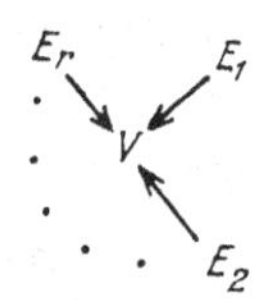

where $E_i = \rho(e_i)$. The diagram of this graph with unoriented edges is called a Dynkin diagram [3]:

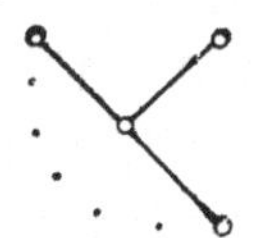

Then $\dim \rho$ is an integer-valued function on the set Γ_0 of vertices of the graph (for D^r, the set Γ_0 consists of $r + 1$ points). Following the methods of [1], it can be shown that the numbers $\dim \rho^+_{t,\,l}$ and $\dim \rho^-_{t,\,l}$ correspond to positive roots of the Dynkin diagram.

6.4. Theorem on the perfectness of the cubicles. As we know already from §3, the l-th upper cubicle $B^+(l)$ is generated by the r elements $h_t(l) = \sum_{j \neq t} e_j(l)$, where $e_j(l) = \sum_{\alpha \in A_j(r,\, l)} e_\alpha = \sum_{i_1, \ldots, i_{l-1}} e_{i_1 \ldots i_{l-1} t}$. We have also proved there that $B^+(l)$ is a Boolean algebra. Every element $v^+(l) \in B^+(l)$ can be written in the form:

$$v_{a,\,l} = \sum_{i \in a} e_i(l) + \sum_{j \in a'} e_j(l)\, h_j(l) = h_0(l) \big(\bigcap_{j \in a'} h_j(l)\big),$$

where a is a subset of $I = \{1, \ldots, r\}$ and $a' = I - a$, and where $h_0(l) = \sum_{j=1}^{r} e_j(l)$. (Note that $h_0(l) \supset h_j(l)$, $j \neq 0$.)

The minimal element of $B^+(l)$, which we denote by $v_{\theta,l}$, corresponds to the empty subset $a = \varnothing$. Thus,

$$v_{\theta,l} = \sum_{j\in I} e_j(l)\,h_j(l) = \bigcap_{j\in I} h_j(l).$$

We now prove a theorem that generalizes Proposition 4.1 to an arbitrary l.

THEOREM 6.1. *Let ρ be a representation of D^r in V, and let $v_{\theta,l}$ be the minimal element of $B^+(l)$. Then ρ splits into a direct sum*

$$\rho = \Big(\bigoplus_{\substack{j\in I\cup\{0\}\\ 1\leqslant k\leqslant l}} \tilde{\rho}_{j,k}\Big) \oplus \tau_{\theta,l},$$

where $\tau_{\theta,l}$ is the restriction of ρ to $\rho(v_{\theta,l}) = V_{\theta,l}$, and each $\tilde{\rho}_{j,k}$ is a multiple of the indecomposable representation $\rho^+_{j,k}$, that is,

$$\tilde{\rho}_{j,k} = \underbrace{\rho^+_{j,k} \oplus \ldots \oplus \rho^+_{j,k}}_{m_{j,k}},\quad m_{j,k} \geqslant 0.$$

The proof is by induction on k. For $k = 1$, the proof was given in Proposition 4.1. We assume that the theorem has been proved for $k = l - 1$ and prove it for $k = l$.

Proposition 4.1 applies to the representation $\rho^{l-1} = (\Phi^+)^{l-1}\rho$. This means that we can choose subspaces $U^{l-1}_{j,1} \subseteq V^{l-1}$ with the following properties: 1) $V^{l-1} = \sum_{j=0} U^{l-1}_{j,1} + V^{l-1}_{\theta,1}$, where $V^{l-1}_{\theta,1} = \rho^{l-1}(v_{\theta,1})$, and this sum is direct; 2) ρ^{l-1} splits into a direct sum: $\rho^{l-1} = \bigoplus_{j=0}^{r} \tilde{\rho}^{l-1}_{j,1} \oplus \tau^{l-1}_{\theta,1}$, where $\tilde{\rho}^{l-1}_{j,1} = \rho^{l-1}|_{U^{l-1}_{j,1}}$, $\tau^{l-1}_{\theta,1} = \rho^{l-1}|_{V^{l-1}_{\theta,1}}$; 3) $\tilde{\rho}^{l-1}_{j,1}$ is a multiple of the atomic representation $\rho^+_{j,1}$. The subspaces $U^{l-1}_{j,1}$ chosen in this way are such that any subspace $\rho^{l-1}(v_{a,1}) \subseteq V^{l-1}$, corresponding to $v_{a,1} \in B^+(1)$, can be expressed as a sum

$$\rho^{l-1}(v_{a,1}) = \sum_{j\in a} U^{l-1}_{j,1} + V^{l-1}_{\theta,1}. \tag{6.11}$$

With the help of the construction in 5.2 we build up subspaces $U_{t,l}$ and $\tilde{V}_{\theta,l}$ such that $U_{t,l} = \sum_{i_1,\ldots,i_{l-1}} \varphi(i_1, \ldots, i_{l-1})\, U^{l-1}_{t,1}$ and $\tilde{V}_{\theta,l} = \sum_{i_1,\ldots,i_{l-1}} \varphi(i_1, \ldots, i_{l-1})\, V^{l-1}_{\theta,1}$. In §5 (Corollary 5.4) we have proved that ρ then splits into the direct sum

$$\rho = \rho(l-1) \oplus \Big(\bigoplus_{j=0}^{r} \tilde{\rho}_{j,l}\Big) \oplus \tilde{\tau}_{\theta,l}. \tag{6.12}$$

where $\tilde{\rho}_{j,l} = \rho|_{U_{j,l}}$ and $\tilde{\tau}_{0,l} = \rho|_{\tilde{V}_{0,l}}$. Also $\tilde{\rho}_{j,l} \cong (\Phi^-)^{l-1}\tilde{\rho}^{l-1}_{j,1}$.

a) We claim that $\tilde{\rho}_{j,l}$ is a multiple of the indecomposable representation $\rho^+_{j,l}$. By construction, $\tilde{\rho}^{l-1}_{j,1} \cong \underbrace{\rho^+_{j,1} \oplus \ldots \oplus \rho^+_{j,1}}_{m_{j,l}}$. Consequently,

$$\tilde{\rho}_{j,l} \cong (\Phi^-)^{l-1}\tilde{\rho}_{j,1}^{l-1} \cong (\Phi^-)^{l-1}\rho_{j,1}^+ \oplus \dots \oplus (\Phi^-)^{l-1}\rho_{j,1}^+ \cong \underbrace{\rho_{j,l}^+ \oplus \dots \oplus \rho_{j,l}^+}_{m_{j,l}}.$$

b) Now we prove that for any $a \subseteq I$ and $v_{a,l} \in B^+(l)$

$$(6.13) \qquad \rho(v_{a,l}) = \sum_{t \in a} U_{t,l} + \tilde{V}_{\theta,l}.$$

We denote the subset $I - \{j\}$ by a_j. Then $v_{a_j,1} = h_j(1) = \sum_{t \neq j} e_t$. Hence, in accordance with (6.11), we can write

$$\rho^{l-1}(h_j(1)) = \sum_{i \in I - \{j\}} U_{i,1}^{l-1} + V_{\theta,1}^{l-1}.$$

We also write $X_j = \sum \varphi(i_1, \dots, i_{l-1})\rho^{l-1}(h_j(1))$. Let us determine X_j. On the other hand,

$$X_j = \sum_{i_1,\dots,i_{l-1}} \varphi(i_1, \dots, i_{l-1}) \Big(\sum_{i \in I - \{j\}} U_{i,1}^{l-1} + V_{\theta,1}^{l-1} \Big) = \sum_{i \in I - \{j\}} U_{i,l} + \tilde{V}_{\theta,l}.$$

But

$$X_j = \sum_{i_1,\dots,i_{l-1}} \varphi(i_1, \dots, i_{l-1}) \rho^{l-1}\Big(\sum_{t \neq j} e_t\Big) =$$
$$= \sum_{i_1,\dots,i_{l-1}} \varphi(i_1, \dots, i_{l-1}) \sum_{t \neq j} \rho^{l-1}(e_t) = \sum_{t \neq j} \sum_{i_1,\dots,i_{l-1}} \varphi(i_1, \dots, i_{l-1}) \rho^{l-1}(e_t).$$

We assume it as proved that $\varphi(i_1, \dots, i_{l-1})\rho^{l-1}(e_t) = \rho(e_{i_1\dots i_{l-1}t})$ if $t \neq i_{l-1}$, and that $\varphi(i_1, \dots, i_{l-1})\rho^{l-1}(e_{i_{l-1}}) = 0$. Consequently,

$$X_j = \sum_{t \neq j} \sum_{\substack{i_1,\dots,i_{l-1} \\ i_{l-1} \neq t}} \rho(e_{i_1\dots i_{l-1}t}) = \rho\Big(\sum_{t \neq j} \sum_{\substack{i_1,\dots,i_{l-1} \\ i_{l-1} \neq t}} e_{i_1\dots i_{l-1}t}\Big) = \rho\Big(\sum_{t \neq j} e_t(l)\Big) = \rho(h_j(l)).$$

So we have proved that

$$(6.14) \qquad \rho(h_j(l)) = \sum_{i \in I - \{j\}} U_{i,l} + \tilde{V}_{\theta,l}.$$

It is known (see §3) that $v_{a,l} = \bigcap_{j \in a'} h_j(l)$ for every $a \neq I$. Therefore, $\rho(v_{a,l}) = \rho\big(\bigcap_{j \in a'} h_j(l)\big) = \bigcap_{j \in a'} \rho(h_j(l))$. Hence, using (6.14), we find that

$$\rho(v_{a,l}) = \bigcap_{j \in a'} \Big(\sum_{i \in I - \{j\}} U_{i,l} + \tilde{V}_{\theta,l} \Big).$$

It follows from (6.12) that the sum $\sum_i U_{i,l} + \tilde{V}_{\theta,l}$ is direct. Consequently, the sublattice of $\mathscr{L}(V)$ generated by $U_{j,l}$ and $\tilde{V}_{\theta,l}$ is distributive. Using these facts, we can show easily that

$$\bigcap_{j\in a'}\Big(\sum_{i\in I-\{j\}} U_{i,l}+\tilde{V}_{\theta,l}\Big)=\sum_{i\in a} U_{i,l}+\tilde{V}_{\theta,l}.$$

Thus, if $a \neq I$, we have proved that

(6.13′) $$\rho(v_{a,l})=\sum_{i\in a} U_{i,l}+\tilde{V}_{\theta,l}.$$

It remains to prove (6.13) when $a = I$. To do this, we have to use the fact that for any $j, s \in I$ with $j \neq s$ we have $v_{I,l} = h_j(l) + h_s(l)$. From this and (6.14) it follows easily that $\rho(v_{I,l})=\sum_{i\in I} U_{i,l}+\tilde{V}_{\theta,l}$. This proves (6.13). In particular, for $a=\varnothing$, we obtain $\rho(v_{\theta,l})=\tilde{V}_{\theta,l}$. By definition,

$$\sum_{i\in a} U_{i,l}+\tilde{V}_{\theta,l}=\sum_{i_1,\dots,i_{l-1}}\varphi(i_1,\dots,i_{l-1})\Big(\sum_{i\in a} U_{i,1}^{l-1}+V_{\theta,1}^{l-1}\Big)=$$
$$=\sum_{i_1,\dots,i_{l-1}}\varphi(i_1,\dots,i_{l-1})\,\rho^{l-1}(v_{a,1}).$$

Thus, we can rewrite (6.13) in the following way:

(6.15) $$\rho(v_{a,l})=\sum_{i_1,\dots,i_{l-1}}\varphi(i_1,\dots,i_{l-1})\,\rho^{l-1}(v_{a,1}).$$

It also follows from what we have proved that every subspace $V_{a,l}=\rho(v_{a,l})$ is such that $\tau_{a,l}=\rho\,|_{V_{a,l}}$ is a direct summand of ρ. For it follows easily from (6.12) and (6.13) that

$$\rho=\rho(l-1)\oplus\Big(\bigoplus_{j\in a'\cup\{0\}}\tilde{\rho}_{j,l}\Big)\oplus\tau_{a,l}.$$

c) We claim that

(6.16) $$\tau_{\theta,l-1}=\Big(\bigoplus_{j=0}^{r}\tilde{\rho}_{j,l}\Big)\oplus\tau_{\theta,l},$$

where $\tau_{\theta,k}=\rho\,|_{V_{\theta,k}}$, $V_{\theta,k}=\rho(v_{\theta,k})$. By definition, $\bigoplus_{j=0}^{r}\tilde{\rho}_{j,l}\oplus\tau_{\theta,l}$ is the restriction of ρ to $\sum_{j=0}^{r} U_{j,l}+\tilde{V}_{\theta,l}$. Clearly,

$$\sum_{j=0}^{r} U_{j,l}+\tilde{V}_{\theta,l}=\sum_{i_1,\dots,i_{l-1}}\varphi(i_1,\dots,i_{l-1})\Big(\sum_{j=0}^{r} U_{j,1}^{l-1}+V_{\theta,1}^{l-1}\Big)=$$
$$=\sum_{i_1,\dots,i_{l-1}}\varphi(i_1,\dots,i_{l-1})\,V^{l-1}=\sum_{i_1,\dots,i_{l-2}}\varphi(i_1,\dots,i_{l-2})\sum_{i_{l-1}}\varphi_{i_{l-1}}V^{l-1}.$$

In §5 we have proved that $\sum_i \varphi_i V^1=\sum_{i=1}^{r}\rho(e_ih_i)=\rho(v_{\theta,1})$. Applying this result to ρ^{l-2} and $\rho^{l-1}\cong\Phi^+\rho^{l-2}$, we find that $\sum_{i_{l-1}}\varphi_{i_{l-1}}V^{l-1}=\rho^{l-2}(v_{\theta,1})$. Thus,

$$\sum_{j=0}^{r} U_{j,l}+\tilde{V}_{\theta,l}=\sum_{i_1,\dots,i_{l-2}}\varphi(i_1,\dots,i_{l-2})\,\rho^{l-2}(v_{\theta,1}).$$

By induction, we assume it as proved that $\sum_{i_1,\dots,i_{l-2}}\varphi(i_1,\dots,i_{l-2})\rho^{l-2}(v_{\theta,1})=\rho(v_{\theta,l-1})$. We write $\tau_{\theta,l-1}=\rho\,|_{V_{\theta,l-1}}$, where

$V_{\theta,\,l-1} = \rho(v_{\theta,\,l-1})$. Now (6.16) is proved. By induction, we assume it as proved that $\rho = \bigoplus_{\substack{j,k\\ \leq l-1}} \tilde{\rho}_{j,k} \oplus \tau_{\theta,\,l-1}$. Comparing this with (6.16), we obtain $\rho = \bigoplus_{\substack{j,k\\ k\leq l}} \tilde{\rho}_{j,k} \oplus \tau_{\theta\, l}$.

MAIN THEOREM I. *In the lattice D^r, $r \geqslant 4$, every element $v \in B^+ = \bigcup_l B^+(l)$ or $v \in B^- = \bigcup_l B^-(l)$ is perfect. This means that for any representation ρ in any space V over a field k, the subspace $\rho(v)$ is such that the restriction $\tau = \rho\,|_{\rho(v)}$ is a direct summand of ρ.*

PROOF. In §3 we have proved that $B^+ = \bigcup_l B^+(l)$ is a lattice. In the course of proving Theorem 6.1 we have shown that every element $v^-_{a,l} \in B^-(l)$ is perfect. The corresponding result for elements $v^-_{a,l} \in B^-(l)$ follows from the duality principle.

The following proposition refines the main theorem.

PROPOSITION 6.3. *Let $v_{a,l} \in B^+(l) \subset D^r$, $r \geqslant 4$, and suppose that $V_{a,\,l} = \rho(v_{a,\,l})$ is different from zero and from V. Then ρ splits into a direct sum*

$$\rho = \Big(\bigoplus_{\substack{j,k\\ k\leq l-1}} \tilde{\rho}_{j,k}\Big) \oplus \Big(\bigoplus_{j\in a'\cup\{0\}} \tilde{\rho}_{j,\,l}\Big) \oplus \tau_{a,\,l},$$

where $\tau_{a,\,l} = \rho\,|\,V_{a,\,l}$, and each representation $\tilde{\rho}_{j,k}$ and $(\tilde{\rho}_{j,\,l})$ is a multiple of the indecomposable representation $\rho^+_{j,\,k}$ and $(\rho^+_{j,\,l})$, respectively.

It is elementary to deduce from our results the following Propositions 6.4 and 6.5.

PROPOSITION 6.4. *Let $\rho^+_{t,\,l}$ be an indecomposable representation of the first kind ($t \in \{1, \ldots, r\}$) in $V_{t,l}$. We set $v_{t,l} = \bigcap_{i\neq t} h_i(l)$ ($v_{t,l} \in B^+(l)$).*

(I) *$\rho^+_{t,\,l}(x) = V_{t,\,l}$ for every element $x \in B^+ = \bigcup_s B^+(s)$ such that $x \supseteq v_{t,l}$. In particular, this identity holds for every $x \in B^+(m)$, $m < l$.*

(II) *$\rho^+_{t,\,l}(y) = 0$ for every element $y \in B^+$ such that $y \subseteq h_t(l)$. In particular, this identity holds for every $y \in B^+(n)$, $n > l$.*

PROPOSITION 6.5. *Let $\rho^+_{0,\,l}$ be an indecomposable representation of the second kind in $V_{0,l}$. Let $v_{\theta,l-1} = \bigcap_i h_i(l-1)$ be the minimal element of $B^+(l-1)$ and $v_{I,\,l} = \sum_{i\in I} e_i(l)$ the maximal element of $B^+(l)$.*

(I) *$\rho^+_{0,\,l}(x) = V_{0,\,l}$ for every element $x \in B^+$ such that $x \supseteq v_{\theta,l-1}$. This means that $\rho^+_{0,\,l}(x) = V_{0,\,l}$ for every $x \in B^+(m)$, $m \leqslant l-1$.*

(II) *$\rho^+_{0,\,l}(y) = 0$ for every element $y \in B^+$ such that $y \subseteq v_{I,l}$. This means that $\rho^+_{0,l}(y) = 0$ for every $y \in B^+(n)$, $\eta \geqslant l$.*

PROPOSITION 6.6. *The element $f_0(l) = \sum_{i_1,\ldots,i_{l-1}} f_{i_1\cdots i_{l-1}0}$ is linearly equivalent to the minimal $v_{\theta,l-1} \in B^+(l-1)$, that is, $\rho(f_0(l)) = \rho(v_{\theta,\,l-1})$ for every representation ρ.*

PROOF. In the course of proving Theorem 6.1 we have shown that $\sum_{i_1,\ldots,i_{l-1}} \varphi(i_1, \ldots, i_{l-1})V^{l-1} = \rho(v_{\theta,\,l-1})$. We assume it as proved (see §7) that

$\varphi(i_1, \ldots, i_{l-1})V^{l-1} = \rho(f_{i_1 \ldots i_{l-1} 0})$. Consequently,

$$\rho(v_{\theta, l-1}) = \sum_{i_1, \ldots, i_{l-1}} \rho(f_{i_1 \ldots i_{l-1} 0}) = \rho(\sum_{i_1, \ldots, i_{l-1}} f_{i_1 \ldots i_{l-1} 0}) = \rho(f_0(l)).$$

PROPOSITION 6.7. *$B^+(l)$ and $B^-(l)$ are Boolean algebras with 2^r elements.*

PROOF. We consider the representation ρ, the direct sum $\bigoplus_{t=1}^{r} \rho^+_{t, l}$ of all indecomposable representations $\rho^+_{t, l}$ with one and the same l. It follows easily from Proposition 6.3 that all the subspaces $\rho(v_{a, l})$ for $a \subseteq I$ are distinct. Thus, $\rho(B^+(l)) \cong \mathscr{B}(I)$. It clearly follows that $B^+(l) \cong \mathscr{B}(I)$.

6.5. Connection between the upper and lower cubicles in D^r, $r \geqslant 4$. We have proved in §3 that the lattices $B^+(l)$ can be ordered in the following way: $v^+_{a,l} \supseteq v^+_{b,l+1}$ for every l and any $a, b \subseteq I$. In particular, the minimal elements $v^+_{\theta,l} \in B^+(l)$ form a descending chain $v^+_{\theta,1} \geqslant v^+_{\theta,2} \geqslant v^+_{\theta,3} \geqslant \ldots$. It follows from Proposition 6.7 that all the elements of this chain are distinct.

Dual to the $v^+_{\theta,l}$ are the elements $v^-_{I,l} \in B^-(l)$. They form (with respect to l) an ascending chain $v^-_{I,1} \subset v^-_{I,2} \subset v^-_{I,3} \subset \ldots$. Let ρ be an arbitrary representation of D^r. As usual, we write $V^+_{\theta, l} = \rho(v^+_{\theta, l})$ and $V^-_{I, l} = \rho(v^-_{I, l})$. In addition, we set $V^+_{\theta,\infty} = \bigcap_l V^+_{\theta,l}$ and $V^-_{I,\infty} = \bigcup_l V^-_{I,l}$.

PROPOSITION 6.8. *$V^-_{I,\infty} \subseteq V^+_{\theta,\infty}$ for every representation ρ of D^r, $r \geqslant 4$.*

PROOF. The representation space V of ρ is finite-dimensional. Therefore, there are integers l and m such that $V^+_{\theta, \infty} = \rho(v^+_{\theta, l}) = V^+_{\theta, l}$ and $V^-_{I, \infty} = \rho(v^-_{I, m}) = V^-_{I, m}$. It follows from the results of this section that ρ splits into a direct sum $\rho = \rho(l) \oplus \tau_{\theta, l}$, where $\tau_{\theta, l} = \rho|_{V^+_{\theta, l}}$ and $(\Phi^+)^l\rho(l) = 0$. We denote the representation space of $\rho(l)$ by U. Since ρ is decomposable with respect to U and $V^+_{\theta,l} = V^+_{\theta,\infty}$, every $x \in D^r$ satisfies $\rho(x) = U\rho(x) + V^+_{\theta, \infty}\rho(x)$. In particular, for the element $x = v^-_{I,m} \in B^-(m)$, we can write $V^-_{I, m} = \rho(v^-_{I, m}) = U\rho(v^-_{I, m}) + V^+_{\theta, \infty}\rho(v^-_{I, m})$.

We claim that $U\rho(v^-_{I, m}) = 0$. Hence it follows at once that $\rho(v^-_{I, m}) = V^-_{I, m} = V^+_{\theta, \infty}V^-_{I, m}$, that is, $V^-_{I,\infty} = V^-_{I,m} \subseteq V^+_{\theta,\infty}$. Suppose the contrary: that $U\rho(v^-_{I, m}) \neq 0$. It follows from the properties of a direct sum that $\rho(l)(x) = U\rho(x)$ for the restriction $\rho|_U = \rho(l)$. Thus, $\rho(l)(v^-_{I, m}) = U\rho(v^-_{I, m})$. If $\rho(l)(v^-_{I, m})$ is non-zero, then, according to the proposition dual to Theorem 6.1, $\rho(l)$ splits into a direct sum:

$\rho(l) = \tau(l) \oplus (\bigoplus_{j, k; k \leqslant m} \tilde{\rho}^-_{j, k})$, where $(\bigoplus_{j, k} \tilde{\rho}^-_{j, k})$ is the restriction of $\rho(l)$ to $\rho(l)(v^-_{I, m})$. Here each $\tilde{\rho}^-_{j, k}$ is a multiple of $\rho^-_{j, k}$. By assumption, $(\Phi^+)^l\rho(l) = 0$. On the other hand, $(\Phi^+)^l\rho(l) = (\Phi^+)^l\tau(l) \oplus (\bigoplus_{j, k} (\Phi^+)^l\tilde{\rho}^-_{j, k})$, and we assume that the sum $\bigoplus_{j, k} \tilde{\rho}^-_{j, k}$ is non-zero. This means that there are j and k such that $\tilde{\rho}^-_{j, k} = \overbrace{\rho^-_{j, k} \oplus \ldots \oplus \rho^-_{j, k}}^{m_{j, k}}$, where $m_{j,k} > 0$. Then

$(\Phi^+)^l\widetilde{\rho_{\bar{j},k}} \supseteq \oplus m_{j,k}(\Phi^+)^l\rho_{\bar{j},k}$. By definition $(\Phi^+)^l\rho_{\bar{j},k} \cong \rho_{\bar{j},k+l}$. Since $\rho_{\bar{j},k}$ is a representation of D^r, $r \geqslant 4$, as we have shown at the beginning of this section, $\rho_{\bar{j},s} \neq 0$ for all $s \geqslant 1$. Consequently, $(\Phi^+)^l\widetilde{\rho_{\bar{j},k}} \neq 0$, and hence $(\Phi^+)^l\rho(l) = 0$. But this contradicts the fact that $(\Phi^+)^l\rho(l) \neq 0$.

PROPOSITION 6.9. *Let* ρ *be a representation of* D^r, $r > 4$, *in* V, *and let* $0 \subset V^-_{I,\infty} \subset V^+_{\theta,\infty} \subset V$ *be a chain of distinct subspaces. Then* ρ *splits into a direct sum* $\rho = \rho^- \oplus \rho_\lambda \oplus \rho^+$, *where* $\rho^- = \rho\,|_{V^-_{I,\infty}}$ *and* $\rho^- \oplus \rho_\lambda = \rho\,|_{V^-_{\theta,\infty}}$.

Note that for ρ^+ and ρ^- there are integers l and m such that $(\Phi^+)^l\rho^+ = 0$ and $(\Phi^-)^m\rho^- = 0$. As regards the ρ_λ, it is not difficult to show that $(\Phi^-)^l\,(\Phi^+)^l\,\rho_\lambda \cong \rho_\lambda$ and $(\Phi^+)^l(\Phi^-)^l\rho_\lambda \cong \rho_\lambda$ for every $l > 0$. For D^4, the representations of type ρ_λ were studied in [5], where they were called regular. Very little is known about the regular representations of the lattices D^r, $r \geqslant 5$. It is clear (see [5], [7]) that the classification of regular indecomposable representations contains as a special case the problem of classifying up to similarity an arbitrary set of linear transformations $A_1, \ldots, A_n$, $n \geqslant 2$, A_i: $V \to V$.

§7. The subspaces $\rho(e_\alpha)$ and the maps φ_i

As usual, let ρ and $\rho^l = (\Phi^+)^l\rho$ be representations of D^r, $r \geqslant 4$, in V and V^l, respectively, and let $\varphi(i_1, \ldots, i_l)$: $V^l \to V$ be the linear map $\varphi_{i_1}\circ\varphi_{i_2}\circ\ldots\circ\varphi_{i_l}$, where φ_{i_k}: $V^k \to V^{k-1}$ are elementary maps (see §5). The basic aim of this section is to prove the formula

$$\rho(e_{i_1\ldots i_l t}) = \varphi(i_1, \ldots, i_l)\,\rho^l\,(e_t), \tag{7.1}$$

where $e_{i_1\ldots i_l t} \in D^r$ is the lattice polynomial defined in §1.

7.1. A lattice in the space $R = \oplus_i \rho(e_i)$. We repeat the construction of the representation $\rho^1 = \Phi^+\rho$ from a representation ρ in V. Let $R = \bigoplus_{i=1}^{r} \rho(e_i)$, and let ∇: $R \to V$ be the linear map defined by the formula $\nabla(\xi_1, \ldots, \xi_r) = \sum_{i=1}^{r} \xi_i$, where $\xi_i \in \rho(e_i)$. We denote by G_i the subspace of R consisting of the vectors $(0, \ldots, 0, \xi_i, 0, \ldots, 0)$ with $\xi_i \in \rho(e_i)$, and let $G'_i = \sum_{t \neq i} G_t$. Then $\rho^1 = \Phi^+\rho$ is defined as the representation in $V^1 = \operatorname{Ker}\nabla$ for which $\rho^1(e_i) = V^1G'_i$.

Let $\mathscr{L}(R)$ be the lattice of all linear subspaces of R. We denote by M the sublattice of $\mathscr{L}(R)$ generated by the subspaces $G_1, \ldots, G_r$ and V^1. It is easy to see that these subspaces satisfy the following relations:

1°. $\sum_{i=1}^{r} G_i = R$.

2°. $G_i \sum_{t \neq i} G_t = 0$ for every i.

3°. $G_iV^1 = 0$ for every i.

From 1° and 2° it is obvious that the sublattice of M generated by $G_1, \ldots, G_r$ is a Boolean algebra, which we denote by $\mathcal{B}$. Clearly, the maximal (unit) element of $\mathcal{B}$ is R.

We define two representations ν^0 and ν^1 of D^r in R by the formulae

$$\text{(7.2)} \qquad \nu^0(e_i) = V^1 + G_i, \quad \nu^1(e_i) = V^1 G_i' = V^1 \sum_{t \neq i} G_t.$$

Thus, $\nu^0(D^r) \in M$ and $\nu^1(D^r) \in M$. It turns out that ν^0 and ν^1 are almost the same as ρ and ρ^1. More precisely, the following is true.

PROPOSITION 7.1. (I) *The map* $\nabla\colon R \to V$ *defines a morphism of representations* $\tilde{\nabla}\colon \nu^0 \to \rho$. *Moreover,* $\rho(x) = \nabla\nu^0(x)$ *for every* $x \in D^r$.

(II) ν^0 *splits into the direct sum* $\nu^0 = \nu^0_0 + \nu^0_1$, *where*

$\nu^0_0 \cong \operatorname{Im} \tilde{\nabla} = \rho_{I,1}$ $\rho_{I,1}$ *is the restriction of* ρ *to* $V_{I,1} = \rho(\sum_{i=1}^r e_i)$, *and* ν^0_1 *is that of* ν^0 *to* V^1. *Here* $\nu^0_1 \cong \operatorname{Ker} \tilde{\nabla}$ *and* ν^0_1 *is a multiple of the atomic representation* $\rho_{\bar{0}1}$.

PROOF. We prove first that ν^0 splits into the direct sum $\nu^0 = \nu^0_0 + \nu^0_1$. Let U be any subspace complementary to $V^1 = \operatorname{Ker} \nabla$ (that is, $UV^1 = 0$ and $U + V^1 = R$). By definition, $\nu^0(e_i) = G_i + V^1 \supseteq V^1$. Consequently, $\nu^0(e_i) = \nu^0(e_i)R = \nu^0(e_i)(U + V^1) = \nu^0(e_i)U + \nu^0(e_i)V^1$. This means (see Proposition 2.3) that ν^0 is decomposable and that V^1 is admissible relative to ν^0 (that is, $V^1(\nu^0(x) + \nu^0(y)) = V^1\nu^0(x) + V^1\nu^0(y)$ for any $x, y \in D^r$).

Since $V^1 = \operatorname{Ker} \nabla$ is admissible, by Proposition 2.2 the correspondence $x \mapsto \nabla\nu^0(x)$, $x \in D^r$, defines a representation in the space $\operatorname{Im} \nabla \cong R/V^1$. Note that $\nabla\nu^0(e_i) = \nabla(G_i + V^1) = \rho(e_i)$. Consequently, $\nabla\nu^0(x) = \rho(x)$ for every $x \in D^r$. Thus ∇ defines a morphism of representations: $\tilde{\nabla}\colon \nu^0 \to \rho$. Obviously, $\operatorname{Im} \tilde{\nabla} = \rho_{I,1}$, where $\rho_{I,1}$ is the restriction of ρ to $\rho(\sum_{i=1}^r e_i) = \sum_{i=1}^r \rho(e_i)$.

Let us verify that $\operatorname{Im} \tilde{\nabla} \cong \nu^0_0$, where ν^0_0 is the restriction of ν^0 to U. First of all we note that the linear map $\nabla_U\colon U \to \operatorname{Im} \nabla$ is an isomorphism. (∇_U denotes the restriction of ∇ to U.) Also, we proved above that

$$\text{(7.3)} \qquad \nu^0(e_i) \overset{\text{def}}{=} G_i + V^1 = \nu^0(e_i)U + \nu^0(e_i)V^1 = \nu^0_0(e_i) + V^1.$$

As we have already mentioned (see 3°), $G_i V^1 = 0$; moreover, it follows from $UV^1 = 0$ that $\nu^0_0(e_i)V^1 = 0$. Consequently,

$$\text{(7.4)} \qquad \dim \rho(e_i) = \dim G_i = \dim \nu^0_0(e_i).$$

Thus, $\operatorname{Im} \tilde{V} = \nu^0_0$.

To finish the proof, we note that $\nu^0_1(e_i) = V^1$ for any $i \in \{1, \ldots, r\}$. Consequently, ν^0_1 is a multiple of the atomic representation $\rho_{\bar{0},1}$ (We recall that $\rho_{\bar{0},1}$ is the representation in the one-dimensional space $W \cong k^1$ for which $\rho_{\bar{0},1}(e_i) = W$ for all i).

We denote by μ the embedding $\mu\colon V^1 \hookrightarrow R$.

PROPOSITION 7.2. (I) *The embedding* $\mu\colon V^1 \to R$ *defines a monomorphism* $\tilde{\mu}\colon \rho^1 \to \nu^1$ *of representations. Also,* $\nu^1(x) = \mu\rho^1(x)$ *for any* $x \in D^r$.

(II) ν^1 *splits into the direct sum* $\nu^1 = \nu^1_1 + \nu^1_0$, *where* ν^1_1 *is the restriction* $\nu^1 \mid_{V^1} \nu_1 \cong \rho^1$, *and* ν^1_0 *is a multiple of the atomic representation* $\rho^+_{0,1}$.

The proof of this is similar to that of the preceding proposition.

7.2. The connection between the maps φ_i and the lattice operations. In this subsection we use as definition of the maps φ_i the formula $\varphi_i = \nabla\pi_i\mu$, where π_i is the projection onto R with kernel G_i' and image G_i. This formula was proved in Proposition 5.2.

It is useful for us to have a lattice definition of a projection.

LEMMA 7.1. *Let R be a finite-dimensional vector space, X a subspace of R, and let $\pi\colon R \to R$ the projection with kernel G' and range of values G. Then $\pi X = G(G' + X)$, where πX is the image of X.*

COROLLARY 7.2. *For every subspace* $Y^1 \subseteq V^1$ *and every* $i \in \{1, \ldots, r\}$

$$\pi_i\mu Y^1 = G_i(G_i' + \mu Y^1).$$

The basic results of this section are formulated below in Theorems 7.1, 7.2, and 7.3.

THEOREM 7.1. *Let ρ be a representation of D^r, $r \geqslant 4$, and $\rho^1 = \Phi^+\rho$. Then*

1°. $\varphi_{i_1}\rho^1(e_{i_1\ldots i_l}) = 0$.

2°. $\varphi_j\rho^1(e_{i_1\ldots i_l}) = \rho(e_{ji_1\ldots i_l})$ *if* $j \neq r_i$.

Here, $e_{i_1\ldots i_l} \in D^r$ are the lattice polynomials defined previously.

We shall show that the proof of 2° follows from the next Proposition 7.3. Its proof is very laborious, and is given in §7.3, 7.4.

PROPOSITION 7.3. *Let $\rho^1 = \Phi^+\rho$, and let ν^0, ν^1 be the representations of D^r, $r \geqslant 4$, in $R = \bigoplus_i \rho(e_i)$ defined by $\nu^0(e_i) = G_i + V^1$, $\nu^1(e_i) = G_i'V^1$.*

Then for every $e_\alpha = e_{i_1\ldots i_l}$ and every $j \neq i_1$,

$$V^1 + G_j(G_j' + \nu^1(e_\alpha)) = \nu^0(e_{j\alpha}). \tag{7.5}$$

PROOF OF THEOREM 7.1. 1°. By definition,

$$\rho^1(e_\alpha) = \rho^1(e_{i_1\ldots i_l}) = \rho^1\Big(e_{i_1}\sum_{\beta\in\Gamma(\alpha)} e_\beta\Big) = \rho^1(e_{i_1})\sum_{\beta\in\Gamma(\alpha)}\rho^1(e_\beta) \subseteq \rho^1(e_{i_1}).$$

Furthermore, $\operatorname{Ker}\varphi_i = \rho^1(e_i)$ (see §5.2). Consequently, $\varphi_{i_1}\rho^1(e_{i_1\ldots i_l}) = 0$.

2°. Suppose that (7.5) has been proved. For any $Y \subset V^1$, according to Corollary 7.1, we have $\varphi_iY = \nabla\pi_j\mu Y = \nabla(G_j(G_j' + \mu Y))$. By definition, $V^1 = \operatorname{Ker}\nabla$. Consequently,

$$\varphi_jY = \nabla(V^1 + G_j(G_j' + \mu Y)). \tag{7.6}$$

Now let $Y = \rho^1(e_\alpha)$. In Proposition 7.2 we have proved that $\mu\rho^1(e_\alpha) = \nu^1(e_\alpha)$. Therefore, $\varphi_j\rho^1(e_\alpha) = \nabla(V^1 + G_j(G_j' + \nu^1(e_\alpha)))$. From (7.5) and Proposition 7.1 ($\nabla\nu^0(\varkappa) = \rho(\varkappa)$) it then follows that

$\varphi_j\rho^1(e_\alpha) = \nabla\nu^0(e_{j\alpha}) = \rho(e_{j\alpha})$.

THEOREM 7.2. *Let $\rho^l = (\Phi^+)^l\rho$ be a representation of D^r. Then for every sequence $i_1, \ldots, i_l$, where $i_j \in \{1, \ldots, r\}$, $i_j \neq i_{j+1}$, and every $t \neq i_l$, we have*

$$\rho(e_{i_1\ldots i_l t}) = \varphi(i_1, \ldots, i_l)\rho^l(e_t).$$

PROOF. By definition, $\rho^{k+1} = \Phi^+\rho^k$. Therefore, applying Theorem 7.1 to ρ^k and ρ^{k+1}, we find that for any $\alpha = (i_1, \ldots, i_m)$ and any $j \neq i$

$$\rho^k(e_{ji_1\ldots i_m}) = \varphi_j^{k+1}\rho^{k+1}(e_{i_1\ldots i_m}),$$

where $\varphi_j^{k+1}\colon V^{k+1} \to V^k$. By definition, $\varphi(i_1, \ldots, i_l) = \varphi_{i_1}\circ \ldots \circ\varphi_{i_l}$. Therefore,

$$\varphi(i_1, \ldots, i_l)\,\rho^l(e_t) = \varphi_{i_1}\circ\ldots\circ\varphi_{i_l}\,\rho^l(e_t) = \varphi_{i_1}\circ\ldots\circ\varphi_{i_{l-1}}\rho^{l-1}(e_{i_l t})$$
$$= \varphi_{i_1}\circ\ldots\circ\varphi_{i_{l-2}}\rho^{l-2}(e_{i_{l-1}i_l t}) = \ldots = \rho(e_{i_1\ldots i_l t}).$$

7.3. Lattice lemmas. Here we prove three lemmas, which will be used in the proof of Proposition 7.3.

LEMMA 7.2. *Let a_1, a_2, b, c be four elements of a modular lattice L such that $a_1 \subseteq a_2$. Then*

(I) $$a_1(b + a_2c) = a_1(a_2b + c);$$

(II) $$a_2 + b(a_1 + c) = a_2 + (a_1 + b)c.$$

PROOF. (I) Since $a_1 \subseteq a_2$, we have $a_1 = a_1a_2$ and hence, $a_1(b + a_2c) = a_1a_2(b + a_2c)$. It follows from Dedekind's axiom that $a_2(b + a_2c) = a_2b + a_2c = a_2(a_2b + c)$. Therefore, $a_1(b + a_2c) = a_1a_2(a_2b + c) = a_1(a_2b + c)$.

(II) Since $a_1 \subseteq a_2$, we have $a_2 = a_1 + a_2$. Therefore

$$a_2 + b(a_1 + c) = a_2 + a_1 + b(a_1 + c) = a_2 + (a_1 + b)(a_1 + c) =$$
$$= a_2 + a_1 + (a_1 + b)c = a_2 + (a_1 + b)c.$$

LEMMA 7.3. *Let $a, a', x_1, \ldots, x_n$ be elements of a modular lattice L such that $aa' = 0$ and $a + a' = 1$, where 0 and 1 are the minimal and maximal elements of L, respectively. Then*

$$a\left(a' + \sum_{i=1}^{n} x_i\right) = \sum_{i=1}^{n} a(a' + x_i).$$

PROOF. Let x be an arbitrary element of L. Then

$$a' + x = 1(a' + x) = (a' + a)(a' + x) = a' + a(a' + x).$$

Using this equation, we can write

$$a\left(a' + \sum_i x_i\right) = a\left(\sum_i (a' + x_i)\right) = a\left(a' + \sum_i a(a' + x_i)\right).$$

Since $a \supseteq \sum_i a(a' + x_i)$, applying Dedekind's axiom we find

$$a\,(a' + \sum_i x_i) = a\,(a' + \sum_i a\,(a' + x_i)) = aa' + \sum_i a\,(a' + x_i) =$$
$$= 0 + \sum_i a\,(a' + x_i) = \sum_i a\,(a' + x_i).$$

LEMMA 7.4. *Let $\mathscr{B}$ be a Boolean algebra with generators $g_1, \ldots, g_r$ such that $\sum_{i=1}^{r} g_i = 1$ and $g_i \sum_{t\neq i} g_t = 0$ for every i. We set $g'_i = \sum_{t\neq i} g_t$. Then $g'_{i_1} g'_{i_2} = \sum_{t\neq i_1, i_2} g_t$.*

7.4. Proof of Proposition 7.3 ($V^1 + G_j(G'_j + \nu^1(e_\alpha)) = \nu^0(e_{j\alpha})$). To begin with we prove a lattice proposition equivalent to the equation $\varphi_{i1}\rho^1(e_{i1} \cdots {}_{i_l}) = 0$.

PROPOSITION 7.4. *Let $\rho^1 = \Phi^+\rho$ and let ν^1 be the representation of D^r in $R = \bigoplus_{i=1}^{r} \rho(e_i)$ defined by $\nu^1(e_i) = V^1 G'_i$. Then*
$V^1 + G_{i_1}(G'_{i_1} + \nu^1(e_{i_1 \ldots i_l})) = V^1$ for every $\alpha = (i_1, \ldots, i_l) \in A(r, l)$.

PROOF. By definition, $\nu^1(e_{i_1 \ldots i_l}) = \nu^1(e_{i_1}) \sum_{\beta\in\Gamma(\alpha)} \nu^1(e_\beta) = V^1 G'_{i_1} \sum_{\beta\in\Gamma(\alpha)} \nu^1(e_\beta) \subseteq G'_{i_1}$. Consequently, $G'_{i_1} + \nu^1(e_{i_1 \ldots i_l}) = G'_{i_1}$, and so

$$V^1 + G_{i_1}(G'_{i_1} + \nu^1(e_{i_1 \ldots i_l})) = V^1 + G_{i_1} G'_{i_1} = V^1 + 0 = V^1.$$

We prove Proposition 7.3 by induction on l, therefore, it will be more convenient for us to prove the following stronger Proposition 7.5, whose first part is equivalent to Proposition 7.3.

PROPOSITION 7.5. (I) *Let ρ be an arbitrary representation of D^r, $r \geqslant 4$, let $\rho^1 = \Phi^+\rho$, and let ν^0, ν^1 be the representations of D^r in $R = \bigoplus_{i=1}^{r} \rho(e_i)$ defined by the equations $\nu^0(e_i) = G_i + V^1$, $\nu^1(e_i) = G'_i V^1$. Then for every $(i_1, \ldots, i_l) \in A(r, l)$ and every $j \neq i_1$, we have*

$$V^1 + G_j(G'_j + \nu^1(e_{i_1 \ldots i_l})) = \nu^0(e_{j i_1 \ldots i_l}). \tag{7.7}$$

(II) *For every $\alpha = (i_1, \ldots, i_{l+1})$, $l \geqslant 1$, and every $t \notin \{i_1, i_2\}$ the elements $e_{i_1} \sum_{\beta\in\Gamma(\alpha)} e_\beta$ and $e_{i_1}(e_t + \sum_{\beta\in\Gamma(\alpha)} e_\beta)$ are linearly equivalent, that is, for every representation τ*

$$\tau(e_{i_1} \sum_{\beta\in\Gamma(\alpha)} e_\beta) = \tau(e_{i_1}(e_t + \sum_{\beta\in\Gamma(\alpha)} e_\beta)), \tag{7.8}$$

where

$$\Gamma(\alpha) = \Gamma(i_1, \ldots, i_{l+1}) =$$
$$= \{\beta = (k_1, \ldots, k_l) \in A(r, l) \mid k_1 \notin \{i_1, i_2\}, k_2 \notin \{i_2, i_3\}, \ldots, k_l \notin \{i_l, i_{l+1}\}\},$$

and $e_{i_1} \sum_{\beta\in\Gamma(\alpha)} e_\beta = e_\alpha$.

REMARK. Apparently, in D^r

(7.9) $$e_{i_1}\sum_{\beta\in\Gamma(\alpha)} e_\beta = e_{i_1}(e_t + \sum_{\beta\in\Gamma(\alpha)} e_\beta)$$

for every $t \notin \{i_1, i_2\}$. However, we can prove this only for $l = 2$ and $l = 3$. A proof for $l = 2$ is given below. Note that (7.9) is a special case of the more general conjecture stated in §3.5 (that $e_\alpha = e'_\alpha$).

Our proof of Proposition 7.5 is by induction on l.

Step 1. We prove (7.7) for $\alpha = (i_1)$, that is, we show that if $j \neq i_1$, then $V^1 + G_j(G'_j + v^1(e_{i_1})) = v^0(e_{ji_1})$.

By definition, $v^1(e_{i_1}) = V^1G'_{i_1} = V^1 \sum_{t\neq i_1} G_t$. It follows from $j \neq i_1$ that $G'_{i_1} \supset G_j$. Hence, applying Lemma 7.2, we can write

$G_j(G'_j + V^1G'_{i_1}) = G_j(G'_jG'_{i_1} + V^1)$. Also by Lemma 7.4, $G'_jG'_{i_1} = \sum_{t\neq j,\, i_1} G_t$.

Using these equations, we find

$$V^1 + G_j(G'_j + v^1(e_{i_1})) = V^1 + G_j(\sum_{t\neq j, i_1} G_t + V^1) =$$
$$= (V^1 + G_j)(V^1 + \sum_{t=j,\, i_1} G_t) = (V^1 + G_j)(\sum_{t\neq j,\, i_1}(V^1 + G_t)) =$$
$$= v^0(e_j)\sum_{t\neq j,\, i_1} v^0(e_t) = v^0(e_j \sum_{t\neq j,\, i_1} e_t) = v^0(e_{ji_1}).$$

Step 2. We prove (7.9) for $l = 1$, that is, we show that if $t \neq i_1, i_2$, then
$$e_\alpha = e_{i_1i_2} = e_{i_1}(e_t + \sum_{\beta\in\Gamma(\alpha)} e_\beta).$$

By definition $\Gamma(i_1, i_2) = \{k \mid k \notin \{i_1, i_2\}\}$. Thus,

$e_{i_1}(e_t + \sum_{\beta\in\Gamma(\alpha)} e_\beta) = e_{i_1}(e_t + \sum_{k\neq i_1,\, i_2} e_k)$. Since $t \neq i_1, i_2$, we have
$e_t + \sum_{k\neq i_1i_2} e_k = \sum_{k\neq i_1i_2} e_k$ hence, $e_{i_1}(e_t + \sum_{\beta\in\Gamma(\alpha)} e_\beta) = e_{i_1}(\sum_{k\neq i_1,\, i_2} e_k) = e_{i_1,\, i_2}$,
as required.

Step 3. Now we suppose that Proposition 7.5 (I) is proved for all $\alpha = (i_1, \ldots, i_\lambda)$, $\lambda \leqslant l - 1$, and Proposition 7.5 (II) for all $\alpha = (i_1, \ldots, i_{\lambda+1})$, $\lambda \leqslant l - 1$. We show that then Proposition 7.5 (I) is also true for any $\alpha = (i_1, \ldots, i_l)$.

Step 3a. We perform some manipulations with the subspace $V^1 + G_j(G'_j + v^1(e_\alpha))$. By definition,

(7.10) $$v^1(e_\alpha) = v^1(e_{i_1}\sum_{\beta\in\Gamma(\alpha)} e_\beta) = v^1(e_{i_1})\sum_{\beta\in\Gamma(\alpha)} v^1(e_\beta) = V^1G'_{i_1}\sum_{\beta\in\Gamma(\alpha)} v^1(e_\beta).$$

We set

(7.11) $$X = \sum_{\beta\in\Gamma(\alpha)} v^1(e_\beta).$$

By definition $v^1(e_\alpha) = V^1G'_i \subseteq V^1$ for every i. Consequently, $v^1(x) \subseteq V^1$ for every element $x \in D^r$. In particular, $\sum_{\beta\in\Gamma(\alpha)} v^1(e_\beta) \subseteq V^1$. Thus, we can write $v^1(e_\alpha) = V^1G'_{i_1}\ \sum_{\beta\in\Gamma(\alpha)} v^1(e_\beta) = G'_{i_1}X$. For brevity, we also set

(7.12) $$V^1 + G_j(G'_j + v^1(e_\alpha)) = F(e_\alpha).$$

Thus, $F(e_\alpha) = V^1 + G_j(G'_j + G'_{i_1}X)$.

By assumption, $j \neq i_1$, hence $G'_{i_1} \overset{\text{def}}{=} \sum_{t\neq i_1} G_t \supseteq G_j$. Using Lemma 7.2 as we did in step 1, we obtain

(7.13) $$F(e_\alpha) = V^1 + G_j(\sum_{t\neq i_1 j} G_t + X).$$

Step 3b. We carry out the proof for $r \geqslant 4$ and, by assumption, $j \neq i_1$. Then the subset $I - \{j, i_1\} = \{1, \ldots, r\} - \{j, i_1\}$ is not empty, and we can write $I - \{j, i_1\} = \{s_3, \ldots, s_r\}$. We show that

(7.14) $$F(e_\alpha) = V^1 + G_j(\sum_{k=3}^{r-1} G_{s_k}(G'_{s_k} + X) + G_{s_r} + X).$$

In (7.13), replacing X by the equal subspace $X + G'_{s_3} X$ and applying Lemma 7.2, we find that

$$F(e_\alpha) = V' + G_j(\sum_{k=3}^{r} G_{s_k} + X + XG'_{s_3}) = V^1 + G_j((\sum_{k=3}^{r} G_{s_k} + X)G'_{s_3} + X) =$$

$$= V^1 + G_j(X + G'_{s_3}(G_{s_3} + \sum_{k=4}^{r} G_{s_k} + X)).$$

Obviously, $G'_{s_3} \supset \sum_{k=4}^{r} G_{s_k}$. Consequently, by Dedekind's axiom,

$G'_{s_3}(G_{s_3} + \sum_{k=4}^{r} G_{s_k} + X) = \sum_{k=4}^{r} G_{s_k} + G'_{s_3}(G_{s_3} + X)$. Thus,

$F(e_\alpha) = V^1 + G_j(X + G'_{s_3}(G_{s_3} + X) + \sum_{k=4}^{r} G_{s_k})$. Note that

$X + G'_{s_3}(G_{s_3} + X) = (X + G'_{s_3})(G_{s_3} + X) = X + G_{s_3}(G'_{s_3} + X)$. Therefore,

(7.15) $$F(e_\alpha) = V^1 + G_j(G_{s_3}(G'_{s_3} + X) + \sum_{k=4}^{r} G_{s_k} + X).$$

Applying transformations similar to those we have used to get from (7.13) to (7.15), we obtain

(7.14′) $$F(e_\alpha) = V^1 + G_j(\sum_{k=3}^{r-1} G_{s_k}(G'_{s_k} + X) + G_{s_r} + X).$$

Step 3c. We show that

(7.16) $$F(e_\alpha) = V^1 + G_j(\sum_{k=3}^{r-1} G_{s_k}(G'_{s_k} + X) + G_{s_r} + X + V^1G'_{s_r}).$$

We assume that Proposition 7.5 (II) has been proved for all $\lambda \leqslant l - 1$, that is, that for every representation τ and every $\alpha = (i_1, \ldots, i_{\lambda+1})$, $\lambda \leqslant l - 1$, and every $t \neq i_1, i_2$, we have

$$\tau(e_{i_1}\sum_{\beta\in\Gamma(\alpha)} e_\beta)=\tau(e_{i_1}(e_t+\sum_{\beta\in\Gamma(\alpha)} e_\beta).$$

Replacing τ by ν^1 in this equation, we see that for all $\alpha = (i_1, \ldots, i_l) \in A(r, l)$

$$\nu^1(e_\alpha)=\nu^1(e_{i_1})\sum_{\beta\in\Gamma(\alpha)}\nu^1(e_\beta)=\nu^1(e_{i_1})(\nu^1(e_t)+\sum_{\beta\in\Gamma(\alpha)}\nu^1(e_\beta)).$$

We have written $\sum_{\beta\in\Gamma(\alpha)}\nu^1(e_\beta) = X$. Now we set

$$\nu^1(e_t)+\sum_{\beta\in\Gamma(\alpha)}\nu^1(e_\beta)=\widetilde{X}.$$

Then we can write $\nu^1(e_\alpha) = \nu^1(e'_{i_1})X = \nu^1(e'_{i_1})\widetilde{X}$. Since $\nu^1(e_{i_1}) = V^1G'_{i_1}$, $X \subseteq V^1$ and $\widetilde{X} \subseteq V^1$, the preceding equation can be rewritten in the following form: $\nu^1(e_\alpha) = G'_{i_1}X = G'_{i_1}\widetilde{X}$. From this it follows easily that $F(e_\alpha) \overset{\text{def}}{=} V^1 + G_j(G'_j + G'_{i_1}X) = V^1 + G_j(G'_j + G'_{i_1}\widetilde{X})$. In the derivation of (7.14), the only property of X we have used is that $X \subseteq V^1$. Consequently, we may replace X by $\widetilde{X}$ and as a result we obtain

$$V^1+G_j(\sum_{k=3}^{r-1} G_{s_k}(G'_{s_k}+X)+G_{s_r}+X)= \tag{7.17}$$

$$=V^1+G_j(\sum_{k=3}^{r-1} G_{s_k}(G'_{s_k}+\widetilde{X})+G_{s_r}+\widetilde{X}).$$

By definition, $X \subseteq \widetilde{X}$. Consequently, we also have

$$F(e_\alpha)=V^1+G_j(\sum_{k=3}^{r-1} G_{s_k}(G'_{s_k}+X)+G_{s_r}+X)\subseteq$$

$$\subseteq V^1+G_j(\sum_{k=3}^{r-1} G_{s_k}(G'_{s_k}+X)+G_{s_r}+\widetilde{X})\subseteq V^1+G_j(\sum_{k=1}^{r-1} G_{s_k}(G'_{s_k}+\widetilde{X})+G_{s_r}+\widetilde{X}).$$

In (7.17) above we have shown that the extreme terms of this inequality are equal. Hence so are the first two terms, that is,

$$F(e_\alpha)=V^1+G_j(\sum_{k=3}^{r} G_{s_k}(G'_{s_k}+X)+G_{s_r}+\widetilde{X}).$$

By definition, $\widetilde{X} = \nu^1(e_t) + X = V^1G'_t + X$, where $t \neq i_1, i_2$. We did not impose any restrictions on s_r except that $s_r \neq j, i_1$. Now we require that $s_r \neq i_2$. Then we can choose $\widetilde{X} = V^1G'_{s_r} + X$, that is, if $s_r \notin \{j, i_1, i_2\}$, we obtain

$$F(e_\alpha)=V^1+G_j(\sum_{k=3}^{r-1} G_{s_k}(G'_{s_k}+X)+G_{s_r}+X+V^1G'_{s_r}). \tag{7.16'}$$

Step 3d. We claim that

$$F(e_\alpha)=V^1+G_j(V^1+\sum_{k=3}^{r} G_{s_k}(G'_{s_k}+X)). \tag{7.18}$$

Since $s_r \neq j$, we have $G'_{s_r} \supseteq G_j$, and so, applying Lemma 7.2 (I) to the right-hand side of (7.16′), we find that

$$F(e_\alpha) = V^1 + G_j(V^1 + G'_{s_r}(G_{s_r} + \sum_{k=3}^{r-1} G_{s_k}(G'_{s_k} + X) + X)).$$

By definition, all the indices $s_3, \ldots, s_r$ are distinct, therefore, $G'_{s_r} \supseteq \sum_{k=3}^{r-1} G_{s_k} \supseteq \sum_{k=3}^{r-1} G_{s_k}(G'_{s_k} + X)$. Using this relation and Dedekind's axiom, we obtain

$$F(e_\alpha) = V^1 + G_j(V^1 + \sum_{k=3}^{r-1} G_{s_k}(G'_{s_k} + X) + G'_{s_r}(G_{s_r} + X)).$$

Since $V^1 \supseteq X$, in accordance with Lemma 7.2 (II) we have $V^1 + G'_{s_r}(G_{s_r} + X) = V^1 + G_{s_r}(G'_{s_r} + X)$. Thus,

$$(7.18') \qquad F(e_\alpha) = V' + G_j(V^1 + \sum_{k=3}^{r} G_{s_k}(G'_{s_k} + X)).$$

Step 3e. Now we prove that $F(e_\alpha) = v^0(e_{j\alpha})$. By definition, G_t and G'_t determine a partition of the identity in M, that is, $G_tG'_t = 0$, $G_t + G'_t = R$. Therefore, applying Lemma 7.3, we find

$$G_t(G'_t + X) = G_t(G'_t + \sum_{\beta \in \Gamma(\alpha)} v^1(e_\beta)) = \sum_{\beta \in \Gamma(\alpha)} G_t(G'_t + v^1(e_\beta)).$$

Using this equation and also the fact that $I - \{j, i_1\} = \{s_3, \ldots, s_r\}$, we can transform (7.18) to the following form:

$$(7.19) \qquad F(e_\alpha) = V^1 + G_j(V^1 + \sum_{k=3}^{r} \sum_{\beta \in \Gamma(\alpha)} G_{s_k}(G'_{s_k} + v^1(e_\beta))) =$$
$$= (V^1 + G_j)(\sum_{t \neq j,\, i_1} \sum_{\beta \in \Gamma(\alpha)} (V^1 + G_t(G'_t + v^1(e_\beta)))).$$

By definition, $\Gamma(\alpha) = \{\beta = (k_1, \ldots, k_{l-1}) \mid k_1 \notin \{i_1, i_2\}, \ldots\}$. Consequently, two cases for the pair $(t,\beta) = (t, k_1, \ldots, t_{l-1})$ are possible: 1) $t = k_1$, or 2) $t \neq k_1$. Earlier, in Proposition 7.4, we have proved that $V^1 + G_{k_1}(G'_{k_1} + v^1(e_{k_1 \ldots k_{l-1}})) = V^1$. If $t \neq k_1$, then by the inductive hypothesis $V^1 + G_t(G'_t + v^1(e_\beta)) = v^0(e_{t\beta})$, where $t\beta = (t_1 k_1, \ldots, k_{l-1})$.

Therefore, we can rewrite (7.19) in the form

$$F(e_\alpha) \stackrel{\text{def}}{=} V^1 + G_j(G'_j + v^1(e_\alpha)) = (V^1 + G_j)(\sum_{t \neq j,\, i_1,\, k_1} \sum_{\beta \in \Gamma(\alpha)} v^0(e_{t\beta})) =$$
$$= v^0(e_j) \sum_{\substack{t \neq j,\, i_1,\, k_1 \\ \beta \in \Gamma(\alpha)}} v^0(e_{t\beta}).$$

Let $j\alpha = (j, i_1, \ldots, i_l)$. Then

$$\Gamma(j\alpha) = \{\gamma = (k_0, k_1, \ldots, k_{l-1}) \mid k_0 \notin \{j, i_1\},\ k_1 \notin \{i_1, i_2\}, \ldots\}.$$

It is not hard to check that

$$(7.7') \qquad V^1 + G_j(G'_j + v^1(e_\alpha)) = v^0(e_j \sum_{\gamma \in \Gamma(j\alpha)} e_\gamma) = v^0(e_{j\alpha}).$$

So we have proved that if (7.7′) $V^1 + G_j(G_j' + \nu^1(e_\alpha)) = \nu^0(e_{j\alpha})$ holds for all $\alpha = (i_1, \ldots, i_\lambda)$, $\lambda \leqslant l - 1$, and (7.8)

$\tau(e_{i_1} \sum_{\beta\in\Gamma(\alpha)} e_\beta) = \tau(e_{i_1}(e_t + \sum_{\beta\in\Gamma(\alpha)} e_\beta))$ for all $\alpha = (i_1, \ldots, i_\lambda)$, $\lambda \leqslant l$, then (7.7′) is true for any $\alpha = (i_1, \ldots, i_l)$.

Step 4. Now we show that if (7.7′) holds for any $\alpha = (i_1, \ldots, i_\lambda)$, $\lambda \leqslant l$, and (7.8) for any $\alpha = (i_1, \ldots, i_\lambda)$, $\lambda \leqslant l$, then (7.8) holds for every $\bar{\alpha} = (j, i_1, \ldots, i_l) \in A(r, l + 1)$; that is, we prove that for every $t \neq j, i_1$

$$(7.8') \qquad \rho(e_{\bar{\alpha}}) \overset{\text{def}}{=} \rho(e_j \sum_{\gamma\in\Gamma(\bar{\alpha})} e_\gamma) = \rho(e_j)(e_t + \sum_{\gamma\in\Gamma(\bar{\alpha})} e_\gamma).$$

To do this we show that

$$(7.20) \qquad V^1 + G_j(G_j' + \nu^1(e_\alpha)) = \nu^0(e_j(e_t + \sum_{\gamma\in\Gamma(j\alpha)} e_\gamma)).$$

From this (7.8) follows almost immediately.

Just as we have proved (7.14′), so we obtain

$$F(e_\alpha) = V^1 + G_j(\sum_{k=3}^{r-2} G_{s_k}(G_{s_k}' + X) + G_{s_{r-1}} + G_{s_r} + X).$$

where $X = \sum_{\beta\in\Gamma(\alpha)} \nu^1(e_\beta)$.

We can number the subset $I - \{j, i_1\} = \{s_3, \ldots, s_{r-1}, s_r\}$ so that $s_{r-1} \neq i_2$, that is, $s_{r-1} \notin \{j, i_1, i_2\}$. Then, by arguments similar to those in step 3c, we obtain

$$(7.21) \qquad F(e_\alpha) = V^1 + G_j(\sum_{k=3}^{r-2} G_{s_k}(G_{s_k}' + X) + G_{s_{r-1}} + G_{s_r} + X + V^1 G_{s_{r-1}}').$$

Using the same techniques as in step 3, we see that (7.21) can be transformed into:

$$(7.22) \quad F(e_\alpha) \overset{\text{def}}{=} V^1 + G_j(G_j' + \nu^1(e_\alpha)) = \nu^0(e_j(e_{s_r} + \sum_{\gamma\in\Gamma(j\alpha)} e_\gamma)),$$

where $j\alpha = (j, i_1, \ldots, i_l)$.

By construction, s_r can be chosen subject only to the condition $s_r \notin \{j, i_1\}$. Therefore, comparing (7.7′) and (7.22), we obtain

$$(7.23) \qquad \nu^0(e_{ji_1\ldots i_l}) = \nu^0(e_j \sum_{\gamma\in\Gamma(j\alpha)} e_\gamma) = \nu^0(e_j(e_t + \sum_{\gamma\in\Gamma(j\alpha)} e_\gamma)),$$

where $t \notin \{j, i_1\}$.

By Proposition 7.1, for every x in D^r we have $\nabla\nu^0(x) = \rho(x)$. Applying the map $\nabla\colon R \to V$ to both sides of (7.23), we obtain

$$\rho(e_j \sum_{\gamma\in\Gamma(j\alpha)} e_\gamma) = \rho(e_j(e_t + \sum_{\gamma\in\Gamma(j\alpha)} e_\gamma)), \quad t \notin \{j, i_1\}.$$

By construction, ρ was an arbitrary representation in V. We have now proved (7.8) and with it Proposition 7.5.

7.5. Proof of the formula $\varphi(i_1, \ldots, i_l)V^l = \rho(f_{i_1\ldots i_l 0})$. The element

$f_{i_1 \ldots i_l 0}$ of D^r is constructed from a sequence $\alpha = (i_1, \ldots, i_l, 0)$ in the following way:

$$f_\alpha = f_{i_1 \ldots i_l 0} = e_{i_1} \sum_{\beta \in \Gamma(\alpha)} e_\beta,$$

where

$$\Gamma(\alpha) = \{\beta = (k_1, \ldots, k_l) \in A(r, l) \mid k_1 \notin \{i_1, i_2\}, k_2 \notin \{i_2, i_3\}, \ldots \ldots, k_{l-1} \notin \{i_{l-1}, i_l\}, k_l \notin \{i_l\}\}.$$

Thus, the polynomial of the second kind $f_{i_1 \ldots i_l 0}$ is defined in terms of the polynomial of the first kind e_β, where $\beta \in (k_1, \ldots, k_l) \in A(r, l)$. Note that by the definition of $A(r, l)$, all $k_i \in I = \{1, \ldots, r\}$. For example,

$e_{i_1 0} = e_{i_1} \sum_{t \neq i_1}$ $e_{i_1 i_2 0} = e_{i_1} \sum_{k_1, k_2} e_{k_1 k_2}$, where $k_1 \neq i_1, i_2$ and $k_2 \neq i_2$.

THEOREM 7.3. *Let V^l be the representation space of $\rho^l = (\Phi^+)^l \rho$. Then $\varphi(i_1, \ldots, i_l) V^l = \rho(f_{i_1 \ldots i_l 0})$.*

The proof of this theorem easily reduces to that of the following formulae:

(I) $\varphi_j^1 V^1 = \rho(f_{j0})$,

(II) $\varphi_j^1(\rho^1(f_{i_1 \ldots i_{l-1} 0})) = \rho(f_{j i_1 \ldots i_{l-1} 0})$, $\quad l \geqslant 2$.

By definition, $f_{j0} = e_j \sum_{t \neq j} e_t$. Consequently, (I) can be rewritten in the following form:

$$\varphi_j^1 V^1 = \rho\left(e_j \sum_{t \neq j} e_t\right) = \rho(e_j h_j).$$

The truth of this equation follows easily from the definition of the elementary map φ_j^1.

Now (II) is obtained from the following assertion, which is analogous to Proposition 7.3.

PROPOSITION 7.6. *Let $\rho^1 = \Phi^+\rho$, and let ν^0, ν^1 be the representations of D^r, $r \geqslant 4$, in $R = \bigoplus_{i=1} \rho(e_i)$ defined by the formulae*

$\nu^0(e_i) = G_i + V^1$, $\nu^1(e_i) = G_i' V^1$. Then for every $f_\alpha = f_{i_1 \ldots i_l 0}$, $l \geqslant 1$, and every $j \neq i_1$

$$V^1 + G_j(G_j' + \nu^1(f_\alpha)) = \nu^0(f_{j\alpha}),$$

where $j\alpha = (j, i_1, \ldots, i_l, 0)$.

The proof of this proposition is, essentially, a repetition of that of Proposition 7.5, with i_l changed to 0 in $\alpha = (i_1, \ldots, i_l)$. A difference occurs only for $l = 1$, when we have to prove that $V^1 + G_j(G_j' + V^1) = \nu^0(f_{j0})$. Here is a proof:

$$V^1 + G_j(G_j' + V^1) = (V^1 + G_j)\left(\sum_{t \neq j} (V' + G_t)\right) =$$
$$= \nu^0(e_j) \sum_{t \neq j} \nu^0(e_t) = \nu^0\left(e_j \sum_{t \neq j} e_t\right) = \nu^0(f_{j0}).$$

All further steps in the proof of Proposition 7.5 remain valid when $i_l = 0$, that is, when $e_{i_1 \ldots i_{l-1} i_l}$ is changed to $f_{i_1 \ldots i_{l-1} 0}$.

§8. Complete irreducibility of the representations $\rho_{t,l}$

8.1. Equivalence of the representations $\rho^+_{t,l}$ and $\rho_{t,l}$. In §6 we have given a functorial definition of the representations $\rho^+_{t,l}$, namely, $\rho^+_{t,l} = (\Phi^-)^{l-1}\rho^+_{t,1}$ for $l \geqslant 2$. In §1 we have given a constructive definition of the representations $\rho_{t,l}$ for $l \geqslant 2$, which we repeat here.

By $W_{t,l}$ ($t \in \{0, 1, \ldots, r\}$, $l = 2, 3, \ldots$) we denote the vector space over k with the basis η_α, $\alpha \in A_t(r, l)$ where

$$A_t(r, l) = \{\alpha = (i_1, \ldots, i_{l-1}, t) \mid i_k \in I = \{1, \ldots, r\} \text{ and } i_1 \neq i_2,\ i_2 \neq i_3, \ldots, i_{l-2} \neq i_{l-1},\ i_{l-1} \neq t\}.$$

By $Z_{t,l}$ we denote the subspace of $W_{t,l}$ spanned by all vectors

$$g_{\alpha;k} = \sum_{\substack{i_k=1 \\ i_k \neq i_{k-1},\, i_{k+1}}}^{r} \eta_{i_1 \ldots i_k \ldots i_{l-1} t}, \text{ where } \alpha = (i_1, \ldots, i_k, \ldots, i_{l-1}, t),$$

and the summation is over all $\alpha' = (i_1, \ldots, i_k, \ldots i_{l-1}, t)$ in which all i_s with $s \neq k$ are fixed.

We set $V_{t,l} = W_{t,l}/Z_{t,l}$, and denote by ϑ the canonical map $\vartheta\colon W_{t,l} \to V_{t,l}$; we denote the vectors $\vartheta\eta_\alpha$ by ξ_α. Thus, $V_{t,l}$ is the space spanned by the vectors ξ_α, $\alpha \in A_t(r, l)$, for which $\sum_{i_k} \xi_{i_1 \ldots i_k \ldots i_{l-1} t} = 0$.

Now let j be a fixed index $j \in I$. By $F_{j,t,l}$ we denote the subspace of $W_{t,l}$ with the basis $\{\eta_\alpha\}$, where α has the form $\alpha = (j, i_2, \ldots, i_{l-1}, t)$. We define a representation $\rho_{t,l}$ in $V_{t,l}$ by setting $\rho_{t,l}(e_j) = \vartheta F_{j,t,l}$.

To establish the isomorphism $\rho_{t,l} \cong \rho^+_{t,l}$, $l \geqslant 2$, it is convenient to introduce auxiliary representations $\nu_{t,l}$ in $W_{t,l}$. We set $\nu_{t,l}(e_j) = F_{j,t,l} + Z_{t,l}$. We now list some of the simplest properties of the $\nu_{t,l}$. Obviously, the map $\vartheta\colon W_{t,l} \to V_{t,l}$ defines a morphism of representations $\tilde{\vartheta}\colon \nu_{t,l} \to \rho_{t,l}$. We introduce the trivial representation τ in $Z_{t,l}$ by setting $\tau(e_i) = Z_{t,l}$ for all $i \in I$. It is easy to see that τ is a multiple of the atomic representation $\rho_{\bar{0},1}$. It turns out that

$$\nu_{t,l} \cong \rho_{t,l} \oplus \tau. \tag{8.1}$$

To establish this isomorphism, it is sufficient to choose in $W_{t,l}$ any subspace U such that $UZ_{t,l} = 0$ and $U + Z_{t,l} = W_{t,l}$. It can be shown elementarily that $\nu_{t,l} = \nu_{t,l}|_U + \nu_{t,l}|_{Z_{t,l}}$ and that $\nu_{t,l}|_U \cong \rho_{t,l}$ and $\nu_{t,l}|_{Z_{t,l}} \cong \tau$. Thus, $\nu_{t,l}$ differs from $\rho_{t,l}$ by the trivial representation τ.

It follows from (8.1) that $\Phi^-\nu_{t,l} \cong \Phi^-\rho_{t,l} \oplus \Phi^-\tau$, and it is clear from the definition of τ that $\Phi^-\tau = 0$. Consequently,

(8.2) $$\Phi^{-}\nu_{t,l} \cong \Phi^{-}\rho_{t,l}.$$

In addition, we set, by definition, $\nu_{t,1} = \rho_{t,1} = \rho^{+}_{t,1}$.

PROPOSITION 8.1. *For every* $l > 1$

$$\Phi^{-}\nu_{t,l-1} \cong \rho_{t,l}.$$

PROOF. We proceed first with an auxiliary construction.

a) We denote by Y_j the subspace of $W_{t,l}$ spanned by the vectors

$g_{j\beta;k} = \sum_{i_k} \eta_{ji_2\ldots i_k\ldots i_{l-1}t}$, where $2 \leqslant k \leqslant l-1$, the index $j \in I$ is fixed and $\beta = (i_2, \ldots, i_k, \ldots, i_{l-1}, t)$. Obviously $Y_j \subseteq F_{j,t,l}Z_{t,l}$. We set $Y = \sum_{j=1}^{r} Y_j$. Clearly, $Y \subseteq Z_{t,l}$. We denote the factor space $W_{t,l}Y$ by Q and the canonical map $W_{t,l} \to Q$ by δ. It follows from the relation $Y \subseteq Z_{t,l}$ that ϑ: $W_{t,l} \to V_{t,l}$ splits into the compositum of the epimorphisms

$$\vartheta = \theta\delta, \quad V_{t,l} \xleftarrow{\theta} Q \xleftarrow{\delta} W_{t,l},$$

where θ is the epimorphism with kernel Ker $\theta = \delta Z_{t,l}$. From the proof, which we give repeat below, we can deduce that $Q \cong \bigoplus_{i=1}^{r} \rho_{t,l}(e_i)$.

b) We write $Q_j = \delta F_{j,t,l}$. By definition, $W_{t,l} = \sum_{j=1}^{r} F_{j,t,l}$ and this sum is direct. Clearly, $Q = \delta W_{t,l} = \delta \sum_j F_{j,t,l} = \sum_j \delta F_{j,t,l} = \sum_{j=1}^{r} Q_j$. We have also defined Ker $\delta = Y = \sum_j Y_j$, where $Y_j \subseteq F_{j,t,l}$. Hence it is easy to deduce that the sum $Q = \sum_{j=1}^{r} Q_j$ is direct.

c) Let j be a fixed index. We define maps μ_j: $W_{t,l-1} \to W_{t,l}$ by means of the system of equations

(8.3) $$\mu_j\eta_{i_2\ldots i_{l-1}t} = \begin{cases} 0 & \text{if} \quad j = i_2, \\ \eta_{ji_2\ldots i_{l-1}t} & \text{if} \quad j \neq i_2. \end{cases}$$

We set $\mu = \sum_{j=1}^{r} \mu_j$. Thus,

$$\mu\eta_{i_2\ldots i_{l-1}t} = \sum_{\substack{s=1 \\ s\neq i_2}}^{r} \eta_{si_2\ldots i_{l-1}t} = g_{\alpha;1},$$

where $\alpha = (i_1, i_2, \ldots, i_{l-1}, t)$. Next we set $\gamma = \delta\mu$ and $\gamma_j = \delta\mu_j$. Thus, the diagram

$$\begin{array}{ccccc} V_{t,l} & \xleftarrow{\theta} & Q & \xleftarrow{\delta} & W_{t,l} \\ & & {\scriptstyle\gamma}\nwarrow & & \uparrow{\scriptstyle\mu} \\ & & & & W_{t,l-1} \end{array}$$

commutes.

LEMMA 8.1. *The maps γ and γ_j defined above have the following properties*:

(I) $\mathrm{Im}\ \gamma = \delta Z_{t,l}$;

(II) $\mathrm{Im}\ \gamma_j = Q_j$;

(III) $\mathrm{Ker}\ \gamma_j = \nu_{t,l-1}(e_j) = F_{j,t,l-1} + Z_{t,l-1}$.

A proof of this lemma is given later; first we complete the proof of Proposition 8.1. From (II) and (III) in Lemma 8.1 it follows that $Q_j \cong W_{t,l-1}/\mathrm{Ker}\ \gamma_j = W_{t,l-1}/\nu_{t,l-1}(e_j)$. Thus, $Q \cong \bigoplus_j W_{t,l-1}/\nu_{t,l-1}(e_j)$. It is also clear that the map γ: $W_{t,l-1} \to Q$ is such that $\gamma = \delta\mu = \delta \sum_j \mu_j = \sum_j \delta\mu_j = \sum_j \gamma_j$. Consequently, $\Phi^-\nu_{t,l-1}$ is a representation in $V = \mathrm{Coker}\ \gamma = Q/\mathrm{Im}\ \gamma$. From (I) in Lemma 8.1 we obtain $V = Q/\mathrm{Im}\ \gamma = Q/\delta Z_{t,l} = Q/\mathrm{Ker}\ \theta = V_{t,l}$. Therefore $\Phi^-\nu_{t,l-1}(e_j) = \theta Q_j = \theta(\delta F_{j,t,l}) = \vartheta F_{j,t,l} = \rho_{t,l}(e_j)$. We have now proved that $\Phi^-\nu_{t,l-1} \cong \rho_{t,l}$.

PROOF OF LEMMA 8.1. From the definition of the maps μ_j and μ it can be verified immediately that

$$(8.4)\qquad \begin{cases} \text{a) } \mathrm{Im}\ \mu_j = F_{j,t,l}, & \text{b) } \mathrm{Ker}\ \mu_j = F_{j,t,l-1}, \\ \text{c) } Z_{t,l} = Y + \mu W_{t,l-1}, & \text{d) } \mu_j Z_{t,l-1} = Y_j. \end{cases}$$

(I) From (8.4) c) we find $\delta Z_{t,l} = \delta(Y + \mu W_{t,l-1}) = \delta(\mathrm{Ker}\ \delta + \mu W_{t,l-1}) = \delta\mu W_{t,l-1} = \gamma W_{t,l-1} = \mathrm{Im}\ \gamma$.

(II) From (8.4) a) we find $\mathrm{Im}\ \gamma_j = \gamma_j W_{t,l-1} = \delta\mu_j W_{t,l-1} = \delta F_{j,t,l} \overset{\mathrm{def}}{=} Q_j$.

(III) Obviously,
$x \in \mathrm{Ker}\ \gamma_j = \mathrm{Ker}\ \delta\mu_j \Leftrightarrow \mu_j(x) \in \mathrm{Ker}\ \delta \Leftrightarrow \mu_j(x) \in (\mathrm{Ker}\ \delta)(\mathrm{Im}\ \mu_j)$. Using the equations $\mathrm{Im}\ \mu_j = F_{j,t,l}$, $\mathrm{Ker}\ \delta = \sum_{j=1}^{r} Y_j$ and the relation $Y_j \subseteq F_{j,t,l}$, it is easy to show that $(\mathrm{Ker}\ \delta)(\mathrm{Im}\ \mu_j) = Y_j$. Thus, we can write

$$(8.5)\qquad x \in \mathrm{Ker}\ \gamma_j \Leftrightarrow \mu_j(x) \in Y_j.$$

As we have mentioned, $\mathrm{Ker}\ \mu_j = F_{j,t,l-1}$ and $Y_j = \mu_j Z_{t,l-1}$. Thus, $Y_j = \mu_j(F_{j,t,l-1} + Z_{t,l-1})$.

From this and (8.5) it is easy to see that $\mathrm{Ker}\ \gamma_j = F_{j,t,l-1} + Z_{t,l-1}$. This proves the Lemma.

PROPOSITION 8.2. $\rho_{t,l} \cong \rho^{+}_{t,l}$.

The proof is by induction on l. By definition, $\rho_{t,1} = \rho^{+}_{t,1}$. Suppose that we have proved that $\rho_{t,k} \cong \rho^{+}_{t,k}$ for all $k \leqslant l - 1$. Then we show that $\rho_{t,l} \cong \rho^{+}_{t,l}$. By (8.2), $\Phi^-\rho_{t,l-1} \cong \Phi^-\nu_{t,l-1}$, and by Lemma 8.1,

$\Phi^{-}\nu_{t,\,l-1} \cong \rho_{t,\,l}$. Consequently, $\Phi^{-}\rho_{t,\,l-1} \cong \rho_{t,\,l}$. By induction, we assume that $\rho_{t,\,l-1} \cong \rho^{+}_{t,\,l-1}$. Hence $\rho_{t,\,l} \cong \Phi^{-}\rho_{t,\,l-1} \cong \Phi^{-}\rho^{+}_{t,\,l-1} \overset{\text{det}}{=} \rho^{+}_{t,\,l}$.

8.2. Complete irreducibility of the representations $\rho_{t,\,l}$. In this subsection we explain the basic steps in the proof of the theorem on complete irreducibility. A full proof of the theorem will be published separately.

THEOREM 8.1. *Let $\rho^{+}_{t,\,l}$ and $(\rho^{-}_{t,\,l})$ be indecomposable representations of D^r, $r \geqslant 4$, in spaces $V_{t,l}$ over a field k of characteristic* 0, *with* dim $V_{t,l} \geqslant 3$. *Then $\rho^{+}_{t,\,l}$ and $(\rho^{-}_{t,\,l})$ are completely irreducible, that is* $\rho^{\pm}_{t,\,l}(D^r) \cong$ **P**(**Q**, m), *where* **P**(**Q**, m) *is the lattice of linear submanifolds of the projective space of dimension m* = dim $V_{t,l} - 1$ *over the field* **Q** *of rational numbers.*

REMARK. The restriction dim $V_{t,l} \geqslant 3$ occurs because the following indecomposable representations are not completely irreducible: 1) all the atomic representations $\rho^{+}_{t,\,1}$ and $\rho^{-}_{t,\,1}$ for $t \in \{0, 1, \ldots, r\}$ for *any lattice* D^r, and 2) the representations $\rho^{+}_{t,\,2}$ and $(\rho^{-}_{t,\,2})$, $t \in \{1, \ldots, r\}$, for D^4. For the latter representations, dim $V_{t,2} = 2$.

We describe the basic steps in the proof of Theorem 8.1. We denote by L the sublattice of $\mathscr{L}(V_{t,\,l})$ generated by the one-dimensional subspaces $k\xi_\alpha$, $\alpha \in A_t(r, l)$. Since $k\xi_\alpha \in \rho^{+}_{t,\,l}(D^r)$, it can be proved elementarily that $L \cong \rho^{+}_{t,\,l}(D^r)$.

The proof of the isomorphism $L \cong$ **P**(**Q**, m) when dim $V_{t,l} \geqslant 3$ is based on the following assertions.

Let $\{\xi_i\}_{i\in A}$ be a set of non-zero vectors in V. We call the set $\{\xi_i\}_{i\in A}$ indecomposable if for every subset B of A $(B \neq \varnothing, B \neq A)$ the intersection of the subspaces $V_B = \sum_{i\in B} k\xi_i$ and $V_{\bar{B}} = \sum_{j\in A-B} k\xi_j$ is non-empty:

$$V_B V_{\bar{B}} \neq 0 \text{ and } V_B + V_{\bar{B}} = V.$$

A set $\{\xi_i\}_{i\in A}$ in a finite-dimensional space V over a field k of characteristic 0 is called rational if we can choose a subset $B \subseteq A$ of linearly independent vectors $\{\xi_i\}_{i\in B}$ such that for any $j \in A$

$$\xi_j = \sum_{i\in B} a_i\xi_i, \text{ where } a_i \in \mathbf{Q}.$$

PROPOSITION 8.3. *Let $\rho^{+}_{t,\,l}$ be an indecomposable representation of D^r $(r \geqslant 4)$ in a space $V_{t,l}$ over a field k of characteristic* 0. *Then for $l > 1$, the set of vectors ξ_α, $\alpha \in A(r, l)$, is indecomposable and rational.*

The proof of the indecomposability of the set $\{\xi_\alpha\}_{\alpha\in A_t(r,\,l)}$ follows easily from the indecomposability of $\rho_{t,\,l}$. The proof of rationality is also elementary.

In establishing the isomorphism $L \cong$ **P**(**Q**, m), the central fact is the following theorem, which is of independent interest.

THEOREM 8.2. *Let $\{\xi_i\}_{i\in A}$ be an indecomposable and rational set* *of vectors in a space V over a field k of characteristic* 0. *If* $\dim V \geqslant 3$, *the lattice L generated by the one-dimensional subspaces $k\xi_i$ is completely irreducible, that is,*

$$L \cong \mathbf{P}(Q, m), \textit{ where } m = \dim V_{t,l} - 1.$$

A proof of this theorem will be published separately.

References

[1] I. N. Bernstein, I. M. Gel'fand and V. A. Ponomarev, Coxeter functors and Gabriel's theorem, Uspekhi Mat. Nauk **28**:2 (1973), 19–33.
= Russian Math. Surveys **28**:2 (1973), 17–32.

[2] G. Birkhoff, Lattice Theory, Amer. Math. Soc., New York 1948.
Translation: *Teoriya struktur*, Izdat. Mir, Moscow 1952.

[3] N. Bourbaki, Éléments de mathématique, XXVI, Groupes et algèbres de Lie, Hermann and Co., Paris 1960. MR **24** # A2641.
Translation: *Gruppy i algebry Li*, Izdat. Mir., Moscow 1972.

[4] P. Gabriel, Unzerlegbare Darstellungen. I, Manuscripta Math. **6** (1972), 71–103.

[5] I. M. Gel'fand and V. A. Ponomarev, Problems of linear algebra and classification of quadruples of subspaces in a finite-dimensional vector space, Coll. Math. Soc. Iános Bolyai **5**, Hilbert space operators, Tihany (Hungary) 1970, 163–237 (in English). (For a brief account, see Dokl. Akad. Nauk SSSR **197** (1971), 762–765.
= Soviet Math. Doklady **12** (1971), 535–539.)

[6] L. A. Nazarova and A. V. Roiter, Representations of partially ordered sets, in the collection "Investigations in the theory of representations", Izdat. Nauka, Leningrad 1972, 5–31.

[7] L. A. Nazarova, Representations of quivers of infinite type, Izv. Akad. Nauk SSSR Ser. Mat. **37** (1973), 752–791.
= Math. USSR – Izv. **7** (1973), 749–792.

Received by the Editors, 10 June 1974

Translated by M. B. Nathanson

Dedicated to P. S. Aleksandrov,
who has done so much for the development
of general ideas in mathematics

LATTICES, REPRESENTATIONS, AND ALGEBRAS CONNECTED WITH THEM I[1]

I. M. Gel'fand and V. A. Ponomarev

In this article the authors have attempted to follow the style which one of them learned from P. S. Aleksandrov in other problems (the descriptive theory of functions and topology).

Let L be a modular lattice. By a representation of L in an A-module M, where A is a ring, we mean a morphism from L into the lattice $\mathscr{L}(A, M)$ of submodules of M. In this article we study representations of finitely generated free modular lattices D^r. We are principally interested in representations in the lattice $\mathscr{L}(K, V)$ of linear subspaces of a space V over a field K ($V = K^n$).

An element a in a modular lattice L is called perfect if a is sent either to O or to V under any indecomposable representation $\rho: L \to \mathscr{L}(K, V)$. The basic method of studying the lattice D^r is to construct in it two sublattices B^+ and B^-, each of which consists of perfect elements.

Certain indecomposable representations $\rho^+_{t,l}$ (respectively, $\rho^-_{t,l}$) are connected with the sublattices B^+ (respectively, B^-). Almost all these representations (except finitely many of small dimension) possess the important property of complete irreducibility. A representation $\rho: L \to \mathscr{L}(K, V)$ is called completely irreducible if the lattice $\rho(L)$ is isomorphic to the lattice of linear subspaces of a projective space over the field Q of rational numbers of dimension $n - 1$, where $n = \dim_K V$. In this paper we construct a certain special K-algebra A^r and study the representations $\rho_A: D^r \to \mathscr{L}_R(A^r)$ of D^r into the lattice of right ideals of A^r. We conjecture that the lattice of right homogeneous ideals of the Q-algebra A^r describes (up to the relation of linear equivalence) the essential part of D^r.

Contents

§1. Basic definitions and statement of results

This article is a further development of the authors' paper [7], but can be read independently.

---

[1] The second part of this article will be published in these Uspekhi **32**:1 (1977), 85–106.

1.1. Lattices. A *lattice* L is a set with two operations: intersection and sum. If $a, b \in L$, then we denote their intersection by ab and their sum[1] by $a + b$. Both these operations are commutative and associative, and, moreover, satisfy the axioms of absorption: $a(a + b) = a$ and $a + ab = a$.

An order relation is defined naturally in a lattice L: $a \subseteq b \Longleftrightarrow a + b = b$. It is easy to deduce that $aa = a$ and $a + a = a$ for every $a \in L$.

A lattice L is called *distributive* if for any $a, b, c \in L$

(1.1) $a(b + c) = ab + ac$,

(1.2) $a + bc = (a + b)(a + c)$.

It can be shown that a lattice is distributive if it satisfies at least one of the equations (1.1) or (1.2).

A lattice L is called *modular* if for all $a, b, c \in L$

$$a \subseteq b \Rightarrow b(a + c) = a + bc$$

EXAMPLE 1. Let A be a ring and let M be a left (right) A-module. Then the set of all submodules of M is a lattice with respect to the operations of intersection ($\cap$) and sum (+). We denote this lattice by $\mathcal{L}(A, M)$.

EXAMPLE 2. In this article we shall most often consider the lattice $\mathcal{L}(K, V)$ of linear subspaces of a finite-dimensional vector space V over a field K. When $\dim V = n$ (that is, $V \cong K^n$), we also denote this lattice by $\mathcal{L}(V)$ or $\mathcal{L}(K^n)$. If U_1 and U_2 are subspaces of V, then we denote their sum and intersection by $U_1 + U_2$ and $U_1 U_2$.

EXAMPLE 3. Let $\mathbf{P}(V) = \mathbf{P}_n(K)$ be the projective space corresponding to $V \cong K^{n+1}$. Then $\mathcal{L}(V)$ is well known to be isomorphic to the lattice of linear submanifolds of the projective space $\mathbf{P}(V)$. We call this lattice a *projective geometry* ($PG(V)$). Thus, if $V = K^3$, then the elements of the corresponding geometry are the points and lines of the projective plane $\mathbf{P}_2(K)$. (If a and b are points in $\mathbf{P}_2(K)$, then $a + b$ is the line passing through a and b. If A and B are lines, then AB is their point of intersection.)

The basic objects of our investigation are the free modular lattices D^r with a finite number of generators $(e_1, \ldots, e_r)$. The lattices D^1 and D^2 are obviously finite. It is not difficult to show that the lattice D^3 is also finite (see Birkhoff [3]). The lattices D^r, $r \geqslant 4$, have a very complicated structure. We are by now only close to an understanding of the structure of $\bar{D}^4$ (the factor lattice of D^4 with respect to linear equivalence, which we define in 1.2).

1.2. Representations. Let L be a modular lattice and $\mathcal{L}(A, M)$ the lattice of submodules of a left A-module M. *A representation* ρ *of* L in M is a morphism[2] $\rho: L \to \mathcal{L}(A, M)$. Here, for any $x, y \in L$ we have $\rho(x + y) = \rho(x) + \rho(y)$ and $\rho(xy) = \rho(x) \cap \rho(y)$, where $\rho(x)$ and $\rho(y)$ are submodules of M.

---

[1] The intersection of the elements a and b is often denoted by $a \cap b$ or $a \wedge b$, and the sum by $a \cup b$ or $a \vee b$.

[2] A *morphism* $\rho: L_1 \to L_2$ of lattices L_1 and L_2 is a mapping such that $\rho(xy) = \rho(x)\rho(y)$ and $\rho(x + y) = \rho(x) + \rho(y)$ for all $x, y \in L_1$.

Throughout this article we are concerned with only two types of representation.

1) A representation of a lattice L in a finite-dimensional vector space V over a field K ($\rho: L \to \mathcal{L}(K, V)$). Such a representation associates with elements $x, y \in L$ subspaces $\rho(x), \rho(y) \subseteq V$.

2) A representation of L in the lattice $\mathcal{L}(A)$ of left ideals of a ring A.

We introduce in L an equivalence relation by setting $x \sim y$ if $\rho(x) = \rho(y)$ for every representation $\rho: L \to \mathcal{L}(K, V)$ in any space V over any field K. It can be shown by examples (even in D^4) that there exist linearly equivalent, but unequal elements.

We consider the set whose elements are the classes of linearly equivalent elements of D^r. The operations $(\cdot)$ and $(+)$ carry over naturally to this set. We denote by $\bar{D}^r$ the lattice obtained in this way. The aim of this paper is the study of the lattices $\bar{D}^r$.

An important technique for the study of D^r is the construction of the sublattice B of perfect elements, to whose definition we now turn.

1.3. Perfect elements in D^r. An element v in a modular lattice L is called *perfect* if for every indecomposable[1] representation $\rho: L \to \mathcal{L}(K, V)$ either $\rho(v) = V$ or $\rho(v) = 0$.

In the free modular lattice D^r with generators $e_1, \ldots, e_r$ we construct two sublattices B^+ and B^-, whose elements, as we shall prove later, are all perfect.

For every integer $l \geqslant 1$ we construct a sublattice $B^+(l)$ consisting of 2^r elements. We shall call this sublattice $B^+(l)$ the l-th *upper cubicle*.

It is quite simple to construct $B^+(1)$. Namely, we set $h_t(1) = \sum_{j \neq t} e_j$. Then the *upper cubicle* $B^+(1)$ is the sublattice of D^r generated by the elements $h_1(1), h_2(1), \ldots, h_r(1)$. It is not difficult to prove (see §3) that $B^+(1)$ is a Boolean algebra with 2^r elements. Thus, $B^+(1)$ is isomorphic to the lattice of vertices of an r-dimensional cube with the natural ordering.

We proceed to the definition of the cubicles $B^+(l)$. The elements of $B^+(l)$ are constructed with the help of polynomials $e_{i_1 \ldots i_l}$, which are of independent interest. We denote by $A(r, l)$ the set whose elements are the sequences $\alpha = (i_1, i_2, \ldots, i_l)$ of integers $1 \leqslant i_j \leqslant r$ such that $i_j \neq i_{j+1}$ for all $1 \leqslant j \leqslant l - 1$. We set, by definition, $A(r, 1) = I = \{1, \ldots, r\}$. The elements $e_\alpha = e_{i_1 \ldots i_l}$ are defined by induction on l as follows. If $l = 1$ and $\alpha = (i_1)$, then $e_\alpha = e_{i_1}$; if $l > 1$ and $\alpha = (i_1, \ldots, i_l)$, then $e_\alpha = e_{i_1} (\sum_{\beta \in \Gamma(\alpha)} e_\beta)$, where $\Gamma(\alpha) \subset A(r, l-1)$ consists of the sequences $\beta = (k_1, \ldots, k_{l-1})$ constructed from a fixed α in the following way:

[1] A representation ρ in a space V is called *decomposable* if there exist non-zero subspaces U_1, U_2 in V that are complementary to each other ($U_1 U_2 = 0$, $U_1 \dotplus U_2 = V$) and such that $\rho(x) = \rho(x) U_1 + \rho(x) U_2$ for every $x \in L$.

$$\Gamma(\alpha) = \{\beta = (k_1, \ldots, k_{l-1}) \mid k_1 \neq i_1, i_2;\ k_2 \neq i_2, i_3, \ldots, k_{l-1} \neq i_{l-1}, i_l\}.$$

For example, if $\alpha = (i_1, i_2)$, then $\Gamma(\alpha) = \{\beta = (k_1) \mid k_1 \neq i_1, i_2\}$, and so $e_{i_1 i_2} = e_{i_1}\left(\sum_{j \neq i_1, i_2} e_j\right)$.

Now we define the elements $h_t(l)$. We set

$$e_t(l) = \sum_{\alpha \in A_t(r, l)} e_\alpha,$$

where $A_t(r, l)$ is the subset of $A(r, l)$ consisting of all sequences $\alpha = (i_1, \ldots, i_{l-1}, t)$ in which the last index is fixed and equal to t. Further, we set

$$h_t(l) = \sum_{j \neq t} e_j(l).$$

Then we define $B^+(l)$ to be the lattice generated by the elements $h_1(l), \ldots, h_r(l)$.

It is not difficult to prove (see §3) that $B^+(l)$ is a Boolean algebra and that elements from different cubicles $B^+(l)$ and $B^+(m)$ can be ordered in the following way: for every $v_l \in B^+(l)$ and $v_m \in B^+(m)$ it follows from $l < m$ that $v_l \supset v_m$. Thus, the set

$$B^+ = \bigcup_{l=1}^{\infty} B^+(l)$$

is itself a lattice.

A second set $B^- = \bigcup_{l=1}^{\infty} B^-(l)$ consists of the elements dual to those of B^+. (We say that a lattice polynomial $g(e_1, \ldots, e_r)$ is dual to a lattice polynomial $f(e_1, \ldots, e_r)$ if it is obtained from f by interchanging the operations $(+)$ and $(\cap)$. Thus, for example, $e_1(e_2 + e_3 + e_4)$ is dual to $e_1 + (e_2 e_3 e_4)$.

One of the main theorems of this article is the following.

THEOREM 1. *The elements of the lattices* $B^+ = \bigcup_{l=1}^{\infty} B^+(l)$ *and* $B^- = \bigcup_{l=1}^{\infty} B^-(l)$ *are perfect.*

1.4. Characteristic functions of an indecomposable representation. Let L be an arbitrary modular lattice. Then the set B of perfect elements in L is a sublattice of L (see §3). If ρ is an arbitrary indecomposable representation of L ($\rho: L \to \mathscr{L}(K, V)$), then every perfect element has, by definition, the following property: either $\rho(v) = 0$ or $\rho(v) = V$ (V is the representation space of ρ). Thus, to every indecomposable representation ρ there corresponds a function χ_ρ on the set B of perfect elements, which is defined as follows:

$$\chi_\rho(v) = \begin{cases} 0 & \text{if } \rho(v) = 0, \\ 1 & \text{if } \rho(v) = V. \end{cases}$$

We call χ_ρ the *characteristic function* of ρ.

In the lattice D^r we have defined two sublattices of perfect elements B^+ and B^-. We claim that $\rho(v^-) \subseteq \rho(v^+)$ for any representation $\rho(\rho: D^r \to \mathcal{L}(K, V))$ $(r \geqslant 4)$ and any $v^- \in B^-$ and $v^+ \in B^+$. We denote by B the sublattice of perfect elements in D^r generated by B^+ and B^-. From $\rho(v^-) \subseteq \rho(v^+)$ it follows that every characteristic function χ_ρ defined on B belongs to one of the following three types: 1) $\chi_\rho(v^+) = 0$ for some $v^+ \in B^+$, hence $\chi_\rho(v^-) = 0$ for all $v^- \in B^-$; 2) $\chi_\rho(v^-) = 1$ for some $v^- \in B^-$, hence, $\chi_\rho(v^+) = 1$ for all $v^+ \in B^+$; 3) $\chi_\rho(v^-) = 0$ for all $v^- \in B^-$ and $\chi_\rho(v^+) = 1$ for all $v^+ \in B^+$. We denote the last function by χ_0^1; In §7 we prove the following theorem.

THEOREM 2. *Let ρ be an indecomposable representation of the lattice D^r $(\rho: D^r \to \mathcal{L}(K, V))$ $(r \geqslant 4)$. If the characteristic function χ_ρ of ρ is of the first or second type, then ρ is defined by its characteristic function uniquely up to isomorphism.*

We shall find all indecomposable representations corresponding to the various characteristic functions of the first or second type. In the following subsection these representations will be constructed explicitly.

As for the indecomposable representations ρ whose characteristic functions are of type 3 ($\chi_\rho = \chi_0^1$), we know at the moment only that there are infinitely many of them. In the case of D^4 the classification of all such representations is known [6]. For the lattices D^r $(r \geqslant 5)$, the classification of the indecomposable representations with $\chi_\rho = \chi_0^1$ contains as special cases such problems as the determination of a canonical form for several linear operators $A_1, \ldots, A_n$ $(A_i: V \to V)$.

1.5. The algebra A^r and the representation ρ_A. Let K be any field. We define the K-algebra A^r as the associative K-algebra with unit element ε generated by $\xi_0, \xi_1, \ldots, \xi_r$ with the relations

$$\xi_i^2 = 0 \quad (i = 1, \ldots, r), \tag{1}$$

$$\xi_0^2 = \xi_0, \tag{2}$$

$$\xi_0 \xi_i = \xi_i \quad (i = 1, \ldots, r), \tag{3}$$

$$\sum_{i=1}^{r} \xi_i \xi_0 = 0. \tag{4}$$

The *standard monomials* in A^r are the products $\xi_{i_1} \ldots \xi_{i_{l-1}} \xi_t$ such that $1 \leqslant i_j \leqslant r$, $i_j \neq i_{j+1}$ for all $1 \leqslant j < l-1$, $i_{l-1} \neq t$, and $0 \leqslant t \leqslant r$. Thus, in a standard monomial ξ_0 can occur only in the last place. It is easy to see that any non-zero monomial can be brought to standard form. The standard monomial $\xi_{i_1} \ldots \xi_{i_{l-1}} \xi_t$ is also denoted by $\xi_\alpha = \xi_{i_1 \ldots i_{l-1} t}$, where $\alpha = (i_1, \ldots, i_{l-1} t)$.

The *degree* of the monomial ξ_α is the number $d(\xi_\alpha)$ defined in the following way: $d(\varepsilon) = d(\xi_0) = 0$, $d(\xi_i) = 1$ for every $i \neq 0$, $d(\xi_\alpha \xi_\beta) = d(\xi_\alpha) + d(\xi_\beta)$ if $\xi_\alpha \xi_\beta \neq 0$. The degree of the element 0 is left undefined.

We denote by V_l ($V_l \subset A^r$) the space of homogeneous polynomials of degree l. It is not difficult to show that this introduces a grading in A^r: $A^r = V_0 \oplus V_1 \oplus \ldots \oplus V_l \oplus \ldots (V_i V_j \subseteq V_{i+j})$, where[1] $V_0 = K\varepsilon \oplus K\xi_0$.

In §8 we shall show that the algebras A^r ($r \geqslant 4$) are infinite-dimensional, and that $\dim V_l > 0$ for all $l \geqslant 0$. The algebras A^1, A^2, A^3 are finite-dimensional, and their dimensions over K are 3, 5, and 11, respectively.

We denote by $\mathscr{L}_R(A^r)$ the lattice of right ideals of A^r (with respect to the operations of intersection $\cap$ and sum $+$).

We define the representation $\rho_A : D^r \to \mathscr{L}_R(A^r)$ by setting $\rho_A(e_i) = \xi_i A^r$, where $\xi_i A^r$ is the right ideal generated by ξ_i, and the $e_i (i = 1, \ldots, r)$ are the generators of D^r.

In §8 we shall prove the following interesting theorem, which establishes a connection between the lattice polynomials e_α (which were defined in 1.3) and the monomials ξ_α in A^r.

THEOREM 3. *For every* $\alpha = (i_1, \ldots, i_{l-1}, t) \in A_t(r, l)$ $(l \geqslant 1)$ *we have* $\rho_A(e_\alpha) = \xi_\alpha A^r$, *where* $\xi_\alpha A^r$ *is the right ideal generated by the monomial* $\xi_\alpha = \xi_{i_1} \ldots \xi_{i_{l-1}} \xi_t$.

This result is due to Gel'fand, Lidskii, and Ponomarev.

1.6. The representations $\rho_{t,l}$. Let $A\xi_t$ be the left ideal generated by the element ξ_t. We introduce the following notation:

$$V_{t,l} = (A\xi_t) \cap V_l \qquad (t = 0, 1, \ldots, r;\ l = 1, 2, \ldots).$$

According to this definition, $V_{i,0} = 0$ if $i \neq 0$ and $V_{0,0} \cong K\xi_0$. For $l = 1$, $V_{i,1} = K\xi_i$ if $i \neq 0$ and $V_{0,1}$ has the dimension $r - 1$ and is the sum of the one-dimensional subspaces $K\xi_i\xi_0$. For $l \geqslant 2$, every subspace $V_{t,l}$ is generated by one-dimensional subspaces $K\xi_\alpha\xi_t$, where $\xi_\alpha\xi_t$ is a monomial of degree l.

We define a representation $\rho_{t,l}$ of D^r in $V_{t,l}$ as follows. We set

$$\rho_{t,l}(e_i) = V_{t,l} \cap (\xi_i A),$$

where $\xi_i A$ is the right ideal generated by ξ_i $(1 \leqslant i \leqslant r)$.

We define a representation ρ_ε of D^r in $K\varepsilon$ (where ε is the unit element of A^r) by setting $\rho_\varepsilon(e_i) = 0$ for every $e_i \in D^r$.

It is elementary to prove (§8) that ρ_A is isomorphic to the direct sum

$$\rho_A \cong \rho_\varepsilon \oplus \rho_{0,0} \oplus \Big(\bigoplus_{l=1}^{\infty} \Big(\bigoplus_{t=0}^{r} \rho_{t,l}\Big)\Big).$$

It turns out that the representations $\rho_{t,l}$ so constructed possess the following remarkable properties.

THEOREM 4. (i) *The representation* $\rho_{0,0}$ *and the representations* $\rho_{t,l}(t = 0, 1, \ldots, r;\ l = 1, 2, \ldots)$ *of* $D^r (r \geqslant 4)$ *are indecomposable. Any two representations* $\rho_{t,l}$ *and* $\rho_{t',l'}$ *such that* $(t, l) \neq (t', l')$ *are not isomorphic.*

(ii) $\rho_{0,0}$ *and* $\rho_{t,l}(t = 0, 1, \ldots, r;\ l = 1, 2, \ldots)$ *are the only representations whose characteristic functions are of the first type (that is,*

[1] By $K\varepsilon$ (respectively $K\xi_0$) we denote the subspace generated by the element ε (respectively, ξ_0).

$\chi_\rho(v^+) = 0$ *for some* $v^+ \in B^+$).

Let V be a linear space over a field K of characteristic 0. A representation ρ of a modular lattice L in V is called completely irreducible if $\rho(L) \subset \mathscr{L}(V, K)$ is isomorphic to the lattice $\mathscr{L}(\mathbf{Q}^n)$ where $n = \dim_K V$ $(n \geqslant 3)$ (that is, $\rho(L)$ is the projective geometry $\mathrm{P}G_{n-1}(\mathbf{Q})$ $(n \geqslant 3)$ over $\mathbf{Q}$.

The following result holds.

THEOREM 5. *All representations* $\rho_{t,l}\colon D^r \to \mathscr{L}(V_{t,l}; K)$ $(r \geqslant 4)$ *over a field* K *of characteristic* 0, *except finitely many, are completely irreducible.*

The only representations that are not completely irreducible are the following:

a) $\rho_{0,0}$ *and* $\rho_{i,1}$ $(i \neq 0)$ *for any* $r \geqslant 4$;

b) $\rho_{i,2}$ $(i \neq 0)$ *for* $r = 4$.

1.8. The lattice F_r^+. We introduce an equivalence relation R in D^r by setting $x \equiv y \pmod R$ if $\rho_{t,l}(x) = \rho_{t,l}(y)$ for any representation $\rho_{t,l}\colon D^r \to \mathscr{L}(V_{t,l}, K)$. We denote the factor lattice D^r/R by F_r^+.

We now state an important conjecture about the structure of the lattices $F_r^+ (r \geqslant 5)$. Let $A_{\mathbf{Q}}^r$ be the $\mathbf{Q}$-algebra A^r, where $\mathbf{Q}$ is the field of rational numbers. A right ideal in $A_{\mathbf{Q}}^r$ is called *homogeneous* if it is equal to a finite sum of ideals $f_{t,l}A_{\mathbf{Q}}^r$, $f_{t,l} \in V_{t,l}$. The lattice of right homogeneous ideals of $A_{\mathbf{Q}}^r$ is denoted by $\mathscr{M}(A_{\mathbf{Q}}^r)$.

CONJECTURE. The factor-lattice F_r^+ of D^r $(r \geqslant 5)$ is isomorphic to the lattice $\mathscr{M}(A_{\mathbf{Q}}^r)$ of right homogeneous ideals of the $\mathbf{Q}$-algebra $A_{\mathbf{Q}}^r$.

We shall soon publish some results obtained jointly with B. V. Lidskii, which bring us close to a proof of this conjecture.

§2. The category $\mathscr{R}(L, K)$

2.1. The category $\mathscr{R}(L, K)$. Let ρ_1 and ρ_2 be two representations of a modular lattice L in spaces V_1 and V_2, respectively. By a morphism $\tilde{u}\colon \rho_1 \to \rho_2$ we mean a linear transformation $u\colon V_1 \to V_2$ such that $u\rho_1(x) \subseteq \rho_2(x)$ for all $x \in L$, where $u\rho_1(x)$ is the image of the subspace $\rho_1(x)$ under the transformation u.

We often denote a morphism $\tilde{u}$ simply by u (the corresponding linear transformation).

We denote by Hom (ρ_1, ρ_2) the set of all morphisms from ρ_1 to ρ_2. It is not difficult to verify that this determines a category $\mathscr{R}(L, K)$ that of finite-dimensional representations over K. In $\mathscr{R}(L, K)$ the direct sum $\rho_1 \oplus \rho_2$ of any pair of objects ρ_1 and ρ_2 is defined in the natural way.

The category $\mathscr{R}(D^r, K)$ is the object of our study. It is easy to show that $\mathscr{R}$ is additive but not Abelian.

Let $\rho \in \mathscr{R}(L, K)$ be a representation of a lattice L in a space V. We denote by V^* the space dual to V. We define the representation $\rho^* \in \mathscr{R}(L, K)$ in V^* by $\rho^*(x) = (\rho(x))^\perp$ for all $x \in L$ (where

$(\rho(x))^\perp \subset V^*$ is the subspace of functionals that vanish on $\rho(x)$). We call ρ^* the representation dual to ρ.

2.2. Decomposable representations and admissible subspaces. Let ρ be a representation of a modular lattice L in a linear space V. A subspace U of V is called *admissible* with respect to a representation ρ if for all $x, y \in L$

$$\text{(I)} \qquad U(\rho(x)+\rho(y)) = U\rho(x) + U\rho(y).$$

It is not difficult to show that (I) is satisfied if and only if

$$\text{(I}'\text{)} \qquad U+\rho(x)\rho(y) = (U+\rho(x))(U+\rho(y)).$$

PROPOSITION 2.1. *Let ρ be a representation of a lattice L in a space V. Let $U \subset V$ be a subspace of V, $U'' = V/U$ be the quotient space, and $\theta: V \to U''$ be the canonical map. Then the following conditions are equivalent*:

1. *U is admissible with respect to ρ.*
2. *The correspondence $x \mapsto U\rho(x)$ defines a representation in U.*
3. *The correspondence $x \mapsto \theta\rho(x)$ defines a representation in U''.*

The proof is elementary (see [3]).■

The representation in an admissible subspace U defined by the correspondence $x \mapsto U\rho(x)$ is called the restriction of ρ to U and is denoted by $\rho|_U$.

PROPOSITION 2.2. *A representation $\rho \in \mathcal{R}(L, K)$ in a space V is decomposable if and only if there exist non-zero subspaces $U_1, \ldots, U_n$ such that $V \cong U_1 \oplus \ldots \oplus U_n$ and for every $x \in L$*

$$\rho(x) = \sum_{i=1}^{n} U_i\rho(x). \quad ■$$

REMARK 1. If a representation $\rho \in \mathcal{R}(L, K)$ is decomposable, with $\rho = \bigoplus_{i=1}^{n} \rho_i$ and if U_i the subspace corresponding to ρ_i, then the U_i are admissible. The converse however, is false, that is, if $V \cong \bigoplus_{i=1}^{n} U_i$ and if each U_i is admissible, then it does not follow that ρ splits into the direct sum of their restrictions.

QUESTION. Is it true that if U is an admissible subspace, then there is a subspace U' complementary to U (that is, $UU' = 0$ and $U + U' = V$) such that U and U' define a splitting of ρ into a direct sum?

PROPOSITION 2.3. *A representation $\rho \in \mathcal{R}(D^r, K)$ splits into a direct sum $\rho = \bigoplus_{i=1}^{n} \rho_i$ of representations ρ_i if and only if there are non-zero subspaces $U_1, \ldots, U_n$ such that $V = \bigoplus_{j=1}^{n} U_j$ and $\rho(e_i) = \sum_{j=1}^{n} \rho(e_i)U_j$ for every $i \in \{1, \ldots, r\}$, where the e_i are the generators of D^r.*■

§3. Perfect elements. Elementary properties of the lattices B^+ and B^-

3.1. In this section we prove that the set B of perfect elements in a modular lattice L is a sublattice.

PROPOSITION 3.1. *Let a and b be perfect elements of a lattice L. Then so are $a + b$ and ab.*

PROOF. If an element a is perfect, then there exists a subspace V' complementary to $\rho(a)$ (that is, $\rho(a)V' = 0$ and $\rho(a) + V' = V$) such that ρ splits into the direct sum of representations $\rho = \rho_q + \rho'$, where $\rho_a = \rho|_{\rho(a)}$ and $\rho' = \rho|_{V'}$. In particular, for $x = b$

$$\rho(b) = \rho(b)\,\rho(a) + \rho(b)\,V'. \tag{3.1}$$

Note that $\rho(b)\rho(a) = \rho(ba)$ and, by Proposition 2.1, $\rho(b)V' = \rho'(b)$. Thus, we can rewrite (3.1) in the following form:

$$\rho(b) = \rho(ab) + \rho'(b). \tag{3.2}$$

Now b is perfect, consequently, in V' there is a subspace V'' such that $V' = \rho'(b) + V''$ and $\rho' = \rho_b' + \rho''$, where $\rho_b' = \rho'|_{\rho'(b)}$ and $\rho'' = \rho'|_{V''}$. Thus, $\rho = \rho_q \oplus \rho' = \rho_a \oplus (\rho_b' \oplus \rho'') = (\rho_a \oplus \rho_b') \oplus \rho''$.

We claim that $\rho_a \oplus \rho_b'$ is the restriction of ρ to the subspace $\rho(a + b)$. Indeed, by definition, $\rho_a \oplus \rho_b' = \rho'|_{\rho(a)+\rho'(b)}$. Moreover, using the fact that $\rho(ab) \subseteq \rho(a)$ and $\rho(b) = \rho(ab) + \rho'(b)$, we can write

$$\rho(a) + \rho'(b) = (\rho(a) + \rho(ab)) + \rho'(b) = \rho(a) + (\rho(ab) + \rho'(b)) = \\ = \rho(a) + \rho(b) = \rho(a+b).$$

Thus, $\rho = \rho_{a+b} \oplus \rho''$.

In other words, $a + b$ is perfect. The fact that ab is perfect can be proved similarly.■

COROLLARY 3.1. *The set S of perfect elements of a modular lattice L is a sublattice of L.*■

PROPOSITION 3.2. *Let L be a modular lattice, and $\rho \in \mathcal{R}(L, K)$ an arbitrary representation. Then the image $\rho(a)$ of a perfect element is a neutral element of the lattice $\rho(L) \subset \mathcal{L}(V, K)$, that is, for any $x_1, x_2 \in L$*

$$(\rho(x_1) + \rho(x_2))\,\rho(a) = \rho(x_1)\,\rho(a) + \rho(x_2)\,\rho(a).$$

PROOF. The element a is perfect, hence there exists a subspace V' such that $V \cong \rho(a) \oplus V'$ and $\rho = \rho_a \oplus \rho' = \rho|_{\rho(a)} \oplus \rho|_{V'}$. Consequently, for any $x_i \in L$ we can write $\rho(x_i) = \rho(x_i)\rho(a) + \rho(x_i)V'$.

Using this identity, we obtain $\rho(a)(\rho(x_1) + \rho(x_2)) = \rho(a)(\rho(x_1)\rho(a) + \rho(x_1)V' + \rho(x_2)\rho(a) + \rho(x_2)V') = \rho(x_1)\rho(a) + \rho(x_2)\rho(a) + \rho(a)(\rho(x_1)V' + \rho(x_2)V')$. By construction $\rho(a)V' = 0$, and a fortiori $\rho(a)(\rho(x_1)V' + \rho(x_2)V') = 0$. Thus, $\rho(a)(\rho(x_1) + \rho(x_2)) = \rho(a)\rho(x_1) + \rho(a)\rho(x_2)$.■

COROLLARY 3.2. *Let S be the sublattice of perfect elements in a lattice L. Then $\rho(S)$ for any representation $\rho \in \mathcal{R}(L, K)$ is a distributive sublattice of neutral elements of $\rho(L)$.*■

CONJECTURE 3.1. *Let L be an arbitrary modular lattice. Then an element* *a is perfect if and only if it is neutral.*

3.2. Elementary properties of the sublattices $B^+(l)$. Now we study some properties of the sublattice B^+ in D^r, which, as we shall prove later, consists of perfect elements. The lattice B^+ is the union of the sublattices $B^+(1), B^+(2), \ldots, B^+(l), \ldots$, which are called *cubicles*. The cubicle $B^+(l)$ is constructed with the help of special elements e_α in the following way. We set

$$e_t(l) = \sum_{\alpha \in A_t(r,\, l)} e_\alpha, \quad h_t(l) = \sum_{i \neq t} e_i(l),$$

where $\alpha = (i_1, \ldots, i_{l-1}, t) \in A_t(r, l)$ is a sequence such that $i_j, t \in \{1, \ldots, r\}$, $i_j \neq i_{j+1}$, $i_{l-1} \neq t$ (the definition of e_α is on page 69).

Now $B^+(l)$ is, by definition, the sublattice generated by the elements $h_1(l), \ldots, h_r(l)$.

THEOREM 3.1. *$B^+(l)$ is a Boolean algebra.*

This theorem is made more precise in the following proposition.

PROPOSITION 3.3. (I) *Every element $v_{a,l}$ of $B^+(l)$ can be written in the following form:*

(i) $$v_{a,\,l} = \sum_{i \in a} e_i(l) + \sum_{j \in I - a} e_j(l)\, h_j(l),$$

where a is an arbitrary subset of $I = \{1, \ldots, r\}$.

If $a \neq I$, this can also be written

(ii) $$v_{c,\,l} = \bigcap_{j \in I - a} h_j(l).$$

(II) *Let $\mathscr{B}(I)$ be the Boolean algebra of all subsets of I. Then the correspondence $a \mapsto v_{a,l}$ defines a morphism $v_l \colon \mathscr{B}(I) \to B^+(l)$ (that is, $v_{(a \cup b),l} = v_{a,l} + v_{b,l}$ u $v_{(a \cap b),l} = v_{a,l} \cap v_{b,l}$ for any $a, b \subset I$).*

REMARK. In §7 (Corollary 7.2) we shall prove that the mapping $v_l \colon \mathscr{B}(I) \to B^+(l)$ is an isomorphism.

The proof of Proposition 3.3 is based on the following lattice-theoretical lemma.

LEMMA 3.1. *Let L be an arbitrary modular lattice, and $\{e_1, \ldots, e_r\}$ a finite set of elements of L. Then the sublattice B generated by the elements $h_j = \sum_{i \neq j} e_i$ $(j = 1, \ldots, r)$ is a Boolean algebra.*

The proof of this lemma reduces to a proof of the formula

$$\sum_{i \in I - b} e_i + \sum_{j \in b} e_j h_j = \bigcap_{j \in b} h_j,$$

where $\emptyset \subset b \subseteq I$. This formula is easily proved (see [7]) by induction on the number of elements in b. ■

3.3. Structure of the lattice B^+. We denote by B^+ the lattice generated by the sublattices $B^+(l)$ $(l = 1, 2, \ldots)$.

THEOREM 3.2. (I) *B^+ is the union of the sets $B^+(l)$.*

(II) *The sublattices $B^+(l)$ in B^+ can be ordered in the following way: let $v^+(l)$ and $v^+(m)$ be any elements of $B^+(l)$ and $B^+(m)$. If $l < m$, then $v^+(l) \supseteq v^+(m)$.*

REMARK. We shall prove in §7 that, in fact, for $l < m$ strict inequality $v^+(l) \supset v^+(m)$ holds.

The proof of this Theorem is based on two lemmas.

We recall that

$$A(r, l) = \{\alpha = (i_1, \ldots, i_l) \mid i_j \in I = \{1, \ldots, r\},$$

where $i_j \neq i_{j+1}$ for all $j < l$.

LEMMA 3.2. *Let $\beta = (i_1, \ldots, i_l) \in A(r, l)$. We set $\beta_j = (i_1, \ldots, i_l, j)$ for $j \neq i_l$. Then $e_\beta \supseteq e_{\beta j}$.*

LEMMA 3.3. *Let $h_t(l)$ be a generator of $B^+(l)$. Then $e_\alpha \subseteq h_t(l)$ for all $\alpha \in A(r, l+1)$.*

The proof of these Lemmas is elementary (see [7]).

Now we prove Theorem 3.2. It follows from Proposition 3.3 that the minimal element in $B^+(l)$ is $v_{\theta,l} = \bigcap_{i=1}^{r} h_i(l)$, and the maximal element in $B^+(l+1)$ is $v_{I,l+1} = \sum_{i=1}^{r} e_i(l+1) = \sum_{\alpha \in A(r,l+1)} e_\alpha$. It follows from Lemma 3.3 that $\bigcap_{i=1}^{r} h_i(l) \supseteq e_\alpha$ for every $\alpha \in A(r, l+1)$. Therefore, $v_{\theta,l} = \bigcap_{i=1}^{r} h_i(l) \supseteq \sum_{\alpha \in A(r,l+1)} e_\alpha = v_{I,l+1}$. Now if $v_{a,l}$ and $v_{b,l+1}$ are arbitrary elements of $B^+(l)$ and $B^+(l+1)$, then $v_{a,l} \supseteq v_{\theta,l} \supseteq v_{I,l+1} \supseteq v_{b,l+1}$, and the theorem is proved.

3.4. The lattice B^-. By definition, the elements of the cubicles $B^-(l)$ are dual to those of the cubicles $B^+(l)$. For example $e_{i_1 i_2} = e_{i_1} \sum_{j \neq i_1, i_2} e_j$, hence, by definition, we set $e^-_{i_1 i_2} = e_{i_1} + (\bigcap_{j \neq i_1, i_2} e_j)$. Similarly,

$$e^-_t(l) = \bigcap_{\alpha \in A_t(r,\, l)} e^-_\alpha, \quad h^-_t(l) = \bigcap_{j \neq t} e^-_j(l).$$

Each cubicle $B^-(l)$ is a Boolean algebra, and the elements $v^-(l)$ and $v^-(m)$ of distinct cubicles $B^-(l)$ and $B^-(m)$ $(l < m)$ are connected by the relation $v^-(l) \subseteq v^-(m)$. Thus, the lattice B^- generated by the $B^-(l)$ is the union $\bigcup_{l=1}^{\infty} B^-(l)$ of the sets $B^-(l)$.

In particular, just as the maximal and minimal elements $v_{I,l}$ and $v_{\theta,l}$ of the cubicles $B^+(l)$ form a chain

$$\sum_{i=1}^{r} e_i = v_{I,1} \supseteq v_{\theta,1} \supseteq v_{I,2} \supseteq v_{\theta,2} \supseteq \ldots \supseteq v_{I,l} \supseteq v_{\theta,l} \supseteq \ldots,$$

so the maximal elements $v^-_{I,l} = \sum_{i=1}^{r} h^-_i(l)$ and minimal elements

$v^-_{\theta,l} = \bigcap_{i=1}^{r} e^-_i(l)$ of the cubicles $B^-(l)$ are ordered dually:

$$\bigcap_{i=1}^{r} e_i = v^-_{\theta,1} \subseteq v^-_{I,1} \subseteq v^-_{\theta,2} \subseteq v^-_{I,2} \subseteq \ldots \subseteq v^-_{\theta,l} \subseteq v^-_I \quad \subseteq \ldots$$

Note that the element $v^-_{\theta,l}$ is dual to $v_{I,l}$ and $v^-_{I,l}$ is dual to $v_{\theta,l}$.

In §7 we shall prove the following proposition.

PROPOSITION 3.4. *Let v^+ and v^- be arbitrary elements of B^+ and B^-. Then $\rho(v^-) \subseteq \rho(v^+)$ for every representation ρ: $D^r \to \mathcal{L}(K, V)$*

We also believe that the following is true.

CONJECTURE. *For every $v^+ \in B^+$ and $v^- \in B^-$*

$$v^- \subset v^+.$$

§4. Proof that $B^+(1)$ and $B^-(1)$ are perfect. Atomic representations

By definition, $B^+(1)$ is the sublattice of D^r generated by the elements $h_1(1), h_2(1), \ldots, h_r(1)$, where $h_j(1) = \sum_{i \neq j} e_i$. The maximal element in the cubicle $B^+(1)$ is $v_{I,1} = \sum_{i=1}^{r} e_i$. We note that $v_{I,1}$ is the maximal element in the entire lattice D^r. The cubicle $B^-(1)$ consists of the elements dual to the elements of $B^+(1)$. (It is generated by the elements $h^-_j(1) = \bigcap_{i \neq j} e_i$. In this section we prove that every element of $B^+(1)$ and $B^-(1)$ is perfect.

4.1. Atomic representations and the perfectness of $B^+(1)$. We define representations $\rho^+_{t,1}$ for $t \in \{0, 1, \ldots, r\}$, which we call (+) *atomic*. By definition, $\rho^+_{t,1}$ is the representation in the one-dimensional space $V^+_{t,1} \cong K$ for which

1) if $t = 0$, then $\rho^+_{0,1}(e_i) = 0$ for all $i = 1, \ldots, r$;

2) if $t \neq 0$, then $\rho^+_{t,1}(e_i) = 0$ for $i \neq t$ and $\rho^+_{t,1}(e_t) = V^+_{t,1}$.

Note that the atomic representations are none other than the representations $\rho_{0,0}$ and $\rho_{i,1}$ defined in §1 on page 72. Namely, $\rho^+_{0,1} \cong \rho_{0,0}$ and $\rho^+_{i,1} \cong \rho_{i,1}$ if $i \neq 0$.

THEOREM 4.1. *Each element $v_{a,1} \in B^+(1)$ is perfect.*

The proof of this Theorem is based on the following lemma.

LEMMA 4.1. *Let ρ be any representation of D^r in a space V. Then ρ is isomorphic to the direct sum $\rho \cong \overline{\rho_{0,1}} + \overline{\rho_{j,1}} + \tau_j$, where $\tau_j = \rho|_{\rho(h_j(1))}$, where $\overline{\rho_{0,1}}$ and $\overline{\rho_{j,1}}$ are multiples of the atomic representations $\rho^+_{0,1}$ and $\rho^+_{j,1}$ (that is, $\overline{\rho_{0,1}} \cong \underbrace{\rho^+_{0,1} + \ldots + \rho^+_{0,1}}_{m_{0,1}}$, where $m_{0,1} \geqslant 0$, and, similarly, $\overline{\rho_{j,1}} \cong \oplus\, m_{j,1}\, \rho^+_{j,1}$, where $m_{j,1} \geqslant 0$).*

PROOF OF LEMMA 4.1. We set $\rho(e_i) = E_i$ and $\rho(h_j(1)) = H_j$. Thus, the subspaces E_i and H_j are such that $H_j = \sum_{i \neq j} E_i$. We also set

$H_0 = \sum_{i=1}^{r} E_i$. Clearly, H_0 has the following property: for every $j \neq 0$

$$E_j + H_j = H_0 \Rightarrow H_j \subseteq H_0. \tag{4.1}$$

We claim that the element $h_j = \sum_{i \neq j} e_i$ of D^r is perfect. We choose subspaces U_0 and U_j in V to satisfy the relations

$$U_0 H_0 = 0, \qquad U_0 + H_0 = V, \tag{4.2}$$

$$U_j H_j = 0, \qquad U_j + E_j H_j = E_j. \tag{4.3}$$

We illustrate the subspaces in the following figure:

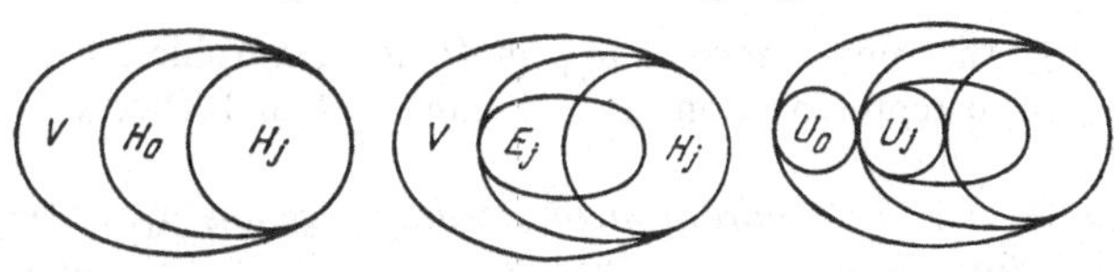

We shall show that $U_j + H_j = H_0$. Clearly, $E_jH_j + H_j = H_j$. Using this and (4.1), (4.3), we see that $U_j + H_j = U_j + E_jH_j + H_j = E_j + H_j = H_0$. Thus, $U_j + H_j = H_0$, and this sum is direct. It easily follows from this equation and the definition of U_0 that $V = U_0 \oplus U_j \oplus H_j$.

We now show that the representation splits into the direct sum $\rho = \rho_0^{-1} \oplus \rho_j^{-1} \oplus \tau_j$, where $\rho_i^{-1} = \rho|_{U_i}$ and $\tau_j = \rho|_{H_j}$.

For this it suffices to show that for every subspace $E_i = \rho(e_i)$ $(i = 1, \ldots, r)$

$$\rho(e_i) = \rho(e_i)U_0 + \rho(e_i)U_j + \rho(e_i)H_j. \tag{4.4}$$

Let us prove this. By construction, $U_0H_0 = U_0(\sum_{i=1}^{r} E_i) = 0$. Consequently, $E_iU_0 = 0$ for every i. By construction, $E_j = U_j + E_jH_j$, hence $E_j = E_j(U_j + E_jH_j) = E_jU_j + E_jH_j$. This proves (4.4) when $i = j$.

If $i \neq j$, it is clear that $E_i \subseteq \sum_{k \neq j} E_k = H_j$, that is, $E_iH_j = E_i$. Next, it follows from $(U_0 + U_j)H_j = 0$ that $(U_0 + U_j)E_i = 0$, and a fortiori $U_0E_i = U_jE_i = 0$. Thus, $E_i = E_iU_0 + E_iU_j + E_iH_j$. This proves (4.4). It means (see Proposition 2.3) that $\rho \cong \rho|_{U_0} \oplus \rho|_{U_j} \oplus \rho|_{H_j}$. Consequently, $h_j(1)$ is perfect.

Thus, the generators $h_1(1), \ldots, h_r(1)$ of $B^+(1)$ are perfect. By Proposition 3.1, this implies that all elements of $B^+(1)$ are perfect, and Theorem 4.1 is proved.

To complete the proof of the lemma, it remains for us to establish that $\rho|_{U_0}$ and $\rho|_{U_j}$ are multiples of the atomic representations $\rho_{0,1}^+$ and $\rho_{j,1}^+$.

By construction of U_0, $U_0H_0 = U_0 \sum_{i=1}^{r} E_i = 0$, and so $U_0E_i = 0$ for every i. Thus, $\rho|_{U_0}(e_i) = \rho(e_i)U_0 = E_iU_0 = 0$ for every i. It follows that $\bar{\rho}_{0,1} = \rho|_{U_0}$ is isomorphic to the direct sum of the atomic representations

$\rho^+_{0,1}$, that is, $\overline{\rho_{0,1}} \cong \underbrace{\rho^+_{0,1} \oplus \ldots \oplus \rho^+_{0,1}}_{m_0}$, where $m_0 = \dim U_0$.

The subspace U_j has the following properties:

a) $U_jE_j = U_j$, and b) $U_jH_j = U_j \sum_{i \neq j} E_i = 0$, that is, $U_jE_i = 0$ for every $i \neq j$. Thus,

$$\bar{\rho}_{j,1}(e_i) = \rho(e_i)U_j = \begin{cases} U_j & \text{if} \quad i = j, \\ 0 & \text{if} \quad i \neq j. \end{cases}$$

Consequently, the representation $\bar{\rho}_{j,1}$ in U_j is isomorphic to the direct sum of the atomic representations $\rho^+_{j,1}$. Lemma 4.1 and Theorem 4.1 are now proved.

COROLLARY 4.1. *The morphism of Boolean algebras $v_1: \mathscr{B}(I) \to B^+(1)$ defined by the formula $a \to v_{a,1}$ (where $a \subseteq I = \{1, \ldots, r\}$ and $v_{a,1} = \sum_{i \in a} e_i + \sum_{j \in I-a} e_jh_j(1) \in B^+(1)$) is an isomorphism.*

PROOF. We construct a representation ρ in $V = K^r$ as follows. Let $\xi_1, \ldots, \xi_r$ be a basis for V. We set $\rho(e_i) = K\xi_i$. It is easy to see that ρ is isomorphic to the direct sum $\bigoplus_{i=1}^{r} \rho^+_{i,1}$ of atomic representations. It is also easy to check that $\rho(h_j(1)) = \sum_{i \neq j} K\xi_i$, that is, $\dim \rho(h_j(1)) = r - 1$. Hence it follows easily that $\rho(v_{a,1}) = \sum_{i \in a} K\xi_i$. Consequently, for any two distinct subsets $a, b \in I$ the corresponding subspaces $\rho(v_{a,1})$ and $\rho(v_{b,1})$ are distinct. This means that $\rho(B^+(1)) \cong \mathscr{B}(I)$ and so $B^+(1) \cong \mathscr{B}(I)$. ■

As we know, any element $v_{a,1} \in B^+(1)$ can be written in the following form: $v_{a,1} = \sum_{i \in a} e_i + \sum_{j \in I-a} e_jh_j(1)$, or, if $a \neq I$, $v_{a,1} = \bigcap_{j \in I-a} h_j(1)$. It follows from Theorem 4.1 that every element $v_{a,1}$ is perfect. So we come to a proposition that refines Theorem 4.1.

PROPOSITION 4.1. *Let $\rho \in \mathscr{R}(D^r, K)$ be any representation. Then $\rho \cong \rho_a(1) \oplus \tau_{a,1}$, where $\tau_{a,1} = \rho|_{\rho(v_{a,1})}$. Here*

(i) *$\rho_a(1) \cong \bigoplus_{j \in (I-a) \cup \{0\}} \bar{\rho}_{j,1}$, and each $\bar{\rho}_{j,1}$ is a multiple of the atomic representation $\bar{\rho}_{j,1} \cong \underbrace{\rho^+_{j,1} \oplus \ldots \oplus \rho^+_{j,1}}_{m_j} (m_j \geqslant 0)$;*

(ii) *if $\tau_{a,1}$ splits into a direct sum $\tau_{a,1} = \oplus \tau_j$ of indecomposable representations τ_j, then none of the τ_j are isomorphic to any of the representations $\rho^+_{i,1}$ for $i \in (I - a) \cup \{0\}$.* ■

We shall use Proposition 4.1 most often when $v_{a,1} = v_{\theta,1}$ is the minimal element of $B^+(1)$. The space $\rho(v_{\theta,1})$ is, as it were, the sum of all subrepresentations that are not (+)-atomic. Namely, $\rho \cong \rho(1) \oplus \tau_{\theta,1}$, where $\tau_{\theta,1} = \rho|_{\rho(v_{\theta,1})}$, and if $\tau_{\theta,1} = \oplus\, \tau_j$, where the τ_j are indecomposable, then none of the τ_j are isomorphic to any of the $\rho^+_{j,1}$ for $i \in \{0, 1, \ldots, r\}$.

4.2. (−)-atomic representations and the cubicle $B^-(1)$. We define representations $\rho^-_{t,1}$ as follows: $\rho^-_{t,1} \overset{\text{def}}{=\!=} (\rho^+_{t,1})^*$. We call them (−)-atomic. It follows from the definition that each $\rho^-_{t,1}$ is a representation in the one-dimensional space $V^-_{t,1} \cong K$, and $\rho^-_{0,1}(e_i) = 0$ for any i; $\rho^-_{j,1}(e_i) = V^-_{j,1}$ if $j \neq i$ and $\rho^-_{j,1}(e_j) = 0$.

The lower cubicle $B^-(1)$ is defined to be the sublattice generated by the elements $h^-_j(1) = \bigcap_{i \neq j} e_i$. Here, $v^-_{\theta,1} = \bigcap_{i=1}^{r} e_i$ is the minimal element in $B^-(1)$. If any element $v^-_{a,1} \in B^-(1)$ with $a \neq \varnothing$ (θ), can be written in the form $v^-_{a,1} = \sum_{j \in a} h^-_j(1)$. Arguments dual to those used in the preceding subsection show that each element $v^-_{a,1} \in B^-(1)$ is perfect. Also, the representation $\rho^-_{a,1} = \rho|_{\rho(v^-_{a,1})}$ is a direct summand of ρ, and $\rho^-_{a,1} \cong \bigoplus_{t \in a \cup \{0\}} m_t \rho^-_{t,1}$, where $m_t \geqslant 0$.

§5. The functors Φ^+ and Φ^-

In this section we define functors Φ^+ and Φ^-: $\mathcal{R} \to \mathcal{R}$, where $\mathcal{R} = \mathcal{R}(D^r, K)$ is the category of representations of D^r in finite-dimensional linear spaces over a field K. These functors play an essential role in proving the theorem that B^+ and B^- are perfect.

Analogues to the functors Φ^+ and Φ^- appeared first in the author's paper [6]. Then a modified form of them, called Coxeter functors, was used effectively to study representations of graphs [2]. A generalization of the Coxeter functors was constructed by Dlab and Ringel [8]. Recent work of Auslander [1] clarified their connection with the classical functors Ext and Tor.

5.1. Definition of the functors Φ^+ and Φ^-. Let ρ be a representation of D^r in a space V. We define a space V^1 and representations ρ^1 in V^1 as follows:

$$V^1 = \{(\xi_1, \ldots, \xi_r) \mid \xi_i \in \rho(e_i), \sum_{i=1}^{r} \xi_i = 0\}, \quad \rho^1(e_i) = \{(\xi_1, \ldots, \xi_{i-1}, 0, \xi_{i+1}, \ldots, \xi_r) \in V^1\},$$

where the e_i are generators of D^r. In other words: we denote by

$\nabla\colon \bigoplus_{i=1}^{r} \rho(e_i) \to V$ the linear map defined by the formula

$\nabla(\xi_1, \ldots, \xi_r) = \sum_{i=1}^{r} \xi_i$. We set $\operatorname{Ker} \nabla = V^1$. Then the following sequence of vector spaces is exact:

$$0 \to V^1 \xrightarrow{\lambda} \bigoplus_{i=1}^{r} \rho(e_i) \xrightarrow{\nabla} V, \quad \text{where } \lambda\colon V^1 \to \bigoplus_i \rho(e_i) \text{ is an embedding.}$$

We denote by π_j the projection onto the space $R = \overset{r}{\underset{i=1}{\oplus}} \rho(e_i)$ (π: $R \to R$) with kernel $\underset{i \neq j}{\oplus} \rho(e_i)$ and range $\rho(e_j)$. We set $\varphi_j = \nabla \pi_j \lambda$. Then $\rho'(e_j) = \operatorname{Ker} \varphi_j$. Thus, from a representation $\rho \in \mathcal{R} = \mathcal{R}(D^r, K)$ we have constructed another representation $\rho' \in \mathcal{R}$. It is easy to check that this correspondence is functorial. We denote by Φ^+ the functor $\rho \to \rho^1$.

A representation ρ^{-1} is constructed from ρ in a dual manner. We set $Q = \overset{r}{\underset{i=1}{\oplus}} (V/\rho(e_i))$. We denote by μ the linear map μ: $V \to Q$ defined by $\mu\xi = (\beta_1 \xi, \ldots, \beta_r \xi)$, where β_i: $V \to V/\rho(e_i)$ is the canonical map. We set $V^{-1} = \operatorname{Coker} \mu = Q/\operatorname{Im} \mu$. Thus, the following sequence is exact:

$$V \xrightarrow{\mu} \overset{r}{\underset{i=1}{\oplus}} (V/\rho(e_i)) \xrightarrow{\theta} V^{-1} \to 0.$$

We set $\psi_j = \theta \pi_j \mu$, where π_j: $Q \to Q$ is the projection into the space $Q = \overset{r}{\underset{i=1}{\oplus}} (V/\rho(e_i))$ with kernel $\underset{i \neq j}{\oplus} (V/\rho(e_i))$ and range $V/\rho(e_j)$. Then $\rho^{-1}(e_j) = \operatorname{Im} \psi_j$. It is not difficult to see that the correspondence $\rho \to \rho^{-1}$ is functorial. We denote this functor by Φ^-.

5.2. **Basic properties of the functors Φ^+ and Φ^-.**

PROPOSITION 5.1. *Let $\rho \in \mathcal{R}(D^r, K)$. Then the following assertions are equivalent*: (i) $\Phi^+\rho = 0$. (ii) *The subspaces $\rho(e_i)$ are linearly independent in V, that is, $\rho(e_j)(\underset{i \neq j}{\Sigma} \rho(e_i)) = \theta$ for every j.* (iii) $\rho \cong \overset{r}{\underset{t=\theta}{\oplus}} \overline{\rho^+_{t,1}}$, *where each $\overline{\rho^+_{t,1}}$ is a multiple of the atomic representation $\rho^+_{t,1}$ (that is, $\overline{\rho^+_{t,1}} \cong \underbrace{\rho^+_{t,1} \oplus \ldots \oplus \rho^+_{t,1}}_{m_{t,1}}$, and $m_{t,1} \geqslant 0$).*

The following proposition describes the dual properties of Φ^-.

PROPOSITION 5.2. *Let $\rho \in \mathcal{R}(D^r, K)$. Then the following assertions are equivalent*: (i) $\Phi^-\rho = 0$. (ii) $\rho(e_j) + (\underset{i \neq j}{\cap} \rho(e_i)) = V$ *for every j (where V is the representation space of ρ).* (iii) $\rho \cong \overset{r}{\underset{t=0}{\oplus}} \overline{\rho^-_{t,1}}$ *where each $\overline{\rho^-_{t,1}}$ is a multiple of the atomic representation $\rho^-_{t,1}$.*

The proofs of these assertions follow immediately from the definitions.■

PROPOSITION 5.3. (i) *If $\rho \cong \overset{n}{\underset{i=1}{\oplus}} \rho_i$, then $\Phi^+\rho \cong \overset{n}{\underset{i=1}{\oplus}} \Phi^+\rho_i$ and $\Phi^-\rho \cong \overset{n}{\underset{i=1}{\oplus}} \Phi^-(\rho_i)$.* (ii) *There exists a natural monomorphism i: $\Phi^-\Phi^+\rho \to \rho$.*
(iii) *There exists a natural epimorphism p: $\rho \to \Phi^+\Phi^-\rho$.*
(iv) $\Phi^+(\rho^*) \cong (\Phi^-\rho)^*$, *where ρ^* is the representation dual to ρ.*

PROOF. Properties (i) and (iv) can be checked directly from the definitions.

Let us prove (ii). The map φ_j: $V^1 \to V$ (where V^1 is the representation space of ρ^1) is such that $\varphi_j(\xi_1, \ldots, \xi_j, \ldots, \xi_r) = \xi_j$, where $\xi_j \in \rho(e_j)$.

Here the condition $(\xi_1, \ldots, \xi_r) \in V^1$ is equivalent to $\sum_{i=1}^{r} \xi_i = 0$, and so $\xi_j \in \rho(e_j)\ (\sum_{i \neq j} \rho(e_i))$. Thus, $\operatorname{Im} \varphi_j \subseteq \rho(e_j)\ (\sum_{i \neq j} \rho(e_i)) = \rho(e_j h_j)$, where $h_j = \sum_{i \neq j} e_i$.

If $\xi_j \in \rho(e_j h_j) = \rho(e_j)\ (\sum_{i \neq j} \rho(e_i))$, this means that $\xi_j = \sum_{i \neq j} \xi_i$, where $\xi_i \in \rho(e_i)$. Then $(\xi_1, \ldots, \xi_{j-1}, -\xi_j, \xi_{j+1}, \ldots, \xi_r) \in V^1$, hence, $\xi_j \in \operatorname{Im} \varphi_j$. Thus, the map φ_j: $V^1 \to V$ has the following properties: $\operatorname{Ker} \varphi_j \stackrel{\text{def}}{=\!=} \rho^1(e_i)$, $\operatorname{Im} \varphi_j = \rho(e_j h_j)$. Consequently,

$\bigoplus_{i=1}^{r} (V^1/\rho^1(e_i)) \cong \bigoplus_{i=1}^{r} \rho(e_i h_i)$. This implies that the following diagram is commutative:

$$
\begin{array}{ccccccc}
0 \to V^1 & \xrightarrow{\lambda} & \bigoplus_{i=1}^{r} \rho(e_i) & \xrightarrow{\nabla} & V & & \\
\| & & \uparrow & & \uparrow i & & \\
0 \to V^1 & \xrightarrow{\lambda'} & \bigoplus_{i=1}^{r} \rho(e_i h_i) & \xrightarrow{\theta} & \operatorname{Coker} \mu' & \to & 0 \\
 & & \uparrow & & & & \\
 & & 0 & & & &
\end{array} \tag{5.1}
$$

Now it is not difficult to check that we can construct a linear map i: $\operatorname{Coker} \mu^1 \to V$ such that the right square of the diagram is commutative. This map i is nothing but the natural isomorphism

$\operatorname{Coker} \mu^1 \cong \sum_{i=1}^{r} \rho(e_i h_i)$. We set $\varphi_j' = \theta \pi_j' \lambda'$, where π_j' is the projection in $\bigoplus_{i=1}^{r} \rho(e_i h_i)$ onto the j-th component. It is easy to see that $\operatorname{Im} \varphi_j' \cong \rho(e_j h_j)$. We define a representation $\widetilde{\rho}$ in $\operatorname{Coker} \mu^1$ by setting $\widetilde{\rho}(e_i) = \operatorname{Im} \varphi_j$. It follows from the definition of Φ^- that $\widetilde{\rho} \cong \Phi^-(\rho^1)$, where $\rho^1 = \Phi^+(\rho)$, is a representation in V^1. Consequently, the embedding i: $\operatorname{Coker} \mu' \to V$ defines a morphism of representations i: $\Phi^-\Phi^+\rho \to \rho$. The proof of (iii) is dual to the one just presented.

We note that $\sum_{i=1}^{r} e_i h_i$ is the minimal element $v_{\theta,1}$ of the cubicle $B^+(1)$. In §4 we have proved that this element is perfect. Thus, ρ splits into the direct sum $\rho \cong \rho(1) \oplus \tau_{\theta,1}$, where $\tau_{\theta,1} = \rho|_{\rho(v_{\theta,1})}$, and $\rho(1)$ is the direct sum $\rho(1) \cong \bigoplus_{t=0}^{r} m_t \rho_{t,1}^+$ $(m_t \geqslant 0)$ of the atomic representations $\rho_{t,1}^+$. We state as a separate corollary the properties of $\Phi^-\Phi^+\rho$, together with the dual assertions for $\Phi^+\Phi^-\rho$.

COROLLARY 5.1. (i) *There exists a natural isomorphism* $\Phi^+\Phi^- \rho \cong \rho|_{\rho(v_{\theta,1})}$, *where* $v_{\theta,1}$ *is the minimal element of* $B^+(1)$. *Here* ρ *is isomorphic to the direct sum* $\rho \cong \Phi^-\Phi^+ \rho \oplus \rho(1)$, *where* $\rho(1) = \bigoplus_{t=0}^{r} m_t \rho^+_{t,1}$ *is a direct sum of atomic representations* $\rho^+_{t,1}$.

(ii) $\Phi^+\Phi^- \rho$ *is isomorphic to the factor representation* $\rho/\rho^-(1)$, *where* $\rho^-(1) = \rho|_{\rho(v^-_{\bar{I},1})}$, *is the restriction of* ρ *to* $\rho(v^-_{\bar{I},1})$, *the image of the maximal element of* $B^-(1)$. *Here* $\rho \cong \rho^-(1) \oplus (\Phi^+\Phi^-\rho)$ *and* $\rho^-(1) \cong \bigoplus_{t=1}^{r} m_t \rho^-_{t,1}$ *(where* $\rho^-_{t,1}$ *is the atomic representation).*∎

We state another proposition in a form convenient for a later application, which is a simple combination of the properties of Φ^+ and Φ^-.

CONSTRUCTION. Let ρ be a representation in V, let τ^1 be a subrepresentation in $\rho^1 = \Phi^+\rho$, and let U^1 be the representation space of τ^1. We set $U = \sum_{i=1}^{r} \varphi_i U^1$ $(U \subseteq V)$, where φ_i: $V^1 \to V$ is the standard map $\varphi_i(\xi_1, \ldots, \xi_i, \ldots, \xi_r) = \xi_i$. We define a representation τ in U by setting $\tau(e_i) = \varphi_i U^1$.

PROPOSITION 5.4. *Suppose that* $\rho^1 = \Phi^+\rho$ *is decomposable*: $\rho^1 = \bigoplus_{j=1}^{n} \tau_j^1$. *Then* ρ *is also decomposable*: $\rho = \rho(1) \oplus (\bigoplus_{j=1}^{n} \tau_j)$, *where* τ_j *is obtained from* τ_j^1 *by the construction described above. Here*

a) $\tau_j \cong \Phi^-\tau_j^1$; б) $\tau_j^1 \cong \Phi^+\tau_j$; в) $\bigoplus_{j=1}^{n} \tau_j = \tau_{\theta,1} \cong \Phi^-\Phi^+\rho$, *where* $\tau_{\theta,1} = \rho|_{\rho(v_{\theta,1})}$; *and* c) $\rho(1)$ *is the direct sum* $\rho(1) = \bigoplus_{t=0}^{r} m_t \rho^+_{t,1}$ $(m_t \geqslant 0)$ *of the atomic representations* $\rho^+_{t,1}$, *and* $\Phi^+\rho(1) = 0$.∎

References

[1] M. Auslander, Representation theory of Artin algebras. I, II, Comm. Algebra **1** (1974), 177–268; 269–310. MR **50** # 2240. – and I. Reiten, III, Comm. Algebra **3** (1975), 239–294. MR **52** # 504.

[2] I. N. Bernstein, I. M. Gel'fand, and V. A. Ponomarev, Coxeter functors and Gabriel's theorem, Uspekhi Mat. Nauk **28**:2 (1973), 19–33.
= Russian Math. Surveys **28**:2 (1973), 17–32.

[3] G. Birkhoff, Lattice Theory, Amer. Math. Soc., New York 1948.
Translation: *Teoriya struktur*, Izdat. Inost. Lit., Moscow 1952.

[4] N. Bourbaki, Eléments de mathématique, XXVI, Groupes et algèbres de Lie, Hermann et Cie., Paris 1960. MR **24** # A2641.
Translation: *Gruppy i algebry Li.* Izdat. Mir, Moscow 1972.

[5] P. Gabriel, Unzerlegbare Darstellungen. I, Manuscripta Math. **6** (1972), 71–103.

[6] I. M. Gel'fand and V. A. Ponomarev, Problems of linear algebra and classification of quadruples of subspaces in a finite-dimensional vector space, Coll. Math. Soc. Iános Bolyai **5**, Hilbert space operators, Tihany (Hungary) 1970, 163–237 (in English). (For a brief account, see Dokl. Akad. Nauk SSSR **197** (1971), 762–765. = Soviet Math. Dokl. **12** (1971), 535–539.)

[7] I. M. Gel'fand and V. A. Ponomarev, Free modular lattices and their representations, Uspekhi Mat. Nauk **29**:6 (1974), 3–58. = Russian Math. Surveys **29**:6 (1974), 1–56.

[8] V. Dlab and C. M. Ringel, Representations of graphs and algebras, Carleton Math. Lect. Notes No. 8 (1974).

[9] L. A. Nazarova and A. V. Roiter, Representations of partially ordered sets, in the coll. "Investigations in the theory of representations", Izdat. Nauka, Leningrad 1972, 5–31.

Received by the Editors, 9 April 1976

Translated by M. B. Nathanson

LATTICES, REPRESENTATIONS, AND ALGEBRAS CONNECTED WITH THEM II[1]

I. M. Gel'fand and V. A. Ponomarev

Contents

§6. The representations $\rho_{t,l}^{+}$ and $\rho_{t,l}^{-}$

We define the representations $\rho_{t,l}^{+}$ ($l = 1, 2, \ldots$) as follows: $\rho_{t,l}^{+}$ is the atomic representation (see §4), and for $l > 1$ we set inductively

$$\rho_{t,l}^{+}=\Phi^{-}\rho_{t,l-1}^{+}.$$

The representations $\rho_{t,l}^{-}$ are, by definition, dual to the $\rho_{t,l}^{+}$, that is,

$$\rho_{t,l}^{-}\overset{\text{def}}{=\!=}(\rho_{t,l}^{+})^{*}.$$

Thus, the $\rho_{t,l}^{-}$ are the (−)-atomic representations, and it follows from properties of the conjugation functor (see Proposition 5.3 (iv)) that

$$\rho_{t,l}^{-}\cong\Phi^{+}\rho_{t,l-1}^{-},\quad l>1.$$

In §8 we shall show that the $\rho_{t,l}^{+}$ are essentially the same as the $\rho_{s,l}$, whose definition in terms of the algebra A was given in §1. More accurately, we shall show that $\rho_{0,l-1} \cong \rho_{0,l}^{+}$ and $\rho_{j,l} \cong \rho_{j,l}^{+}$ if $j \neq 0$.

The functorial definition of $\rho_{t,l}^{+}$ is more convenient when we are interested in such categorical properties as decomposability and when there is no need to investigate the intrinsic structure of the representation.

DEFINITION 6.1. *The dimension of a representation $\rho \in \mathcal{R}(D^{r}, K)$ in the*

[1] The first part of this article was published in these Uspekhi **31**:5 (1976), 71–88 = Russian Math. Surveys **31**:5 (1976), 67–85.

space V is the sequence of integers: $\dim \rho = (n; m_1, \ldots, m_r)$, *where* $n = \dim V$ *and* $m_i = \dim \rho(e_i)$.

For example, the atomic representations $\rho^+_{0,1}$ and $\rho^+_{\alpha,1}|_{t\neq 0}$ have $\dim \rho^+_{0,1} = (1; 0, \ldots, 0)$ and $\dim \rho^+_{t,1} = (1; 0, \ldots, 0, 1, 0, \ldots 0)$, where the 1 stands in the $(t+1)$st place.

PROPOSITION 6.1. *Let* $\rho \in \mathcal{R}(D^r, K)$ *be an indecomposable representation.*

(I) *Then there are the following possibilities for* $\Phi^+\rho$:

a) $\Phi^+\rho = 0 \iff \rho \cong \rho^+_{t,l}$ *for some* $t \in \{0, 1, \ldots r\}$,

b) $\Phi^+\rho \neq 0 \iff (\Phi^-\Phi^+\rho \cong \rho)$;

here, the representation $\Phi^+\rho = \rho^1$ *is also indecomposable, and its dimension* $\dim \rho^1 = (n^1; m^1_1, \ldots, m^1_r)$ *can be computed from* $\dim \rho = (n; m_1, \ldots, m_r)$ *by the formula*:

$$n^1 = \sum_{i=1}^{r} m_i - n, \quad m^1_j = \sum_{i\neq j} m_i - n.$$

(II) *There are the following possibilities for* $\Phi^-\rho$:

a) $(\Phi^-\rho = 0) \iff (\rho \cong \rho^-_{t,1}$ *for some* $t \in \{0, 1, \ldots, r\})$,

b) $(\Phi^-\rho \neq 0) \iff (\Phi^+\Phi^-\rho = \rho)$; *here, the representation* $\Phi^-\rho$ *is also indecomposable, and its dimension* $\dim \Phi^-\rho = (n^-; m^-_1, \ldots, m^-_r)$ *can be computed from* $\dim \rho = (n; m_1, \ldots, m_r)$ *by the formula*

$$n^- = (r-1)n - \sum_{i=1}^{r} m_i, \qquad m^-_i = n - m_i.$$

PROOF. (I) a) and b), except for the assertions about the dimensions, clearly follow from Proposition 5.4. From the same proposition we find that $\Phi^-\Phi^+\rho \cong \rho$ if and only if $\rho(v_{\theta,1}) = V$. By definition

$v_{\theta,1} = \bigcap_{i=1}^{r} h_i(1)$. Hence, $\rho(v_{\theta,1}) = \rho(\bigcap_i h_i(1)) = V$ implies that $\rho(h_i(1)) = V$ for all i, and so $\rho(e_i h_i(1)) = \rho(e_i)V = \rho(e_i)$.

Therefore, we can rewrite the diagram (5.1) in the following way:

$$\begin{array}{ccccccccc} 0 & \longrightarrow & V^1 & \longrightarrow & \bigoplus_i \rho(e_i) & \longrightarrow & V = \sum_{i=1}^{r} \rho(e_i) & & \\ & & \| & & \| & & \| & & \\ 0 & \longrightarrow & V^1 & \xrightarrow{\mu'} & \bigoplus_i \rho(e_i h_i) & \xrightarrow{\theta'} & \operatorname{Coker} \mu' & \longrightarrow & 0. \end{array}$$

From this we find that $\dim V^1 = n^1 = \sum_i \dim \rho(e_i) - \dim V = \sum_{i=1}^{r} m_i - n$. The formula $m^1_j = \sum_{i\neq j} m_i - n$ is also easily proved. Part (II) of Proposition 6.1 follows by duality.

THEOREM 6.1. *Let* ρ *be an indecomposable representation of the lattice* D^r, *where the number* r *of generators of* D^r *is at least* 4. *Then there are the following three mutually incompatible possibilities*:

(I) $((\Phi^+)^{l-1}\rho \neq 0$ *and* $(\Phi^+)^l \rho = 0) \Longleftrightarrow$ (*there is a* $t \in \{0, 1, \ldots, r\} \mid \rho \cong \rho^+_{t,l}$),

(II) $((\Phi^-)^{m-1}\rho \neq 0$ *and* $(\Phi^-)^m \rho = 0) \Longleftrightarrow$ (*there is an* $s \in \{0, 1, \ldots, r\} \mid \rho \neq \rho^-_{s,m}$).

(III) $\forall_{l,m}((\Phi^+)^l\rho \neq 0$ *and* $(\Phi^-)^m \rho \neq 0)$.

PROOF. Let ρ be an arbitrary indecomposable representation. There are the following possibilities:

1) there is an $l > 0$ such that $(\Phi^+)^{l-1}\rho \neq 0, (\Phi^+)^l\rho = 0$;

2) there is an $m > 0$ such that $(\Phi^-)^{m-1}\rho \neq 0, (\Phi^-)^m\rho = 0$;

3) $(\Phi^+)^l\rho \neq 0$ and $(\Phi^-)^l\rho \neq 0$ for every $l > 0$.

We consider these cases separately.

1) $(\Phi^+)^l\rho \neq 0, (\Phi^+)^l\rho = 0$. We write $(\Phi^+)^{l-1}\rho = \rho^{l-1}$. It follows from Proposition 6.1 that ρ^{l-1}, as well as ρ, is indecomposable, and that $\rho \cong (\Phi^-)^{l-1}\rho^{l-1} = (\Phi^-)^{l-1}(\Phi^+)^{l-1}\rho$. Since $\Phi^+\rho^{l-1} = 0$, it follows from Proposition 6.1 that $\rho^{l-1} \cong \rho^+_{t,1}$. Consequently, $\rho \cong (\Phi^-)^{l-1}\rho^+_{t,1} = \rho^+_{t,l}$.

2) $(\Phi^-)^{m-1}\rho \neq 0, (\Phi^-)^m\rho = 0$. Arguments similar to those used in 1) show that $(\Phi^-)^{m-1}\rho \cong \rho^-_{s,1}$ and $\rho \cong (\Phi^+)^{m-1}\rho^-_{s,1} = \rho^-_{s,m}$.

For the proof of the theorem it remains to show that if $\rho \in \mathcal{R}(D^r, K)$ and $r \geqslant 4$, then 1) and 2) are mutually exclusive.

Let $\rho \cong \rho^+_{t,l}$. We claim that $(\Phi^-)^m\rho^+_{t,l} \neq 0$ for every m. By definition, $(\Phi^-)^m\rho^+_{t,l} = \rho^+_{t,l+m}$. Thus, we must prove that $\rho^+_{t,k} \neq 0$ for every $k > 0$. By Proposition 6.1, $\dim \rho^+_{t,k} = (n_{t,k}; m^1_{t,k}, \ldots, m^r_{t,k})$ can be computed recursively from the formulae

$$n_{t,k} = (r-1)n_{t,k-1} - \sum_{i=1}^{r} m^i_{t,k-1}; \qquad m^i_{t,k} = n_{t,k-1} - m^i_{t,k-1}.$$

It is not difficult to deduce from them that for $r \geqslant 4$ the terms of the sequence $\{n_{t,l}\}$ $(l = 1, 2, \ldots)$ can be found from the recurrence relation

$$n_{t,l} = (r-2)n_{t,l-1} - n_{t,l-2}, \qquad l \geqslant 3,$$

and the initial conditions $n_{t,1} = 1$, $n_{t,2} = r-2$ for $t \neq 0$, and $n_{0,1} = 1$, $n_{0,2} = r-1$.

For $r \geqslant 4$ the terms of $\{n_{t,l}\}$ increase monotonically with l, and so all the $\rho^+_{t,l}$ are different from zero. Thus, if $(\Phi^+)^{l-1}\rho \neq 0$ and $(\Phi^+)^l\rho = 0$, then $(\Phi^-)^m\rho \neq 0$ for every m, that is, 1) and 2) are mutually exclusive.

REMARK. The lattices D^1, D^2, D^3 are finite and each has only finitely many indecomposable representations (up to isomorphism). The numbers of these representations of D^1, D^2, and D^3 are 2, 4, and 9, respectively.

If $r \geqslant 3$ and ρ is an indecomposable representation of D^r, then there are positive l and m such that $(\Phi^+)^l\rho = 0$ and $(\Phi^-)^m\rho = 0$. Therefore, each indecomposable representation of D^r, $r \leqslant 3$, can be described both in the form $\rho^+_{t,l}$ and $\rho^-_{t,m}$. For example, in D^3 the following isomorphisms hold: $\rho^+_{0,1} \cong \rho^-_{0,3}$, $\rho^+_{0,2} \cong \rho^-_{0,2}$, $\rho^+_{0,3} \cong \rho^-_{0,1}$. The dimensions of these representations are, respectively, $(1; 0, 0, 0)$, $(2; 1, 1, 1)$, and $(1; 1, 1, 1)$.

§7. The perfectness of the lattices B^+ and B^-. Characteristic functions

7.1. Proof of the theorem on the perfectness of the sublattices B^+ and B^-. This proof is based on the following proposition. (As usual, we write $\rho^1 = \Phi^+\rho$, where V^1 and V are the representation spaces of ρ^1 and ρ; $\rho_j\colon V^1 \to V$ is the standard map $\rho_j(\xi_1, \ldots, \xi_j, \ldots, \xi_r) = \xi_j$; and $e_{i_1 \ldots i_l}$ are the lattice polynomials in $D^r (r \geqslant 4)$ (see §1, p. 70).)

PROPOSITION 7.1. *For every $\rho \in \mathcal{R}(D^r, K)$, $r \geqslant 4$,*

$$\varphi_j\rho^1(e_{i_1\ldots i_l}) = \begin{cases} 0, & \text{when } j = i_1, \\ \rho(e_{ji_1\ldots i_l}), & \text{when } j \neq i_1. \end{cases}$$

We omit the proof of this proposition. It is the central and most complicated part of [7].

We recall that the generators $h_t(l)$ of the sublattice $B^+(l)$ are defined in the following way:

$$e_t(l) = \sum_{i_1, \ldots, i_{l-1}} e_{i_1, \ldots, i_{l-1}t}, \qquad h_t(l) = \sum_{i \neq t} e_i(l),$$

where $e_t(l)$ is the sum of all possible $e_{i_1 \ldots i_{l-1}t}$ in which the last index is t.

COROLLARY 7.1. *For every $\rho \in \mathcal{R}(D^r, K)$, $r \geqslant 4$,*

$$\rho(h_t(l+1)) = \sum_{j=1}^{r} \varphi_j\rho^1(h_t(l)).$$

The proof obviously follows from Proposition 7.1 and the definition of $h_t(l)$.■

FUNDAMENTAL THEOREM 7.1. *All the elements of the lattices B^+ and B^- are perfect.*

Before proving this theorem, we prove the following proposition.

PROPOSITION 7.2. (I). *Every element $h_j(l) \in B^+(l)$ $(j = 1, \ldots, r)$ is perfect, that is, for every representation $\rho \in \mathcal{R}(D^r, K)$ in V*

$$\rho \cong \rho_u \oplus \tau_{j,l},$$

where $\tau_{j,l} = \rho\,|_{\rho(h_j(l))}$ и $\rho_u = \rho\,|_U$, and U is a space complementary to $\rho(h_j(l))$ in V.

(II). *The representation ρ_u satisfies the relation*

$$\rho_u = \Big(\bigoplus_{\substack{t,k \\ 0<k<l}} \bar{\rho}_{t,k}\Big) \oplus \bar{\rho}_{0,l} \oplus \bar{\rho}_{j,l},$$

where each representation $\bar{\rho}_{t,k}$ and $\bar{\rho}_{s,l}$ is a multiple of the indecomposable representation $\rho^+_{t,k}$ and $\rho^+_{s,l}$, that is, $\bar{\rho}_{t,k} \cong \underbrace{\rho^+_{t,k} \oplus \ldots \oplus \rho^+_{t,k}}_{m_{t,k}}$, where $m_{t,k} \geqslant 0$.

(III). *If $\tau_{j,l}$ splits into a direct sum $\tau_{j,l} = \bigoplus_{i=1}^{n} \tau_i$ of indecomposable representa-*

tions τ_i, then none of the τ_i are isomorphic to any of the representations $\rho^+_{t,k}$ $(t = 0, 1, \ldots, r; 0 < k < l)$, $\rho^+_{0,l}$, or $\rho^+_{j,l}$.

The proof is by induction on l. For $l = 1$ the corresponding assertion was proved earlier in Lemma 4.1. Now we assume that the proposition has been proved for the elements $h_j(l-1)$. Then the representation $\rho^1 = \Phi^+\rho$ can be written as

$$\rho^1 = \rho^1_u \oplus \tau^1_{j,\, l-1},$$

where $\rho^1_u = \rho^1|_{U^1}$, U^1 is a complement to $\rho^1(h_j(l-1))$ in V^1, and $\tau^1_{j,l-1} = \rho^1|_{\rho^1(h_j(l-1))}$.

We apply the construction of §5 to ρ^1. By Proposition 5.4, ρ is isomorphic to a direct sum $\rho = \rho(1) \oplus \rho_u \oplus \tau_{j,l}$, where $\rho(1) \cong \bigoplus_{t=0}^{r} m_t \rho^+_{t,1}$ $(m_t \geqslant 0)$ and the representations ρ_u and $\tau_{j,l}$ are constructed from ρ^1_u and $\tau^1_{j,l-1}$ as in §5. Then $\tau_{j,l}$ is the restriction of ρ to the subspace $\sum_{i=r}^{r} \varphi_i(\rho^1(h_j(l-1)))$. From Corollary 7.1 it is clear that this subspace is nothing but $\rho(h_j(l))$. Thus, $\tau_{j,l}$ is a direct summand of ρ. This proves part (I) of Proposition 7.2 (that is, we have shown that the element $h_{j,l}$ is perfect).

Before proving parts (II) and (III) of Proposition 7.2, we now prove Theorem 7.1.

PROOF OF THEOREM 7.1. Since the elements $h_1(l), \ldots, h_r(l)$ are generators of the lattice $B^+(l)$, and since each of them is perfect, the entire sublattice $B^+(l)$ is perfect (that is, consists of perfect elements). We have proved in §3 that B^+ is the union of the sets $B^+(l)$. Therefore, B^+ is also perfect. The perfectness of B^- follows by duality.

COMPLETION OF THE PROOF OF PROPOSITION 7.2 (II). We may assume by induction that we have proved that

$$\rho^1_u = \bigoplus_{\substack{t,k \\ 0<k<l-1}} \tilde{\rho}_{t,k} \oplus \tilde{\rho}_{0,\, l-1} \oplus \tilde{\rho}_{j,\, l-1},$$

where each of the $\tilde{\rho}_{t,k}$ and $\tilde{\rho}_{s,l-1}$ is a multiple of the indecomposable representation $\rho^+_{t,k}$ and $\rho^+_{s,l-1}$. By Proposition 5.4, $\rho_u \cong \Phi^- \rho^1_u$, where ρ_u is the representation constructed from ρ^1_u. Consequently,

$$\rho_u \cong \Big(\bigoplus_{\substack{t,k \\ 0<k<l-1}} \Phi^-\tilde{\rho}_{t,k}\Big) \oplus (\Phi^-\tilde{\rho}_{0,\, l-1}) \oplus (\Phi^-\tilde{\rho}_{j,\, l-1}).$$

By assumption, $\tilde{\rho}_{t,k} = \underbrace{\rho^+_{t,k} \oplus \ldots \oplus \rho^+_{t,k}}_{m_{t,k}}$, $m_{r,k} \geqslant 0$. Moreover, $\Phi^-\rho^+_{t,k} = \rho^+_{t,k+1}$ by definition. Consequently,

$\Phi^- \tilde{\rho}_{t,k} \cong \underbrace{\rho^+_{t,k+1} \oplus \ldots \oplus \rho^+_{t,k+1}}_{m_{t,k}}$. Thus $\rho_u \cong \bigoplus_{\substack{t,k \\ 1<k<l}} \bar{\rho}_{t,k} \oplus \bar{\rho}_{0,l} \oplus \bar{\rho}_{j,l}$, where $\bar{\rho}_{t,k} = \Phi^- \tilde{\rho}_{t,k-1}$, and each such representation is a multiple of the indecomposable representation $\rho^+_{t,k}$. Finally, bearing in mind that $\rho(1) = \bigoplus_{t=0}^{r} m_t \rho^+_{t,1}$, we can write

$$\rho = \rho(1) \oplus \Big(\bigoplus_{\substack{t,k \\ 1<k<l}} \bar{\rho}_{t,k}\Big) \oplus \bar{\rho}_{0,l} \oplus \bar{\rho}_{j,l} \oplus \tau_{j,l} = \bigoplus_{\substack{t,k \\ 0<k<l}} \bar{\rho}_{t,k} \oplus \bar{\rho}_{0,l} \oplus \bar{\rho}_{j,l} \oplus \tau_{j,l}.$$

This proves part (II) of Proposition 7.2. Part (III) is proved similarly (here we have to use the isomorphisms $\tau_{j,l} \cong \Phi^- \tau^1_{j,l-1}$ and $\tau^1_{j,l-1} \cong \Phi^+ \tau_{j,l}$).

As we know, any element of $B^+(l)$ can be written in the form $v_{a,l} = \sum_{i\in a} e_i(l) + \sum_{j\in I-a} e_i(l)h_j(l)$ or, if $a \neq I$, $v_{a,l} = \bigcap_{j\in I-a} h_j(l)$, where a is a subset of $I = \{1, \ldots, r\}$. The following proposition makes more precise the properties of a perfect element.

PROPOSITION 7.3. *Let $\rho \in \mathscr{R}(D^r, K)$ by an arbitrary representation. Then*

$$\rho \cong \Big(\bigoplus_{\substack{t,k \\ 0<k<l}} \bar{\rho}_{t,k}\Big) \oplus \Big(\bigoplus_{s\in(I-a)\cup\{0\}} \bar{\rho}_{s,l}\Big) \oplus \tau_{a,l},$$

where $\tau_{a,l} = \rho|_{\rho(v_{a,l})}$ and each $\bar{\rho}_{t,k}$ and $\bar{\rho}_{s,l}$ is a multiple of the indecomposable representation $\rho^+_{t,k}$ and $\rho^+_{s,l}$. Here, if $\tau_{a,l}$ splits as $\tau_{a,l} = \bigoplus_{i=1}^{m} \tau_i$ into a direct sum of indecomposable representations τ_i, then none of the τ_i are isomorphic to any of the representations $\rho^+_{t,k}$ $(t = 0, 1, \ldots, r;\ k = 1, \ldots, l-1)$ or $\rho^+_{s,l} (s \in (I-a) \cup \{0\})$.

We omit the simple proof of this proposition.■

COROLLARY 7.2. (i). *The correspondence $a \mapsto v_{a,l}$ defines an isomorphism of the Boolean algebras $v_l\colon \mathscr{B}(I) \to B^+(l)$.*

(ii) *Let $v_{a,l}$ and $v_{b,m}$ be arbitrary elements of $B^+(l)$ and $B^+(m)$. If $l < m$, then $v_{a,l} \supset v_{b,m}$.*

PROOF. We consider the representation $\rho = \rho^+_{0,l} \oplus \rho^+_{1,l} \oplus \ldots \oplus \rho^+_{r,l}$. By Proposition 7.3, if a and $b \subseteq I$ and $a \neq b$, then the subspaces $\rho(v_{a,l})$ and $\rho(v_{b,l})$ are distinct. Hence, if $a \neq b$, then $v_{a,l} \neq v_{b,l}$.

Let $v_{\theta,l-1} = \bigcap_{i=1}^{r} h_j(l)$ be the minimal element in $B^+(l-1)$, and $v_{I,l} = \sum_{i=1}^{r} e_i(l)$ the maximal element in $B^+(l)$. It is not difficult to check that the subspaces $\rho(v_{\theta,l-1})$ and $\rho(v_{I,l})$ can be described as follows:

$$\rho(v_{\theta,\, l-1}) \cong \bigoplus_{t=0}^{r} V_{t,\, l}^{+} \quad \text{и} \quad \rho(v_{I,\, l}) \cong \bigoplus_{t=1}^{r} V_{t,\, l}^{+},$$

where $V_{t,l}^{+}$ is the representation space of $\rho_{t,l}^{+}$. Thus, $\rho(v_{\theta,l-1}) \supset \rho(v_{I,l})$, and so $v_{\theta,l-1}$ is strictly larger than $v_{I,l}$ $(v_{\theta,l-1} \supset v_{I,l})$. Now if $v_{a,l-1} \in B^{+}(l-1)$ and $v_{b,l} \in B^{+}(l)$ are arbitrary then $v_{a,l-1} \supseteq v_{\theta,l-1} \supset v_{I,l} \supseteq v_{b,l}$. Consequently, $v_{a,l-1} \supset v_{b,l}$.■

7.2. The connection between the representations of B^{+} and B^{-}. We have proved that the elements $v_{\theta,l}$ (where $v_{\theta,l}$ is the minimal element of $B^{+}(l)$) form a decreasing chain: $v_{\theta,1} \supset v_{\theta,2} \supset \ldots \supset v_{\theta,l} \supset \ldots$. Dual to the elements $v_{\theta,l}$ of D^{r} are the maximal elements $v_{I,l}^{-}$ of the lower cubicles $B^{-}(l)$ $(v_{I,l}^{-} = \sum_{i=1}^{r} h_{j}^{-}(l)$, where $h_{j}^{-}(l) = \bigcap_{i \neq j} e_{i}^{-}(l))$. These elements also form a chain: $v_{I,1}^{-} \subset v_{I,2}^{-} \subset \ldots \subset v_{I,l}^{-} \subset \ldots$.

Let $\rho \in \mathscr{R}(D^{r}, K)$. We write $V_{\theta,l} = \rho(v_{\theta,l})$ and $V_{I,l}^{-} = \rho(v_{I,l}^{-})$, and we set

$V_{\infty}^{+} = \bigcap_{l=1}^{\infty} V_{\theta,l}$ and $V_{\infty}^{-} = \bigcap_{l=1}^{\infty} V_{I,l}^{-}$.

PROPOSITION 7.4. *$V_{\infty}^{-} \subseteq V_{\infty}^{+}$ for every representation $\rho \in \mathscr{R}(D^{r}, K)$, $r \geqslant 4$.*

PROPOSITION 7.5. *Every representation $\rho \in \mathscr{R}(D^{r}, K)$, $r \geqslant 4$, is isomorphic to a direct sum $\rho \equiv \rho^{-} \oplus \rho \oplus \rho^{+}$, where $\rho^{-} = \rho|_{V_{\infty}^{-}}$, $(\rho^{-} \oplus \rho_{\lambda}) = \rho|_{V_{\infty}^{+}}$. Here $(\Phi^{+})^{l}\rho^{+} = 0$ for some $l > 0$, $(\Phi^{-})^{l}\rho^{-} = 0$ for some $l > 0$, and for every $l > 0$*

$$(\Phi^{-})^{l}(\Phi^{+})^{l}\rho_{\lambda} \cong \rho_{\lambda} \text{ and } (\Phi^{+})^{l}(\Phi^{-})^{l}\rho_{\lambda} \cong \rho_{\lambda}.$$

The proof of these propositions follows easily from Theorem 6.1 and Proposition 7.3.■

Note that in Proposition 7.5 $\rho^{+} \cong \bigoplus_{t,\, l} \bar{\rho}_{t,l}^{+}$, where each $\bar{\rho}_{t,l}^{+}$ is a multiple of the indecomposable representation $\bar{\rho}_{t,l}^{+} \cong \oplus m_{t,l}\, \rho_{t,l}^{+}$, $m_{t,l} \geqslant 0$, similarly, $\rho^{-} \cong \bigoplus_{t,\, l} \bar{\rho}_{t,l}^{-}$, where each $\bar{\rho}_{t,l}^{-}$ is a multiple of the indecomposable representation $\rho_{t,l}^{-}$. A classification of the ρ_{λ} is known at present only for the lattice D^{4} [6]. The classification of the representations ρ_{λ} of D^{r}, $r \geqslant 5$, seems at present a hopeless problem.

7.3. Indecomposable representations and characteristic functions. Let B be a sublattice of perfect elements in a modular lattice L. With each indecomposable representation ρ: $L \to \mathscr{L}(K, V)$ we associate a function χ_{ρ} on the set B in the following way:

$$\chi_{\rho}(v) = \begin{cases} 0 & \text{if } \rho(v) = 0, \\ 1 & \text{if } \rho(v) = V, \end{cases}$$

where $v \in B$ and V is the representation space of ρ. This χ_{ρ} is called the *characteristic function.* The set with two elements $\{0, 1\}$ can be regarded as a lattice (a Boolean algebra), which we denote by 2. The characteristic function χ_ρ then becomes a lattice morphism $\chi_\rho : B \to 2$. It is clear from the definition of the characteristic function that if two elements v_1 and v_2 of B are linearly equivalent, then $\chi_\rho(v_1) = \chi_\rho(v_2)$. Thus, χ_ρ is completely determined once its corresponding map $\chi_\rho : \bar{B} \to 2$ is specified.

We do not know the entire lattice of perfect elements in D^r $(r \geqslant 4)$. We know only its two sublattices B^+ and $B^- \subset D^r$. We denote by B the lattice generated by B^+ and B^-, and by $\bar{B}$ the factor lattice of B by the relation of linear equivalence. We know that $\rho(v^-) \subseteq \rho(v^+)$ for any representation $\rho \in \mathscr{R}(D^r, K)$ $(r \geqslant 4)$ and any $v^+ \in B^+$ and $v^- \in B^-$. Consequently, $\bar{B} = B^+ \cup B^-$ (that is, if $v \in \bar{B}$, then either $v \in B^+$ or $v \in B^-$), and if $v^+ \in B^+$ and $v^- \in B^-$, then $v^- \subset v^+$ in $\bar{B}$.

We shall study the characteristic functions $\chi_\rho : \bar{B} \to 2$ $(\bar{B} \subseteq \bar{D}^r)$. It follows from the ordering of $\bar{B}$ that there are three kinds of morphisms $\chi: \bar{B} \to 2$: Either (1) there is an element $v^+ \in B^+$ such that $\chi(v^+) = 0$. Then $\chi(v^-) = 0$ for every $v^- \in B^-$; or (2) there is a $v^- \in B^-$ such that $\chi(v^-) = 1$. Then $\chi(v^+) = 1$ for every $v^+ \in B^+$; or (3) $\chi(v^+) = 1$ for every $v^+ \in B^+$ and $\chi(v^-) = 0$ for every $v^- \in B^-$. We denote the last function by χ_0^1. Characteristic functions of the first kind are denoted by χ^+, and those of the second kind denoted by χ^-.

THEOREM 7.2. *Let ρ be an indecomposable representation of D^r, and suppose that its characteristic function χ_ρ is of the first or second kind. Then ρ is determined by its characteristic function uniquely up to isomorphism. Moreover, if χ_ρ is a function of the first kind, then $\rho \cong \rho^+_{t,l}$ for some $t \in \{0, 1, \ldots, r\}$ and $l \in \{1, 2, \ldots\}$; if χ_ρ is a function of the second kind, then $\rho \cong \rho^-_{s,m}$ for some $s \in \{0, 1, \ldots, r\}$ and $m \in \{1, 2, \ldots\}$.*

Before proving this theorem, we define morphisms $\chi^+_{t,l}: \bar{B} \to 2$ and $\chi^-_{t,l}: \bar{B} \to 2$. For every $l = 1, 2, \ldots$ we define

$$\chi^+_{0,\,l}(v^+_m) = \begin{cases} 1 \text{ if } & m < l, \\ 0 \text{ if } & l \leqslant m, \end{cases}$$

where v^+_m is an arbitrary element of $B^+(m)$.

$$\chi^+_{t,\,l}(v^+_m)\,|_{t \neq 0} = \begin{cases} 1 \text{ if } & m < l, \text{ or if } m = l \text{ and } v^+_m \supseteq h_s(m), \text{ where } s \neq t; \\ 0 \text{ if } & l < m \text{ or if } m = l \text{ and } v^+_m \subseteq h_t(m), \end{cases}$$

where $h_s(m)$ $(s = 0, 1, \ldots r)$ are generators of the cubicle $B^+(m)$.

The characteristic functions $\chi^-_{t,l}: \bar{B} \to 2$ are defined similarly, namely,

$$\chi^-_{0,l}(v^-_m) = \begin{cases} 1 \text{ if } & l \leqslant m, \\ 0 \text{ if } & m < l, \end{cases}$$

$$\chi^-_{t,l}(v^-_m)|_{t \neq 0} = \begin{cases} 1 \text{ if } l < m \text{ or if } m = l \text{ and } h^-_t(m) \subseteq v^-_m, \\ 0 \text{ if } m < l \text{ or if } m = l \text{ and } v^-_m \subseteq h^-_s(m), \text{ where } s \neq t. \end{cases}$$

LEMMA 7.1. *Let $\chi: \bar{B} \to 2$ be an arbitrary morphism. Then there are the following three possibilities:*

1) *χ is a morphism of the first kind (that is, $\exists v^+ \in B^+ | \chi(v^+) = 0$). Then $\chi = \chi^+_{t,l}$ for some $t \in \{0, 1, \ldots, r\}$ and $l = \{1, 2, \ldots\}$;*

2) *χ is a morphism of the second kind (that is, $\exists v^- \in B^- | \chi(v^-) = 1$). Then $\chi = \chi^-_{s,m}$ for some $s \in \{0, 1, \ldots, m\}$ and $l = \{1, 2, \ldots\}$;*

3) *$\chi = \chi^1_0$, that is, $\chi(v^-) = 0$ for every $v^- \in B^-$ and $\chi(v^+) = 1$ for every $v^+ \in B^+$.*

The proof follows easily from the description of the lattices B^+ and B^-.■

PROOF OF THEOREM 7.2. We claim that each morphism $\chi^+_{t,l}: \bar{B} \to 2$ has one and only one indecomposable representation ρ such that $\chi_\rho = \chi^+_{t,l}$ and $\rho \cong \rho^+_{t,l}$. We first consider the case $t \neq 0$. The equality $\chi_\rho = \chi^+_{t,l}$ indicates that $\rho(h_t(l)) = 0$.

It follows from this and from Proposition 7.4 that

$$\rho \cong \Big(\bigoplus_{\substack{s,k \\ 0<k<l}} \bar{\rho}_{s,k}\Big) \oplus \bar{\rho}_{0,l} \oplus \bar{\rho}_{t,l},$$

where each $\bar{\rho}_{s,k}$ and $\bar{\rho}_{s,l}$ is a multiple of the indecomposable representation $\rho^+_{s,k}$ and $\rho^+_{s,l}$. But ρ is indecomposable, by hypothesis, and so $\rho \cong \rho^+_{s,k}$, where for the pair (s, k) either $k < l$ and s is arbitrary, or $(s, k) = (0, l)$, or $(s, k) = (t, l)$.

The equality $\chi_\rho = \chi^+_{t,l}$ also indicates that $\rho(h_i(l)) = V$ for any $i \neq t$. It follows from $\rho(h_i(l)) = V$, from Proposition 7.2, and from the indecomposability of ρ that ρ cannot be isomorphic to any $\rho^+_{s,k}$ with $k < l$ or $(s, k) = (0, l)$ or $(s, k) = (i, l)$. This reasoning applies to all i $(i \neq t)$.

Therefore, the only possibility is that $\rho \cong \rho^+_{t,l}$.

This proves that $(\chi_\rho = \chi^+_{t,l}) \Rightarrow (\rho \cong \rho^+_{t,l})$.

Together with Lemma 7.1 this means that $(\chi_\rho = \chi^+_{t,l}) \Longleftrightarrow (\rho \cong \rho^+_{t,l})$. The proof for the case $t = 0$ proceeds similarly. The proof for characteristic functions of the second kind $(\chi_\rho = \chi^-_{t,l})$ proceeds dually.■

Little can be said about indecomposable representations ρ for which $\chi_\rho = \chi^1_0$ (that is, $\chi_\rho(v^+) = 1$ and $\chi_\rho(v^-) = 0$ for all $v^+ \in B^+$ and $v^- \in B^-$). They are precisely the representations that are not annihilated by any of the functions $(\Phi^+)^l$ or $(\Phi^-)^l$. Their classification is known only for the lattice D^4 [6]. There are infinitely many such indecomposable representations $\rho \in \mathcal{R}(D^4, K)$ and each of them has not only integer invariants, but also a continuous invariant $\lambda \in K$ (similar to an eigenvalue of a linear transformation).

One might suppose that the equality $\chi_\rho = \chi_{\rho'} = \chi_0^1$, where ρ and ρ' are non-isomorphic indecomposable representations, is a consequence of the fact that we do not know the entire lattice B of perfect elements in D^4.

However, we believe that this is not the case. We offer the following conjecture.

CONJECTURE. *The lattice B of perfect elements in D^r $(r \geqslant 4)$ is the union of B^+ and B^-.*

§8. The algebras A^r and the representations $\rho_{t,l}$

8.1. The algebras A^r. Let K be a field. We define the K-algebra A^r as the associative K-algebra with unit ε and with the generators $\xi_0, \xi_1, \ldots, \xi_r$ satisfying the following relations:

(i) $$\xi_i^2 = 0 \quad (i = 1, \ldots, r),$$

(ii) $$\xi_0^2 = \xi_0,$$

(iii) $$\xi_0 \xi_i = \xi_i \quad (i = 1, \ldots, r),$$

(iv) $$\sum_{i=1}^{r} \xi_i \xi_0 = 0.$$

We write A instead of A^r when this cannot lead to misunderstandings.

By a standard monomial in A^r we mean a product $\xi_{i_1} \ldots \xi_{i_{l-1}} \xi_t$, where for every $1 \leqslant j < l - 1$

$$i_j \in \{1, \ldots, r\}, \ i_j \neq i_{j+1} \text{ and } t \in \{0, 1, \ldots, r\}.$$

We denote this standard monomial by $\xi_{i_1} \ldots \xi_{i_{l-1} t} = \xi_\alpha$, where $\alpha = (i_1, \ldots, i_{l-1}, t)$ is a sequence of indices. Clearly, every non-zero monomial in A^r can be put in standard form.

The degree of ξ_α is the integer $d(\xi_\alpha)$ defined as follows: $d(\xi_0) = d(\varepsilon) = 0$; $d(\xi_i) = 1$ for every $i \neq 0$; and $d(\xi_\alpha \xi_\beta) = d(\xi_\alpha) + d(\xi_\beta)$ if $\xi_\alpha \xi_\beta \neq 0$. The degree of zero is undefined.

We denote by V_l $(V_l \subseteq A^r)$ the space of homogeneous polynomials of degree l. For example, V_1 is the space that contains the monomials $\xi_1, \ldots, \xi_r, \xi_1 \xi_0, \ldots, \xi_r \xi_0$ of degree 1, and $\dim V_1 = 2r - 1$.

It is easy to see that in this way A^r becomes a graded algebra: $A^r = V_0 \oplus V_1 \oplus \ldots \oplus V_l \oplus \ldots; V_i V_j \subseteq V_{i+j}$, where $V_0 = K\varepsilon \oplus K\xi_0$.

Direct calculation shows that the algebras A^1, A^2, and A^3 are finite-dimensional over K of dimensions 3, 5, and 11, respectively.

Let us show, for example, that $A^2 = V_0 \oplus V_1$, and that $\dim_K A^2 = 5$. Note that $\xi_1 \xi_2 = \xi_2^2 + \xi_1 \xi_2 = (\sum_{i=1}^{2} \xi_i) \xi_2 = (\sum_{i=1}^{2} \xi_i)(\xi_0 \xi_2) = (\sum_{i=1}^{2} \xi_i \xi_0) \xi_2 = 0$. Similarly, $\xi_2 \xi_1 = 0$, and so $V_2 = 0$. Thus, $V_l = 0$ for every $l \geqslant 2$. In every algebra A^r we have $\dim V_0 = 2$ and $\dim V_1 = 2r-1$. Therefore, $\dim A^2 = \dim V_0 + \dim V_1 = 5$.

In the same way it can be shown that $A^3 = V_0 \oplus V_1 \oplus V_2$, and that $\dim V_0 = 2$, $\dim V_1 = 5$, $\dim V_3 = 4$.

As regards the algebras A^r, $r \geqslant 4$, they are all infinite-dimensional, as we shall show later, and the numbers $d_l = \dim V_l$ form a sequence strictly increasing with l.

8.2. The representations ρ_A and $\rho_{t,l}$. We define a representation $\rho_A : D^r \to \mathcal{L}_R(A^r))$ of D^r in $\mathcal{L}_R(A^r)$ of right ideals of $A = A^r$ in the following way. We set

$$\rho_A(e_i) = \xi_i A,$$

where $\xi_i A$ is the right ideal generated by ξ_i, and e_i is a generator of D^r ($i = 1, \ldots, r$). If x and y are lattice polynomials in D^r, then, by definition, $\rho_A(x \cap y) = \rho_A(x) \cap \rho_A(y)$ and $\rho_A(x+y) = \rho_A(x) + \rho_A(y)$, where $\rho_A(x)$ and $\rho_A(y)$ are right ideals of A.

Let $A\xi_t$ be the left ideal generated by ξ_t. For every $l \geqslant 0$ and $t = 0, 1, \ldots, r$, we set

$$V_{t,l} = (A\xi_t) \cap V_l. \tag{8.1}$$

It is immediately clear from this formula that $V_{i,0} = 0$ for $i \neq 0$ and $V_{0,0} = K\xi_0$. It is also easy to see that all the subspaces $V_{i,1}$ ($i \neq 0$) are one-dimensional: $V_{i,1} = K\xi_i$. Any space $V_{t,l}$, where $(t, l) = (0, 1)$ or $l > 1$, is the sum $\Sigma\, K\xi_\alpha\xi_t$ of all subspaces $K\xi_\alpha\xi_t$ such that $d(\xi_\alpha\xi_t) = l$.

We define a linear representation $\rho_{t,l}$: $D^r \to \mathcal{L}(K, V_{t,l})$ in $V_{t,l}$ by setting

$$\rho_{t,l}(e_j) = V_{t,l} \cap (\xi_j A) \qquad (j = 1, \ldots, r), \tag{8.2}$$

where $\xi_j A$ is the right ideal generated by ξ_j.

We also define the representation ρ_ε in the one-dimensional space $K\varepsilon$ so that $\rho_\varepsilon(e_j) = 0$ for all $j = 1, \ldots, r$.

EXAMPLE 1. $\rho_{i,1}$: $D^r \to \mathcal{L}(K, V_{i,1})$ ($i \neq 1$) is the representation in the space $V_{i,1} = K\xi_i$ such that $\rho_{i,1}(e_i) = K\xi_i$ and $\rho_{i,1}(e_j) = 0$ if $i \neq j$.

EXAMPLE 2. $\rho_{0,1}$: $D^r \to \mathcal{L}(K, V_{0,1})$ is the representation in $V_{0,1}$ that is generated by the vectors $\xi_1\xi_0, \ldots, \xi_r\xi_0$ ($\sum_{i=0}^{r} \xi_i\xi_0 = 0$). Clearly, $\dim V_{0,1} = r-1$. By definition, for this representation $\rho_{0,1}(e_j) = K\xi_j\xi_0$.

PROPOSITION 8.1. *For every algebra A^r ($r \geqslant 4$)*

$$\rho_A \cong \rho_\varepsilon \oplus \rho_{0,0} \oplus \left(\bigoplus_{l=1}^{\infty} \left(\bigoplus_{t=0}^{r} \rho_{t,l}\right)\right). \tag{8.3}$$

The proof reduces to a verification of two elementary assertions.

1) For every $l \geqslant 1$ the space V_l is isomorphic to a direct sum:

$$V_l \cong V_{0,l} \oplus V_{1,l} \oplus \ldots \oplus V_{r,l}.$$

2) Clearly, every ideal $\xi_i A = \rho(e_i)$ is homogeneous with respect to the grading $A = V_0 \oplus V_1 \oplus \ldots$, that is, $\xi_i A = \overset{r}{\underset{l=1}{\oplus}} (V_l \cap \xi_i A)$. To complete the proof of the proposition it remains to check that every subspace $V_l \cap \xi_i$ splits into a direct sum

$$(V_l \cap \xi_i A) \cong \overset{r}{\underset{t=0}{\oplus}} \rho_{t,l}(e_i) = \overset{r}{\underset{t=0}{\oplus}} (V_{t,l} \cap \xi_i A) = \overset{r}{\underset{t=0}{\oplus}} (A\xi_t \cap V_l \cap \xi_i A). \quad \blacksquare$$

REMARK. (8.3) also holds for the algebras A^1, A^2, and A^3. True, in these cases the sum on the right contains only finitely many terms. For example, $A^3 = \rho \oplus \rho_{0,0} \oplus (\overset{2}{\underset{l=1}{\oplus}} (\overset{3}{\underset{t=0}{\oplus}} \rho_{t,l}))$.

8.3. The connection between the representations $\rho_{t,l}$ and $\rho^+_{t,l}$. By definition, the atomic representation $\rho^+_{t,l}$ in the one-dimensional space $V^+_{t,l} \cong K$ is such that: a) $\rho^+_{0,1}(e_i) = 0$ for every $i = 1, \ldots, r$; b) $\rho^+_{j,1}(e_j) = V^+_{j,1}$ and $\rho^+_{j,1}(e_i) = 0$ if $i \neq j$. Thus,

$$\rho^+_{0,1} \cong \rho_{0,0} \text{ and } \rho^+_{i,1} \cong \rho_{i,1} \quad (i \neq 0).$$

The representations $\rho^+_{t,l}$, $l > 1$, were constructed from the atomic representations by means of the functors Φ^-, namely, $\rho^+_{t,l} \overset{\text{def}}{=} (\Phi^-)^{l-1} \rho^+_{t,1}$ for every $t = 0, 1, \ldots, r$.

PROPOSITION 8.2. *For every $l \geqslant 1$,*

$$\rho_{0,l-1} \cong \rho^+_{0,l}, \quad \rho_{t,l} \cong \rho^+_{t,l}, \quad t = 1, \ldots, r.$$

For $l = 1$, as we have already mentioned, the proposition is evident. For $l > 1$ the proof is by induction. We assume that the isomorphism $\rho_{j,l} \cong \rho^+_{j,l}$ (or $\rho_{0,l-1} \cong \rho^+_{0,l}$) has already been proved and we claim that similar isomorphisms hold for $l \neq 1$. To see this, obviously, it is enough to prove that

$$\Phi^- \rho_{j,l} \cong \rho_{j,l+1}, \tag{8.4}$$

$$\Phi^- \rho_{0,l-1} \cong \rho_{0,l}. \tag{8.5}$$

Let us prove (8.5), say. We denote by ξ_{jL} the linear map of A into itself defined by the formula $\xi_{jL}: x \to \xi_j x$ for every $x \in A$. The map ξ_{jL} ($j \neq 0$) acts on the standard monomial $\xi_\alpha \in V_{t,l-1}$ by the formula

$$\xi_{jL}(\xi_\alpha) = \xi_{jL}(\xi_{i_1} \ldots {}_{i_{l-1}t}) = \begin{cases} 0 & \text{if } j = i_1, \\ \xi_{ji_1 \ldots i_{l-1}t} & \text{if } j \neq i_1. \end{cases}$$

Hence $\xi_{jL} V_{t,l-1} \subseteq V_{t,l}$ for $j \neq 0$; moreover, it is easy to see that

$$\xi_{jL}V_{t,\,l-1}=\rho_{t,\,l}(e_j)\qquad(j\neq 0).$$

Let μ denote the map $V_{0,l-1} \to \bigoplus_{j=1}^{r} \rho_{0,l}(e_j)$ defined by the formula

$$\mu x=(\xi_1 x,\ \ldots,\ \xi_r x).$$

Let θ denote the map $\theta\colon \bigoplus_{j=1}^{r} \rho_{0,l}(e_j) \to V_{0,l}$ such that $\theta(x_1, \ldots, x_r) = \sum_{i=1}^{r} x_i$, where $x_i \in \rho_{0,l}(e_i)$. Thus, we have a sequence

$$V_{0,\,l-1} \xrightarrow{\mu} \bigoplus_{j=1}^{r} \rho_{0,\,l}(e_j) \xrightarrow{\theta} V_{0,\,l} \tag{8.6}$$

for which $\theta\mu = 0$. It is clear from the definition of Φ^- that to establish the isomorphism $\Phi^-\rho_{0,l-1} = \rho_{0,l}$ we have to prove the following:
a) $\rho_{0,l}(e_i) \cong V_{0,l-1}/\rho_{0,l-1}(e_i)$; b) the sequence (8.6) is exact at the middle term, that is, Ker θ = Im μ; c) $V_{0,l}$ = Coker μ (however, if Ker θ = Im μ, then $V_{0,l} \cong$ Coker μ, because θ is an epimorphism).

We omit the simple proof of these assertions.■

COROLLARY 8.1. *The algebras A^r, $r \geqslant 4$, are infinite-dimensional, and the numbers n_l = dim V_l form an increasing sequence $n_0 < n_1 < n_2 < \ldots$*

The proof follows from the isomorphism $\rho_{t,l} = \Phi^-\rho_{t,l-1}$ and the formulae of Proposition (6.1), in which the dimension of $\Phi^-\rho$ can be found from that of ρ.■

8.4. The lattice polynomials $e_{i_1\ldots i_l}$ and the representation ρ_A. The lattice polynomials $e_\alpha = e_{i_1\ldots i_l}$ in D^r were defined as follows. Let $A(r, l)$ be the set whose elements are sequences $\alpha = (i_1, \ldots, i_l)$ of integers $1 \leqslant i_j \leqslant r$ such that $i_j \neq i_{j+1}$ for every $1 \leqslant j \leqslant l-1$. We set $A(r, 1) = \{1, \ldots, r\}$. The elements e_α are defined by induction on l: if $l = 1$ and $(\alpha) = (i_1)$, then $e_\alpha = e_{i_1}$; if $l > 1$ and $\alpha = (i_1, \ldots, i_l)$, then $e_\alpha = e_{i_1}(\sum_{\beta\in\Gamma(\alpha)} e_\beta)$, where $\Gamma(\alpha) \subset A(r, l-1)$ and

$$\Gamma(\alpha)=\{\beta=(k_1,\ \ldots,\ k_{l-1})\mid k_1\neq i_1,\ i_2;\ k_2\neq i_2,\ i_3;\ \ldots;\ k_{l-1}\neq i_{l-1},\ i_l\}.$$

We denote by $A_t(r, l)$, $t \in \{0, 1, \ldots, r\}$, the set whose elements are sequences $\beta = (i_1, \ldots, i_{l-1}, t)$ such that $i_j \in \{1, \ldots, r\}$, $i_j \neq i_{j+1}$, $i_{l-1} \neq t$. For every $\beta \in A_0(r, l)$, $l > 1$, we define

$$e_\beta=e_{i_1\ldots i_{l-1}0}=e_{i_1}\sum_{\gamma\in\Gamma(\beta)} e_\gamma,$$

where $\Gamma(\beta) \subset A(r, l-1)$ and $\Gamma(\beta)$ is constructed from the sequence $\beta = (i_1, \ldots, i_{l-1}, 0)$ in the following way:

$$\Gamma(\beta)=\{\gamma=(k_1,\ \ldots,\ k_{l-1})\mid k_1\neq i_1,\ i_2;\ k_2\neq i_2,\ i_3;\ \ldots;\ k_{l-2}\neq i_{l-2},\ i_{l-1},\ k_{l-1}\neq i_{l-1}\}.$$

For example,

$$e_{i_1 0} = e_{i_1} \sum_{t \neq i_1} e_t, \quad e_{i_1 i_2 0} = e_{i_1} \Big(\sum_{\substack{k_1 \neq i_1 i_2 \\ k_2 \neq i_2}} e_{k_1 k_2} \Big).$$

The next theorem establishes a remarkable connection between the monomials e_α and the representation $\rho_A: D^r \to \mathscr{L}_R(A)$ $(\rho_A(e_i) = \xi_i A)$.

THEOREM 8.1. *Let ρ_A be the representation of D^r defined above in the lattice of right ideals of A. Then for every $\alpha = (i_1, \ldots, i_{l-1}, t) \in A_t(r, l)$: $\rho_A(e_\alpha) = \xi_\alpha A$, where $\xi_\alpha A$ is the right ideal generated by the element $\xi_\alpha = \xi_{i_1} \ldots \xi_{i_{l-1}} \xi_t$.*

This theorem is based on the following proposition, which is proved in [7]. Note that this proposition refines Proposition 7.1.

PROPOSITION 8.3. *Let ρ be an arbitrary representation of D^r $(r \geqslant 4)$, and $\rho^1 = \Phi^+\rho$. Let φ_j: $V^1 \to V$ be the standard map from the representation space V^1 of ρ^1 into the representation space V of ρ, defined by the formula $\varphi_j(\xi_1, \ldots, \xi_j, \ldots, \xi_r) = \xi_j$, where $(\xi_1, \ldots, \xi_r) \in V^1 \Longleftrightarrow \sum_{i=1}^{r} \xi_i = 0$.*

Then

$$\varphi_j V^1 = \rho(e_{j,0}), \quad \varphi_{i_1}\rho^1(e_{i_1 \ldots i_{l-1} t}) = 0,$$

$$\varphi_j \rho^1(e_{i_1 \ldots i_{l-1} t}) = \rho(e_{j i_1 \ldots i_{l-1} t}), \quad \textit{if} \quad j \neq i. \quad \blacksquare$$

We have proved that the representations $\rho_{t,l}$ satisfy $\Phi^+\rho_{t,l} \cong \rho_{t,l-1}$, and that the map $\xi_{jL}: V_{t,l-1} \to V_{t,l}$ is the one we have denoted by φ_j: $V^1 \to V$. Consequently, Proposition 8.3 can be restated for $s \in \{0, 1, \ldots, r\}$ in the following way.

LEMMA 8.1. *For every $m \geqslant 1$ and every $\alpha = (i_1, \ldots, i_{l-1}, t) \in A_t(r, l)$*

$$\rho_{s, m}(e_{i_1 0}) = \xi_{i_1} V_{s, m-1},$$

$$\rho_{s, m}(e_{i_1 i_2 \ldots i_{l-1} t}) = \xi_{i_1} \rho_{s, m-1}(e_{i_2 \ldots i_{l-1} t}),$$

where $\xi_{i_1} \rho(e_{\beta_t})$ is the image of the subspace $\rho(e_{\beta_t}) \subseteq V_{s,m-1}$ under $\xi_{i_1 L}$. $\blacksquare$

PROOF OF THEOREM 8.1. Since ρ_A splits into a direct sum of representations $\rho_{s,m}$, we can write

$$\rho_A(e_\alpha) = \sum_{m=0}^{\infty} \sum_{s=0}^{r} \rho_{s, m}(e_\alpha). \tag{8.7}$$

Let us find the subspaces $\rho_{s,m}(e_\alpha) = \rho_{s,m}(e_{i_1 \ldots i_{l-1} t})$ for various s and m. For this we have to analyze the cases $t \neq 0$ and $t = 0$. We first consider $t \neq 0$.

I. a) Let $t \neq 0$. We determine $\rho_{s,m}(e_{i_1 \ldots i_{l-1} t})$ when $m < l$. Applying Lemma 8.1 repeatedly, we obtain

(8.8) $$\rho_{s,m}(e_{i_1\ldots i_{l-1}t}) = \xi_{i_1}\rho_{s,m-1}(e_{i_2\ldots i_{l-1}t}) =$$
$$= \xi_{i_1}(\xi_{i_2}\rho_{s,m-2}(e_{i_3\ldots i_{l-1}t})) = \xi_{i_1i_2}\rho_{s,m-2}(e_{i_3\ldots i_{l-1}t}) = \ldots$$
$$\ldots = \xi_{i_1i_2\ldots i_{m-1}}\rho_{s,1}(e_{i_m\ldots i_{l-1}t}).$$

By definition $e_{i_m\ldots i_{l-1}t} \in e_t\,(l-m+1) \subset h_p(l-m+1)$ for any $p \neq t$. Consequently, $\rho_{s,1}(e_{i_m\ldots i_{l-1}t}) \subseteq \rho_{s,1}(h_p(l-m+1))$. As we know, the element $h_p(l-m+1)$ is perfect, and $\rho_{s,1} \cong \rho^+_{s,1}$ if $s \neq 0$, and $\rho_{0,1} \cong \rho^+_{0,2}$. Therefore, if $l > m$, then $\rho_{s,1}(h_p(l-m+1)) = 0$, and a fortiori $\rho_{s,e}(e_{i_m\ldots i_{l-1}t}) = 0$. Thus, if $m < l$, then $\rho_{s,m}(e_{i_1\ldots i_{l-1}t}) = 0$.

b) We now determine $\rho_{s,m}(e_{i_1\ldots i_{l-1}t})$ for $m = l$. By analogy to the chain of equalities (8.8), we find

$$\rho_{s,l}(e_{i_1\ldots i_{l-1}t}) = \xi_{i_1\ldots i_{l-1}}\rho_{s1}(e_t).$$

If $s = 0$, then, by definition, $\rho_{0,1}(e_t) = K\xi_t\xi_0 = \xi_t(K\xi_0)$, and if $s = t$, then $\rho_{s,1}(e_t) = K\xi_t = \xi_t K$ and $\rho_{s,1}(e_t) = 0$ for $s \neq t$. Thus, we can write

$$\rho_{s,l}(e_{i_1\ldots i_l t}) = \begin{cases} \xi_{i_1\ldots i_{l-1}t}K\xi_0 & \text{if } s = 0, \\ \xi_{i_1\ldots i_{l-1}t}K\varepsilon & \text{if } s = t, \\ 0 & \text{if } s \neq t. \end{cases}$$

c) For $m > l$ we obtain

(8.9) $$\rho_{s,m}(e_{i_1\ldots i_{l-1}t}) = \xi_{i_1\ldots i_{l-1}}\rho_{s,m-l+1}(e_t).$$

By definition, for $n > 1$ the subspaces $\rho_{s,n}(e_t)$ satisfy $\rho_{s,n}(e_t) = \xi_t V_{s,n-1}$. Consequently, $\rho_{s,m}(e_{i_1\ldots i_{l-1}t}) = \xi_{i_1\ldots i_{l-1}t}V_{s,m-l}$. We can therefore rewrite (8.9) as follows:

$$\rho_{s,m}(e_\alpha) = \rho_{s,m}(e_{i_1\ldots i_{l-1}t}) = \xi_{i_1\ldots i_{l-1}t}V_{s,m-l} = \xi_\alpha V_{s,m-l} \qquad (m > l).$$

d) Now we insert all the relevant expressions for $\rho_{s,m}(e_\alpha)$ in (8.7). This gives

$$\rho_A(e_\alpha) = \rho_A(e_{i_1\ldots i_{l-1}t}) =$$
$$= \sum_{s=0}^{r}\rho_{s,l}(e_\alpha) + \sum_{s=0}^{r}\rho_{s,l+1}(e_\alpha) + \ldots + \sum_{s=0}^{r}\rho_{s,l+n}(e_\alpha) + \ldots =$$
$$= \xi_\alpha K\varepsilon + \xi_\alpha K\xi_0 + \sum_{s=0}^{r}\xi_\alpha V_{s,1} + \ldots + \sum_{s=0}^{r}\xi_\alpha V_{s,n} + \ldots =$$
$$= \xi_\alpha\Big(K\varepsilon + K\xi_0 + \sum_{n=1}^{\infty}\sum_{s=0}^{r}V_{s,n}\Big) = \xi_\alpha A.$$

The proof that $e_\alpha = e_{i_1} \dots e_{i_{l-1}0}$ proceeds similarly, with the help of the formula $\rho_{0,m}(e_{i,0}) = \xi_{i_1} V_{0,m-1}$. ∎

§9. Complete irreducibility of the representations $\rho^+_{t,l}$ and $\rho^-_{t,l}$

In this section all vector spaces are finite-dimensional over a field K of characteristic zero.

9.1. Systems of vectors. Let $R \subset V$ be a finite set of elements of a finite-dimensional space V. Then R is called a system of vectors in V if all elements $\alpha \in R$ are non-zero and R generates V. A subset $R' \subset R$ is called a subsystem of R if R' generates V.

EXAMPLE 1. A root system (for the definition, see [4]) is a system of vectors.

EXAMPLE 2. Let $V_{t,l} \subseteq A^r$ be the representation space of $\rho_{t,l}$. Then the set of $\xi_\alpha = \xi_{i_1 \dots i_{l-1} t}$ (monomials of degree l) is a system of vectors.

A subset $B \subset R$ is called a *basis of* R if B is a basis of V.

A system R is called *indecomposable* if for every subset $R_1 \subset R$ the intersection of the subspaces $V_1 = \sum_{\alpha \in R_1} K\alpha$ and $V_2 = \sum_{\alpha \in R - R_1} K\alpha$ is non-empty. ($K\alpha$ denotes the one-dimensional subspace generated by α.)

We introduce several concepts, which allow us to give a convenient criterion for the indecomposability of a system R. Let $B = \{\alpha_1, \dots, \alpha_n\}$ be a fixed basis of R. We associate with each vector $\beta \in R$ a subset (chamber) $C_\beta \subseteq B$ in the following way:

$$(\alpha_i \in C_\beta) \Longleftrightarrow (\beta = b_i\alpha_i + \sum_{j \neq i} b_j\alpha_j \quad \text{and} \quad b_i \neq 0).$$

Two vectors α_i and α_j in B are called *simply-connected* if they belong to the same chamber (that is, if there exists a $\beta \in R$ such that $\beta = b_i\alpha_i + b_j\alpha_j + \sum_{k \neq i,j} b_k\alpha_k$, where $b_i \neq 0$ and $b_j \neq 0$). Two basis vectors α_i and α_k are called *connected*, in symbols $\alpha_i \sim \alpha_k$, if there is a sequence of vectors $\alpha_i = \alpha^{(0)}, \alpha^{(1)}, \dots, \alpha^{(m)} = \alpha_k$ such that every pair $\alpha^{(j)}, \alpha^{(j+1)}$ is simply-connected. If we take every vector $\alpha \in B$ to be connected to itself, then it is not difficult to check that connectedness is an equivalence relation. A basis B is called *connected* if every pair of vectors $\alpha_i, \alpha_j \in B$ is connected.

PROPOSITION 9.1. *The following assertions are equivalent*:

(i) *Every basis B of R is connected.*

(ii) *The system R is indecomposable.*

We claim that a basis is not connected if and only if the system is decomposable.

1) Let B be a disconnected basis of R and $B_1 \subset B$ be a non-trivial connected component. Then for every vector $\beta \in R$ with chamber C_β

either $C_\beta \subset B_1$ or $C_\beta \subset (B - B_1)$. We set $V_1 = \sum_{\alpha \in B_1} K\alpha$ and $\bar{V}_1 = \sum_{\alpha \in B - B_1} K\alpha$. It follows easily from the definition of a chamber that $\beta \in \sum_{\alpha \in C_\beta} K\alpha$. Thus, for every vector $\beta \in R$, either $\beta \in V_1$ or $\beta \in \bar{V}_1$. This means that R is decomposable.

2) If R is decomposable, then there exist two non-empty complementary subsets R_1 and R_2 ($R_1 \cap R_2 = 0$, $R_1 + R_2 = R$) such that the subspaces $V_i = \sum_{\beta \in R_i} K\beta$ ($i = 1, 2$) are disjoint. Let B be any basis of R. Clearly, each basis vector belongs to one of the subspaces V_1 or V_2. Thus, the set B splits into two complementary subsets B_1 and B_2, where $\alpha \in B_j \Longleftrightarrow \alpha \in V_j$. Here B_j is a basis of $V_i (i = 1, 2)$. Let $\beta \in R$ be an arbitrary vector. Then the condition of decomposability implies that either $\beta \in V_1$ or $\beta \in V_2$. Now $\beta \in V_i$ clearly implies that $C_\beta \subseteq B_i$. Consequently, B is not connected.■

An indecomposable system R is called minimal if for every $\beta \in R$ the subsystem $R' = R - \{\beta\}$ is decomposable.

PROPOSITION 9.2. *Let R be a minimal indecomposable system and $B = \{\alpha_1, \ldots, \alpha_n\}$ a basis of R. Then:*

a) *the number m of elements in $R - B$ satisfies $1 \leqslant m < n$;*

b) *for each $\beta \in R - B$ the chamber $C_\beta \subseteq B$ contains at least two elements;*

c) *the set $R - B$ can be numbered in such a way that $R - B = \{\beta_1, \ldots, \beta_m\}$ and for each $j > 1$ the chamber $C_j \stackrel{\text{def}}{=} C_{\beta_j}$ satisfies $\varnothing \subset C_j \cap (\sum_{i=1}^{j-1} C_i) \subset \sum_{i=1}^{j-1} C_i$ (where $\subset$ denotes strict inclusion) and the subsets $\sum_{i=1}^{k} C_i$ form a strictly increasing chain:*

$$C_1 \subset \sum_{i=1}^{2} C_i \subset \sum_{i=1}^{3} C_i \subset \ldots \subset \sum_{i=1}^{m} C_i = B.$$

We omit the simple proof of this proposition.■

Let R be a system in $V = K^n$. The lattice generated by the one-dimensional subspace $K\alpha$, $\alpha \in R$, is denoted by $\mathscr{M}(R)$. A system R in V is called *completely irreducible* if $\mathscr{M}(R) \cong \mathscr{L}(\mathbf{Q}^n)$, (where $\mathscr{L}(\mathbf{Q}^n)$ is the lattice of linear subspaces of $\mathbf{Q}^n$).

A system R is called *rational* if there is a basis $B = \{\alpha_1, \ldots, \alpha_n\} \subset R$ such that each vector $\beta \in R$ can be written as a sum $\beta = \sum_{i=1}^{n} \lambda_i \alpha_i$ with rational coefficients λ_i.

It is easy to show that if R is rational, then every vector $\beta \in R$ for any basis $B = \{\alpha_i'\} \subseteq R$ can be written as a sum $\beta = \sum_{i=1}^{n} \lambda_i' \alpha_i'$ with $\lambda_i' \in \mathbf{Q}$.

THEOREM 9.1. *Let R be an indecomposable rational system in $V \cong K^n$ over a field K of characteristic 0. If $\dim V = n \geqslant 3$, then R is completely irreducible, that is $\mathcal{M}(R) \cong \mathcal{L}(Q^n)$.*

The proof requires several Lemmas.

MAIN LEMMA 9.1. *Let $V = K^n$, where K is a field of characteristic 0, and let $R = \{\alpha_0, \alpha_1, \ldots, \alpha_n\}$ be a system in V such that $\sum_{i=0}^{n} \alpha_i = 0$. If $n \geqslant 3$, then R is completely irreducible, that is, $\mathcal{M}(R) \cong \mathcal{L}(Q^n)$).*

We do not give the rather tedious proof of this lemma. We only remark that it is very similar to the standard procedure for introducing coordinates in a projective space.

REMARK It is easily seen that any n vectors $\{\alpha_i\}$ of the system R in Lemma 9.1. form a basis of V.

LEMMA 9.2. *Let R be a system as in Lemma 9.1 (that is, $R \subset K^n$, $R = \{\alpha_0, \ldots, \alpha_n\}$, $\sum_{i=0}^{n} \alpha_i = 0$), and let $\mathcal{M}(R)$ be the lattice generated by the subspaces $K\alpha_i$. Let $x_0, x_1, \ldots, x_n$ be non-zero elements of $\mathcal{M}(R)$ such that any n of them are linearly independent (that is $\forall_{i,j|i\neq j} x_i \sum_{k \neq i,j} x_k = 0$). Then $x_0, x_1, \ldots, x_n$ are generators of $\mathcal{M}(R)$.*

PROOF. It follows easily from the lemma that the x_i are one-dimensional subspaces of V. Therefore, by applying the fundamental lemma, we see that each element x_i can be represented in the form $x_i = Kf_i$, where $f_i = \sum_{j=1}^{n} \lambda_{ij}\alpha_j$ and $\lambda_{ij} \in Q$.

Since $\dim V = n$ and all $f_i \neq 0$, clearly, the $n + 1$ vectors $f_0, f_1, \ldots, f_n$ are linearly dependent, consequently, $f_0 = \sum_{i=1}^{n} \gamma_i f_i$ with $\gamma_i \in Q$. By assumption, any n of the vectors $f_0, f_1, \ldots, f_n$ are linearly independent. Starting from this it is not hard to show that all the coefficients γ_i in the expansion $f_0 = \sum_{i=1}^{n} \gamma_i f_i$ are non-zero.

We set $f_0' = -f_0$ and $f_i' = \gamma_i f_i$ if $i \neq 0$. Then the system $R' = \{f_0', f_1', \ldots, f_n'\}$ satisfies the conditions of Lemma 9.1, hence, $\mathcal{M}(R') \cong \mathcal{L}(Q^n)$. Since $Kf_i' = x_i$ and $x_i \in \mathcal{M}(R)$, it follows that $\mathcal{M}(R') \subseteq \mathcal{M}(R)$.

It follows from the isomorphism $\mathcal{M}(R') \cong \mathcal{L}(Q^n)$ that $\mathcal{M}(R')$ contains every one-dimensional space of the form Ky, where $y = \sum_{i=1}^{n} \mu_i f_i'$ with $\mu_i \in Q$. By definition, $f_i' = \gamma_i \sum_{j=1}^{n} \lambda_{ij}\alpha_j$. Hence it is easy to show that the

α_j can be expressed rationally in terms of the f_j'. This means that $K\alpha_j \in \mathcal{M}(R')$ and so $\mathcal{M}(R') \subseteq \mathcal{M}(R)$. We have shown earlier that $\mathcal{M}(R') \subseteq \mathcal{M}(R)$. Therefore, $\mathcal{M}(R') = \mathcal{M}(R)$. ∎

LEMMA 9.3. *Let $R = \{\alpha_0, \alpha_1, \ldots, \alpha_{n+1}\}$ be a system of $n + 2$ vectors in $V = K^n$ such that $\sum_{i=0}^{m} \alpha_i = 0$, $\sum_{j=k}^{n+1} \alpha_j = 0$, where $1 \leqslant k \leqslant m \leqslant n$. If $n \geqslant 3$, then R is completely irreducible.*

PROOF. By assumption, the vectors α_i generate V. The conditions of the lemma then imply that $B = \{\alpha_1, \ldots, \alpha_n\}$ is a basis of R. We consider separately the cases $m = 1$, $k = n$, and $1 < m$, $k < n$.

1) Let $m = 1$. This means that $\alpha_0 + \alpha_1 = 0$ and $\sum_{j=1}^{n+1} \alpha_j = 0$. Since $K\alpha_0 = K\alpha_1$, the subspaces $K\alpha_i$ $(i = 1, 2, \ldots, n + 1)$ are generators of $\mathcal{M}(R)$. Thus, this case reduces to the main lemma.

2) Let $k = n$. This means that $\sum_{i=1}^{n} \alpha_i = 0$ and $\alpha_n + \alpha_{n+1} = 0$. This is clearly another application of the main lemma.

3) Let $1 < m$ and $k < n$. We claim that $\mathcal{M}(R)$ contains the element Kf, where $f = \sum_{i=1}^{n} \alpha_i$. We write $V_0 = \sum_{i=1}^{m} K\alpha_i$, $V_1 = \sum_{j=k}^{n} K\alpha_j$ (that is, $\dim V_0 = m$, $\dim V_1 = n - k + 1$). The condition $\sum_{i=0}^{m} \alpha_i = 0$ implies that the subspace $y_0 = K\alpha_0$ is defined on the basis $\{\alpha_1, \ldots, \alpha_m\}$ of V_0 by the system of equations

$$\begin{cases} x_1 = x_2, \\ \cdot\ \cdot\ \cdot\ \cdot\ \cdot\ \cdot \\ x_{m-1} = x_m. \end{cases}$$

Here it is clear that each individual equation $x_i = x_{i+1}$ $(1 \leqslant i < m)$ defines in V_0 an $(m - 1)$-dimensional subspace, which we denote by W_i. We write y_j for the line $K\alpha_j$. It is easy to check that $W_i = (y_0 + \ldots + y_{i-1} + \hat{y}_i + \hat{y}_{i+1} + y_{i+2} + \ldots + y_m)$ $(1 \leqslant i < m)$, where the hats over y_i and y_{i+1} denote that the corresponding summands are omitted.

Thus, $W_i \in \mathcal{M}(R)$ because $y_i \in \mathcal{M}(R)$.

Similarly, the line $y_{n+1} = K\alpha_{n+1}$ is defined in the basis $\{\alpha_k, \ldots, \alpha_n\}$ of V_1 by the system of equations

$$\begin{cases} x_k = x_{k+1}, \\ \cdot\ \cdot\ \cdot\ \cdot\ \cdot\ \cdot \\ x_{n-1} = x_n, \end{cases}$$

Here each individual equation $x_j = x_{j+1}$ $(k \leqslant j < n)$ is that of the subspace

$$W_j = (y_k + \ldots + y_{j-1} + \hat{y}_j + \hat{y}_{j+1} + y_{j+2} + \ldots + y_{n+1}).$$

We define elements L_i in $\mathcal{M}(R)$ as follows:

$$L_i = \begin{cases} (y_0 + y_{i-1} + \hat{y}_i + \hat{y}_{i+1} + y_{i+2} + \ldots + y_n) & \text{for} \quad 1 \leqslant i < m, \\ (y_1 + y_{i-1} + \hat{y}_i + \hat{y}_{i+1} + y_{i+2} + \ldots + y_{n+1}) & \text{for} \quad m \leqslant i < n. \end{cases}$$

It is not difficult to verify that in the coordinate system $\{\alpha_1, \ldots, \alpha_n\}$ the hyperplane $L_i\,(1 \leqslant i < n)$ is defined by the equation $x_i = x_{i+1}$. Consequently, the system of equations

$$\begin{cases} x_1 = x_2, \\ \quad \cdots \\ x_{n-1} = x_n \end{cases}$$

defines a one-dimensional subspace Kf, where f has the coordinates $(1, 1, \ldots, 1)$ in the basis $\{\alpha_1, \ldots, \alpha_n\}$. It is also clear that $Kf = \bigcap_{i=1}^{n-1} L_i$. Since the L_i are elements of $\mathcal{M}(R)$ we see that $Kf \in \mathcal{M}(R)$. We denote the system $\{\alpha_1, \ldots, \alpha_n, -f\}$ by $\widetilde{R}$. Obviously, this system satisfies the conditions of the main lemma 9.1, hence $\mathcal{M}(\widetilde{R}) \cong \mathcal{L}(\mathbf{Q}^n)$. It follows from $Kf \in \mathcal{M}(R)$ that $\mathcal{M}(\widetilde{R}) \subseteq \mathcal{M}(R)$.

Owing to the isomorphism $\mathcal{M}(\widetilde{R}) \cong \mathcal{L}(\mathbf{Q}^n)$ the one-dimensional subspaces $y_0 = K\alpha_0 = K(\sum_{i=1}^{m} \alpha_i)$ and $y_{n+1} = K\alpha_{n+1} = K(\sum_{j=k}^{n} \alpha_j)$ belong to $\mathcal{M}(\widetilde{R})$. Consequently, $\mathcal{M}(R) \subseteq \mathcal{M}(\widetilde{R})$. We have proved above that $\mathcal{M}(\widetilde{R}) \subseteq \mathcal{M}(R)$. Therefore $\mathcal{M}(R) = \mathcal{M}(\widetilde{R}) \cong \mathcal{L}(\mathbf{Q}^n)$. ∎

PROOF OF THEOREM 9.1. We choose in R any minimal indecomposable subsystem R' and claim that R' is completely irreducible, that is $\mathcal{M}(R') \cong \mathcal{L}(\mathbf{Q}^n)$).

Let $B = \{\alpha_1, \ldots, \alpha_n\}$ be a basis of R'. We denote by m the number of elements in $R' - B$. It follows from Proposition 9.2 that $1 \leqslant m < n$. We recall that with each element $\beta \in R'$ we associate a chamber C_β (a subset of B) by the following rule: $(\alpha_i \in C_\beta) \Longleftrightarrow (\beta = \lambda_i \alpha_i + \sum_{j \neq i} \lambda_j \alpha_j$, with $\lambda_i \neq 0)$. It follows from Proposition 9.2 that if $\beta \in R' - B$, then the number of elements $d(\beta)$ in C_β is at least 2 $(d(\beta) \geqslant 2)$.

We break the proof into several steps and use induction on the number m of elements of $R' - B$.

Step 1. Let $m = 1$, that is, $R - B = \{\beta\}$. Since R' is indecomposable it evidently follows that $C_\beta = B$, that is, $\beta = \sum_{i=1}^{n} \lambda_i \alpha_i$, where all $\lambda_i \neq 0$. We set $\alpha_0' = -\beta$ and $\alpha_i' = \lambda_i \alpha_i$ for $i \geqslant 1$. Since $\sum_{i=0}^{n} \alpha_i' = 0$, the system

$R_1 = \{\alpha'_0, \ldots, \alpha'_n\}$ clearly satisfies the conditions of the main Lemma 9.1, and so $\mathcal{M}(R') = \mathcal{M}(R_1) \cong \mathcal{L}(Q^n)$.

Step 2. Let $R' - B = \{\beta_1, \beta_2\}$. It follows from Proposition 9.2 that the chambers C_1 and C_2 of β_1 and β_2 satisfy $C_1 \cup C_2 = B$, $C_1 \cap C_2 \neq 0$, $C_i \neq B$. Hence it is easy to see that the basis B can be numbered in the following way: $B = \{\alpha_1, \ldots, \alpha_n\}$, where $\beta_1 = \sum_{i=1}^{s_1} \lambda_i \alpha_i$ and $\beta_2 = \sum_{j=s_0}^{n} \mu_j \alpha_j$, with $1 < s_0 \leqslant s_1 < n$, and all of the coefficients λ_i and μ_j are non-zero. (Note that $s_1 = d(C_1)$ and $n - s_0 + 1 = d(C_2)$, where $d(C_i)$ is the number of elements in C_i. Here, $s_1 - s_0 + 1 = d(C_1 \cap C_2)$.) We consider separately the cases $d(C_1 \cap C_2) = 1$ and $d(C_1 \cap C_2) > 1$.

Step 2a). Let $d(C_1 \cap C_2) = 1$. This means in other words that $s_1 = s_0 = s$, that is, $\beta_1 = \sum_{i=1}^{s} \lambda_i \alpha_i$ and $\beta_2 = \sum_{j=s}^{n} \mu_j \alpha_j$, where $1 < s < n$. We define vectors α'_i in the following way. We set $\alpha'_0 = -\frac{1}{\lambda_s}\beta_1$; $\alpha'_i = \frac{\lambda_i}{\lambda_s}\alpha_i$, if $1 \leqslant i \leqslant s$; $\alpha'_j = \frac{\mu_j}{\mu_s}\alpha_j$, if $s < j \leqslant n$; $\alpha'_{n+1} = \frac{-1}{\mu_s}\beta_2$. Then $\sum_{i=0}^{s} \alpha'_i = 0$ and $\sum_{j=s}^{n+1} \alpha'_j = 0$, and $\{\alpha'_1, \ldots, \alpha'_n\}$ is a basis of V. Let R_1 denote the system $(\alpha'_0, \ldots, \alpha'_{n+1})$. Clearly, R_1 satisfies the conditions of Lemma 9.3, and so $\mathcal{M}(R_1) \cong \mathcal{L}(Q^n)$. From the construction it is clear that $\mathcal{M}(R') = \mathcal{M}(R_1)$ thus, $\mathcal{M}(R') \cong \mathcal{L}(Q^n)$.

Step 2b). Let $d(C_1 \cap C_2) > 1$. In other words, we can write $\beta_1 = \sum_{i=1}^{s_1} \lambda_i \alpha_i$, $\beta_2 = \sum_{j=s_0}^{n} \mu_j \alpha_j$, where $1 < s_0 < s_1 < n$. We use the notation $V_1 = \sum_{i=1}^{s_1} K\alpha_i$ and $V_2 = \sum_{j=s_0}^{n} K\alpha_j$. From $1 < s_0 < s_1 < n$ it follows that $\dim V_1 = s_1 \geqslant 3$ and $\dim V_2 = n - s_0 + 1 \geqslant 3$. Let $R_1 = \{\alpha_1, \ldots, \alpha_s, \beta_1\}$ and $R_2 = \{\alpha_{s_0}, \ldots, \alpha_n, \beta_2\}$. Thus, R_i is a system in V_i ($i = 1, 2$). The same arguments as in step 1 show that R_1 and R_2 are completely irreducible, that is,

$$\mathcal{M}(R_1) \cong \mathcal{L}(Q^{s_1}), \quad \mathcal{M}(R_2) \cong \mathcal{L}(Q^{n-s_0+1}).$$

Since $\mathcal{M}(R_1) \cong \mathcal{L}(Q^{s_1})$, the lattice $\mathcal{M}(R_1)$ contains every one-dimensional subspace of the form Kx, where $x = \sum_{i=1}^{s} \gamma_i \alpha_i$ with $\alpha_i \in Q$. In particular, if $\alpha_0 = -\sum_{i=1}^{s_1} \alpha_i$, then $K\alpha_0 \in \mathcal{M}(R_1)$.

It follows from Lemma 9.2 that the subspaces $K\alpha_0, K\alpha_1, \ldots, K\alpha_{s_1}$ can be chosen as new generators of $\mathcal{M}(R_1)$ that is, $\mathcal{M}(R_1) = \mathcal{M}_1(K\alpha_0, K\alpha_1, \ldots, K\alpha_{s_1})$, where $\mathcal{M}_1(K\alpha_0, \ldots, K\alpha_{s_1})$ denotes the lattice generated by the one-dimensional subspaces $K\alpha_0, \ldots, K\alpha_{s_1}$.

The same statements can be made about the lattice $\mathcal{M}(R_2)$, generated by the system $R_2 = \{\alpha_{s_0}, \ldots, \alpha_n, \beta\}$ in V_2. Namely, $\mathcal{M}(R_2)$ contains $K\alpha_{n+1}$, where $\alpha_{n+1} \overset{\text{def}}{=} -\sum_{j=s_0}^{n} \alpha_j$ and $\mathcal{M}(R_2) = \mathcal{M}_2(K\alpha_{s_0}, \ldots, K\alpha_n, K\alpha_{n+1})$.

We write $\mathcal{M}(R') = \mathcal{M}(K\beta_1, K\alpha_1, \ldots, K\alpha_n, K\beta_2)$. Now it is easy to see that $\mathcal{M}(R') = \mathcal{M}(K\alpha_0, K\alpha_1, \ldots, K\alpha_n, K\alpha_{n+1})$. New generators for $\mathcal{M}(R')$ can be chosen so that $\sum_{i=0}^{s_1} \alpha_i = 0$ and $\sum_{j=s_0}^{n+1} \alpha_j = 0$. Consequently, the conditions of Lemma 9.1 are satisfied and $\mathcal{M}(R') \cong \mathcal{L}(\mathrm{Q}^n)$.

Step 3. We now turn to the general case $R' - B = \{\beta_1, \ldots, \beta_{m+1}\}$. From Proposition 9.2 it follows that we can renumber B so that in the new numbering the vectors $\beta_j \in R' - B$ can be written in the following form:

$$\beta_1 = \sum_{i=1}^{s_1} \lambda_{1i}\alpha_i,$$

$$\beta_2 = \sum_{\alpha_i \in C_2 \cap C_1} \lambda_{2i}\alpha_i + \sum_{j=s_1+1}^{s_2} \mu_{2j}\alpha_j,$$

$$\cdots\cdots\cdots\cdots\cdots\cdots\cdots\cdots$$

$$\beta_k = \sum_{\alpha_i \in C_k \cap L_{k-1}} \lambda_{ki}\alpha_i + \sum_{j=s_{k-1}+1}^{s_k} \mu_{kj}\alpha_j,$$

$$\beta_{m+1} = \sum_{\alpha_i \in C_{m+1} \cap L_m} \lambda_{m+1,\,i}\alpha_i + \sum_{j=s_m+1}^{n} \mu_{m+1,\,j}\alpha_j,$$

where C_j is the chamber of the element β_j, and $L_k \overset{\text{def}}{=} \{1, 2, \ldots, s_k\} = \sum_{j=1}^{k} C_j$. From Proposition 9.2 it also follows that $1 < s_1 < s_2 < \ldots < s_m < s_{m+1} = n$. Since $m > 1$, we have $s_m \geqslant 3$, and so by induction we may assume that the proposition is proved for the system $R_m = \{\alpha_1, \ldots, \alpha_{s_m}, \beta_1, \ldots, \beta_m\}$ in $V_m \overset{\text{def}}{=} \sum_{i=1}^{s_m} K\alpha_i$. This means that the lattice $\mathcal{M}_m \overset{\text{def}}{=} \mathcal{M}(K\alpha_1, \ldots, K\alpha_{s_m}, K\beta_1, \ldots, K\beta_m)$ is isomorphic to $\mathcal{L}(\mathrm{Q}^{s_m})$. Consequently, there is an element $y = K\alpha_0$ in $\mathcal{M}_m$ where $\alpha_0 \overset{\text{def}}{=} -\sum_{j=1}^{s_m} \alpha_j$. Here, the elements $K\alpha_0, K\alpha_1, \ldots, K\alpha_{s_m}$ are generators of $\mathcal{M}_m$. Hence we conclude that $\mathcal{M}(R') = \mathcal{M}(K\alpha_1, \ldots, K\alpha_n, K\beta_1, \ldots, K\beta_{m+1}) = \mathcal{M}(K\alpha_1, \ldots, K\alpha_n, K\alpha_0, K\beta_{m+1})$. Consequently, the arguments of step 2 apply to $\mathscr{M}(R')$ and so $\mathscr{M}(R') \cong \mathscr{L}(\mathbf{Q}^n)$.

Step 4. Now we return to the case of an arbitrary (non-minimal) system R. Let $R' = \{\alpha_1, \ldots, \alpha_n, \beta_1, \ldots, \beta_l\}$ be a minimal subsystem. It is clear that a basis $B = \{\alpha_1, \ldots, \alpha_n\}$ of R' is also a basis of R. Let $R - R' = \{\gamma_1, \ldots, \gamma_k\}$. Since R is rational, each vector γ_i can be represented in the form $\gamma_i = \sum_{j=1}^{n} \lambda_{ij}\alpha_j$, where $\lambda_{ij} \in \mathbf{Q}$. Since $\mathscr{M}(R') \cong \mathscr{L}(\mathbf{Q}^n)$, we see that $K\gamma_1 \in \mathscr{M}(R')$. Consequently, $\mathscr{M}(R) = \mathscr{M}(R') \cong \mathscr{L}(\mathbf{Q}^n)$. ■

THEOREM 9.2. *All but a finite number of the representations* $\rho_{t,l}: D^r \to \mathscr{L}(K, V_{t,l})$ *over a field K of characteristic* 0 *are completely irreducible.*

Only the following representations are not completely irreducible:

a) $\rho_{0,0}$ *and* $\rho_{i,1} (i \neq 0)$ *for any* $r \geqslant 4$,

b) $\rho_{i,2} (i \neq 0)$ *for* $r = 4$.

PROOF. The representations $\rho_{0,0}$ and $\rho_{i,1}$ are not completely irreducible, since $\dim V_{0,0} = \dim V_{i,1} = 1$. It is also easy to find that $\dim V_{i,2} = 2$ for $r = 4$ and $i \neq 0$, so that the systems $\rho_{i,2}$ $(i \neq 0,\ r = 4)$ are not completely irreducible.

For all other representations $\rho_{t,l}$ we can show that $\dim V_{t,l} \geqslant 3$.

To prove the complete irreducibility of the other representations $\rho_{t,l}$, we have to verify that: 1) the system of vectors $R_{t,l} \overset{\text{def}}{=\!=} \{\xi_\alpha\}$, $\alpha \in A_t(r, l)$, is rational, and that 2) the system $R_{t,l}$ is indecomposable.

The rationality of $R_{t,l}$ follows easily from the fact that the complete system of equations satisfied by the vectors ξ_α consists of $\sum_s \xi_{i_1 \ldots i_{j-1} s i_{j+1} \ldots i_{l-1} t} = 0$, where the summation is over all vectors ξ_α in which the indices i_k $(k \neq j)$ are fixed.

The indecomposability of $R_{t,l}$ clearly follows from that of the representations $\rho_{t,l}$.

References

[1] M. Auslander, Representation theory of Artin algebras. I, II. Comm. Algebra **1** (1974), 177–268; 269–310. MR **50** # 2240. – and I. Reiten, III, Comm. Algebra **3** (1975), 239–294. MR **52** # 504.

[2] I. N. Bernstein, I. M. Gel'fand, and V. A. Ponomarev, Coxeter functors and Gabriel's theorem, Uspekhi Mat. Nauk **28**:2 (1973), 19–33.
= Russian Math. Surveys **28**:2 (1973), 17–32.

[3] G. Birkhoff, Lattice Theory, Amer. Math. Soc., New York 1948.
Translation: *Teoriya struktur*, Izdat. Inost. Lit., Moscow 1952.

[4] N. Bourbaki, Eléments de mathématique, XXVI, Groupes et algèbres de Lie, Hermann et Cie., Paris 1960. MR **24** # A2641.
Translation: *Gruppy i algebry Li*, Izdat. Mir, Moscow 1972.

[5] P. Gabriel, Unzerlegbare Darstellungen. I, Manuscripta Math. **6** (1972), 71–103.

[6] I. M. Gel'fand and V. A. Ponomarev, Problems of linear algebra and classification of quadruples of subspaces in a finite-dimensional vector space, Coll. Math. Soc. Iános Bolyai **5**, Hilbert space operators, Tihany (Hungary) 1970, 163–237 (in English). (For a brief account, see Dokl. Akad. Nauk SSSR **197** (1971), 762–765. = Soviet Math. Dokl. **12** (1971), 535–539.

[7] I. M. Gel'fand and V. A. Ponomarev, Free modular lattices and their representations, Uspekhi Mat. Nauk **29**:6 (1974), 3–58. = Russian Math. Surveys **29**:6 (1974), 1–56.

[8] V. Dlab and C. M. Ringel, Representations of graphs and algebras, Carleton Math. Lect. Notes No. 8 (1974).

[9] L. A. Nazarova and A. V. Roiter, Representations of partially ordered sets, in the coll. "Investigations in the theory of representations", Izdat. Nauka, Leningrad 1972, 5–31.

Received by the Editors 9 April 1976

Translated by M. B. Nathanson